Math
Problems in
Plasticity

Mathematical Problems in Plasticity

Roger Temam

Director, Institute for Scientific Computing and Applied Mathematics
Indiana University

Translated by L. S. Orde

DOVER PUBLICATIONS, INC.
Mineola, New York

Copyright

Bibliographical Note

This Dover edition, first published in 2018, is an unabridged republication of the English translation by L. S. Orde, as printed by Gauthier-Villars, Paris, in 1985. It was originally offered in French by Bordas, Paris, in 1983, under the title *Problèmes mathématiques en plasticité*. This volume includes a new preface and a new appendix by Jean-François Babadjian and the author.

Library of Congress Cataloging-in-Publication Data

Names: Temam, Roger, author.
Title: Mathematical problems in plasticity / Roger Temam, director, Institute for Scientific Computing and Applied Mathematics, Indiana University ; translated by L.S. Orde.
Other titles: Problemes mathematiques en plasticite. English
Description: Dover edition. | Mineola, New York : Dover Publications, Inc., 2018. | English translation originally published: Paris : Gauthier-Villars, 1985.
Identifiers: LCCN 2018030839| ISBN 9780486828275 | ISBN 0486828271
Subjects: LCSH: Plasticity. | Plasticity—Mathematical models. | Equilibrium.
Classification: LCC QA931 .T3813 2018 | DDC 531/.385—dc23
LC record available at https://lccn.loc.gov/2018030839

Manufactured in the United States by LSC Communications
82827101 2018
www.doverpublications.com

Table of Contents

Preface 2018

We are pleased that Dover Publications, Inc. has decided to offer a new version of *Mathematical Problems in Plasticity*, which was first published in French in 1983 and in English in 1985, by Gauthier-Villars.

As far as we know, no other titles on this subject have been written since then, although a number of books have appeared on the computational side or closer to the engineering side. Hence, this volume remains, we hope, useful.

I have not been extensively involved in such mathematical problems after the publication of this book, but many important developments have occurred, which are reviewed in the new appendix written in collaboration with Jean-François Babadjian. In brief, the new developments listed in this appendix include:

- plasticity as a variational problem;
- softening phenomena;
- nonassociative plasticity;
- multiscale plasticity problems;
- dynamic models;
- regularity of solutions;
- finite plasticity; and
- other problems.

I would like to thank Jean-François for preparing the appendix and for gathering an extensive bibliography. I thank Elisa Davoli and Gilbert Strang for their comments. And, finally, I thank Dover for producing this book.

Roger Temam

Foreword

The object of this book is to study, from the mathematical standpoint the problem of the equilibrium of a perfectly plastic body, under certain conditions. Although the present study is devoted to this problem which will be described in detail below, it seems highly probable that the tools and methods used should serve in other fields and in particular in solving problems relating to the 'evolution' of plastic phenomena, the mechanics of fracture, and in a quite different sphere of ideas, in certain optimal control problems.

The plasticity problem considered here is that of determining the equilibrium of a three-dimensional body occupying a volume Ω of $\mathbb{R}^3$, when that body is in equilibrium under a set of certain body forces of volume density f, and certain surface forces of surface density g, applied on a part Γ_1 of its boundary Γ. The unknowns of the problem are, on the one hand, the field $u = u(x)$ which defines the displacement from the unconstrained equilibrium position of the body at the point x, and, on the other hand the stress field $\sigma=\sigma(x)$ (which is a field of symmetric tensors of order 2 defined over Ω); lastly the strain or displacement field u is given to be equal to u_0 on a portion Γ_0 of Γ, complementary to Γ_1. Under certain hypotheses, and in particular on the hypothesis that the material behaves as a perfect plastic, the fields u and σ are respectively the solutions of two well-known variational problems, which we shall call *the strain problem* (or displacement problem) and *the stress problem*; the strain problem is also known as the Hencky problem.

These two problems constitute *two problems in the calculus of variations* whose solution has been brought to a successful conclusion only very recently ([1]); the purpose of this volume is to give an up-to-date account of these developments.

The study of the stress problem presents few difficulties in itself: it can easily be shown that it has an unique solution except in the case where the set of admissible stresses is empty (this can happen, depending on the forces f and g, under conditions to be explained and more precisely defined later). The strain problem, in marked contrast, presents formidable difficulties which have only recently been overcome. In particular:

- The infimum of the strain problem, which we denote by $\mathcal{P}$, can be equal to $-\infty$.
- When the infimum is finite $(> -\infty)$, the minimising sequences of $\mathcal{P}$ are, in the best case, bounded in a space which is non-reflexive and it is not possible to extract subsequences which converge for convenient topologies.
- When plastification occurs on Γ_0, one must expect, on mechanical grounds, that the *boundary condition* $u = u_0$ *will not in fact be satisfied on* Γ_0; it can also be foreseen, from the mechanical standpoint, that sliding or slippages can occur in the interior, which would correspond mathematically to the displacement field u presenting discontinuities on some surface interior to Ω. This would exclude the possibility that u should belong to a space of continuous functions, and even rules out, because of the trace theorems, a Sobolev space. The construction of a suitable function space is therefore one of our main objectives.

Our work is essentially concerned with these two areas. Firstly we have to pinpoint the relationship between the stress problem and the strain problem; they are dual to each other and the duality is easily described with the help of the general methods available in convex analysis (cf. W. Fenchel [1][2][3], R.T. Rockafellar [1][2], I. Ekeland-R. Temam [1]). Secondly we make a study in depth of the

---

([1]) Numerous bibliographical references will be given at appropriate points throughout the book and in the Bibliography at the end.

strain problem $\mathcal{P}$; this is done by introducing a generalised (or relaxed) problem $\mathcal{Q}$ which is related to $\mathcal{P}$ in the following way:

1) $\mathcal{Q}$ is an extension of $\mathcal{P}$ ($\mathcal{Q} \supset \mathcal{P}$), that is to say it corresponds to minimising the same functional, but over a larger space.

2) The infimum of $\mathcal{Q}$ is the same as that of $\mathcal{P}$ (it is obviously less than or equal to it and we prove that strict inequality cannot hold).

3) The problem $\mathcal{Q}$ always has solutions (save in an exceptional limiting case), and the solutions of $\mathcal{Q}$ are in particular the limits or cluster points, for certain topologies, of the minimising sequences of $\mathcal{P}$.

To define the problem $\mathcal{Q}$, it has been necessary to introduce and study a new function space, the space $BD(\Omega)$ which is the space of summable (i.e. Lebesgue-integrable) functions from Ω into $\mathbb{R}^3$ whose deformation is a bounded measure on Ω (1), that is to say $BD(\Omega)$ is the space of functions v such that:

$v = (v_1, v_2, v_3) \in L^1(\Omega)^3$, and

$\varepsilon_{ij}(v) = \frac{1}{2}\left(\frac{\partial v_i}{\partial x_j} + \frac{\partial v_j}{\partial x_i}\right)$ is a bounded measure on Ω for $i,j = 1,2,3$.

This space has been introduced and studied in particular by H. Matthies-G. Strang-E. Christiansen [1], P. Suquet [1]-[7], G. Strang-R. Temam [1][3], R. Temam [9][11]. It allows us to define and characterise the *finite-energy spaces*, those for which the energy of the plastic deformation has a finite value. Let us suppose, for simplicity, that the plastic deformation energy associated with a displacement field v is given by

$$\int_\Omega |\varepsilon(v)(x)|dx = \int_\Omega \left\{\sum_{i,j=1}^{3} |\varepsilon_{ij}(v)(x)|^2\right\}^{1/2} dx .$$

A simplified form of the problem of characterising the associated finite-energy space is embodied in the following question:

---

(1) BD for bounded deformation by analogy with BV for bounded variation.

What can be said about a sequence of regular functions v_m *from* Ω *into* $\mathbb{R}^3$, *which converge uniformly (or in* $L^1(\Omega)^3$*) to a limit* v, *and which are such that the associated 'energy of deformation' remains bounded:*

$$\int_\Omega |\varepsilon(v_m)|dx \leq \textit{constant}$$

The introduction of the space $BD(\Omega)$ enables us to answer this question easily (cf. Chapter II where the question is posed and answered in another form better suited to the problem being dealt with)

* * *

We now turn to a description of the contents in sequential order. The book is divided into three Chapters. Chapter I begins by recalling the necessary parts of the theory in functional analysis, convex analysis, and mechanics and gives a precise definition of the strain problem and the stress problem. The remainder of the chapter contains *a systematic study of these two problems using the methods of convex analysis*. The duality of these problems is made clear; an auxiliary variational problem is introduced, which we call the *Limit Analysis* problem, which enables us to tell whether or not the infimum of $\mathscr{P}$ is finite $(> -\infty)$, and by dual considerations, whether there are any admissible stress fields for the associated stress problem. Chapter I finishes with the introduction and study of a partially relaxed version of $\mathscr{P}$, the problem denoted by $\mathscr{PR}$, in which the boundary constraint $u = u_0$ on Γ_0 is partially suppressed.

Chapter II has, as its object, the definition and study of the strai problem $\mathscr{Q}$, which generalises the initial problem $\mathscr{P}$. It contains a rather full investigation of the function space $BD(\Omega)$ and certain other allied function spaces. It also introduces and studies the concept of convex functions of measures. The object here is to generalise the measures $|\mu|$, μ^+, μ^- associated with a given bounded measure μ : if f is a suitable convex function defined on $\mathbb{R}^m$ and μ a bounded measure with values in $\mathbb{R}^m$, it is possible, under certain conditions, to define $f(\mu)$ as a bounded measure with values in $\mathbb{R}$; this applies for example when $f(s) = \sqrt{1 + s^2}$ and $\langle\langle f(\mu) = \sqrt{1 + \mu^2}\rangle\rangle$

Once these preliminary tools have been constructed we give in Chapter II, the definition of problem $\mathcal{Q}$. We study its relations with $\mathcal{P}$ and the stress problem; finally we prove that $\mathcal{Q}$ has a solution.

Chapter III gives further developments of two kinds. First we consider models of 'imperfectly plastic' type and study their convergence towards the perfectly plastic model: this involves therefore the study of perturbed variational problems, in respect of both stress and strain, and the study of their limit as a perturbation parameter tends to zero. Secondly we give a study, analogous to that made in Chapters I and II, of a model of perfectly plastic plates. The study which we present (following the treatment by F. Demengel [1]-[3]), is based on the introduction of a modified function space $HB(\Omega)$, the space of functions with bounded Hessian; it is the analogue here of the space $BD(\Omega)$. This allows one to deal with, for example, the deformation of *plates which bend* (forming one or more folds) without breaking.

* * *

The author is well aware of the limitations, from the mechanical point of view, of the model studied: the Hencky model which has more in common with a model of nonlinear elasticity with threshold (cf. P.Germain [3]) provides a very unsatisfactory representation of the phenomena involved in plasticity. It would be preferable to start studying the evolutionary model of Prandtl-Reuss in the form of the quasi-static approximation or in the general form. We think however that at least some of the research and techniques developed here represents an indispensable first step towards a study of the Prandtl-Reuss model. Moreover the introduction of the space $BD(\Omega)$ and certain of the developments which flow from it appear to provide a natural framework for problems in solid mechanics where discontinuities appear; it could therefore well prove useful in the theoretical study of problems in the mechanics of fracture. *From the numerical viewpoint*, it seems that our approach could be useful: in calculations up to the breaking point the use of convex analysis allows more precise results than usual to be obtained (with conditions that are both necessary and sufficient) and the formulations in Chapter II can take account of surfaces of discontinuity; the first numerical calculations based on these ideas were done by D. Clement [1] and F. Crépel [1]. Finally, as we mentioned earlier, *certain optimal control problems* appear to be amenable to analogous methods in so far as the investigating and study of necessary conditions for optimality is concerned; this subject will be developed elsewhere but the interested reader may already care to consult J.L. Lions [3] and R. Temam [12].

In conclusion I should like to thank all those who, in one way or another, have contributed to this work or have encouraged it. I thank Monsieur Paul Germain for the interest which he has shown in this work, in particular on the occasion of the publication of certain articles which form the underlying basis for this book, and also in the reference P. Germain [3]. I thank Monsieur Jacques-Louis Lions for his interest and for having accepted this book in the series for which he is responsible.

I should also like to thank Gilbert Strang for a friendly collaboration which has been both very agreeable and fruitful, and for his interest during the preparation of this book; I likewise thank Robert Kohn who collaborated in an article which plays a very useful role at the end of Chapter II. Françoise Demengel and Pierre Suquet have been kind enough to read parts of the manuscript and I thank them for their comments. Finally I thank Madame Le Meur and Madame Maynard who kindly dealt with the typing of the text.

Chapter I
Variational problems in plasticity theory

Introduction

In this first chapter we recall various subjects in functional analysis, convex analysis, and in mechanics which will constantly be required in what follows. Next we formulate the variational problems relating to perfect plasticity which form the main theme of this book. Finally the rest of the chapter is devoted to a study of these problems which makes systematic use of the tools of convex analysis.

In Section 1 we remind the reader of some of the basic concepts of functional analysis and function spaces of the Sobolev type and of a few additional results. Section 2 recalls some results from convex analysis: properties of conjugate convex functions, variational problems in their dual aspects. Examples taken from the calculus of variations are given and in particular a study is made of the dual variational problems of three-dimensional linear elasticity theory, namely the problems whose solutions give respectively the strain field and the stress field. Section 3 recalls some fundamental ideas in the mechanics of continuous media, and in particular those involved in writing down the equations of equilibrium and in formulating the constitutive laws.

From Section 4 onwards and throughout this chapter and the next, we concentrate on studying the variational problems for the strain, and for the stress, in the Hencky plasticity model *(the static problem)*. Section 4 gives a systematic study of the duality between the strain problem and the stress problem. Section 5 deals with Limit Analysis. An auxiliary variational problem is introduced, the Limit Analysis problem, which gives a criterion which can be used to decide whether the infimum of the strain problem is finite or not, and dually, whether the stress problem possesses admissible states of stress. Although the general idea of limit analysis has long been known, and is routinely used in mechanics in plastic analysis of structures, fracture mechanics and yield calculations, it seems that the limit analysis problem described here (introduced by G. Strang and the author) is new (at least in the form given here). It should also be pointed out incidentally that our study, based on convex analysis, gives conditions which are both necessary *and* sufficient, not merely necessary *or* sufficient conditions, as is often the case in limit analysis.

Finally Section 6, which ties in with the rest of the Chapter from the point of view of the *methods* (the systematic use of duality and convex analysis), is a first step towards the realisation of the goals of Chapter II; the variational problem, as formulated in Section 4, has, in general, no solution and it is a generalised form of the problem which will be solved in Chapter II. In Section 6 we consider a first generalisation of the problem, corresponding to a partial relaxation of the boundary conditions; we describe in Section 6 this relaxed problem and study its relationship with the initial strain problem.

1. Function spaces

1.1 Elementary concepts

We denote by $x = (x_1,x_2,\ldots,x_n)$, $y = (y_1,y_2,\ldots,y_n)$, $z = (z_1, z_2,\ldots,z_n)$, ... , the points of the Euclidean space $\mathbb{R}^n$, and call $e_1 = (1,0,\ldots,0)$, $e_2 = (0,1,\ldots,0)$, ... , $e_n = (0,\ldots,0,1)$, the canonical basis of $\mathbb{R}^n$.

The differential operator $\partial/\partial x_i$ $(1\leq i\leq n)$ will be denoted by D_i and if $j = (j_1,\ldots,j_n)$ is a set of non-negative integers, D^j denotes the differential operator

$$D^j = D_1^{j_1}\ldots D_n^{j_n} = \frac{\partial^{[j]}}{\partial x_1^{j_1}\ldots\partial x_n^{j_n}} ,$$

of order

$$[j] = j_1+\ldots+j_n . \qquad (1.1)$$

When $j_i = 0$, $D_i^{j_i}$ represents the identity operator.

When X is a subset of $\mathbb{R}^n$, we write $\mathscr{C}(X)$ for the space of real continuous functions defined on X . When Ω is an open subset of $\mathbb{R}^n$, we shall need to consider in addition to the spaces $\mathscr{C}(\Omega)$ and $\mathscr{C}(\bar{\Omega})$ the spaces $\mathscr{C}^k(\Omega)$ and $\mathscr{C}^k(\bar{\Omega})$, that is the spaces of real continuous functions, defined on Ω and $\bar{\Omega}$ respectively, which are k times continuously differentiable; $\mathscr{C}_0^k(\Omega)$ and $\mathscr{C}_0(\Omega)$ are the subspaces of $\mathscr{C}^k(\Omega)$ and $\mathscr{C}(\Omega)$ respectively, comprising the functions with compact support in Ω ; the modification required to these definitions when $k = \infty$ is obvious; when k is finite and Ω

bounded, it is elementary that $\mathscr{C}^k(\bar{\Omega})$ is a Banach space for the norm defined by

$$\|u\|_{\mathscr{C}^k(\bar{\Omega})} = \sum_{[j]\leq k} \sup_{x\in\bar{\Omega}} |D^j u(x)| .$$

We shall sometimes also write $\mathscr{D}(\Omega)$ for the space $\mathscr{C}_0^\infty(\Omega)$, as in the Theory of Distributions by L. Schwartz [1].

If Ω is any open subspace of $\mathbb{R}^n$ and $1\leq p<\infty$, we call $L^p(\Omega)$ the space of (classes of) real functions on Ω which are L^p for the Lebesgue measure $dx = dx_1 \ldots dx_n$. It is a Banach space for the norm

$$(1.2) \qquad \|u\|_{L^p(\Omega)} = \left(\int_\Omega |u(x)|^p dx\right)^{1/p} .$$

For $p = \infty$, $L^\infty(\Omega)$ is the space of (classes of) real functions on Ω which are measurable and essentially bounded; it is also a Banach space for the norm

$$(1.3) \qquad \|u\|_{L^\infty(\Omega)} = \underset{x\in\Omega}{\text{sup.ess.}} |u(x)| .$$

When $p = 2$, $L^2(\Omega)$ is a Hilbert space for the scalar product

$$(1.4) \qquad (u,v) = \int_\Omega u(x)\, v(x)\, dx ,$$

the corresponding norm being then denoted by $|u|$,

$$|u| = \|u\|_{L^2(\Omega)} = \{(u,u)\}^{1/2} .$$

We shall now very briefly recall a few standard techniques and concepts.

Duality and weak convergence

If X is a Banach space, X' denotes its dual which is also a

Banach space, and X'' denotes the bidual of X, that is to say the dual of the dual of X (also known as the second conjugate of X). The space X can be identified with a subspace of X'', and the Banach space X is said to be reflexive when $X = X''$. Thus, for $1 \leq p < \infty$, the dual of $L^p(\Omega)$ is $L^{p'}(\Omega)$ where $p' = p/(p-1)$ and for $1 < p < \infty$, $L^p(\Omega)$ is a reflexive space; $L^1(\Omega)$ and $L^\infty(\Omega)$ are however not reflexive spaces and neither are the spaces $\mathscr{C}^k(\bar{\Omega})$.

When X is a reflexive Banach space, it is possible to extract, from any sequence ϕ_m bounded in X a weakly convergent subsequence ϕ_{m_i}, or in other words there exists a $\phi \in X$ such that, for all $\psi \in X'$:

$$\langle \phi_{m_i}, \psi \rangle \to \langle \phi, \psi \rangle, \quad \text{for } m_i \to \infty. \tag{1.5}$$

Similarly if X is a Banach space, not necessarily reflexive, whose dual is X', then any sequence ψ_m bounded in X' contains a weak-star (weak*) convergent subsequence ψ_{m_i}, or in other words there exists a $\psi \in X'$ such that, for all ϕ belonging to X

$$\langle \psi_{m_i}, \phi \rangle \to \langle \psi, \phi \rangle, \quad \text{for } m_i \to \infty. \tag{1.6}$$

Thus if u_m is a bounded sequence in $L^p(\Omega)$, and $1 < p \leq \infty$ there exists a $u \in L^p(\Omega)$ and a subsequence u_{m_i} such that, for all $v \in L^{p'}(\Omega)$,

$$\int_\Omega u_{m_i}(x)\, v(x)\, dx \to \int_\Omega u(x)\, v(x)\, dx, \quad \text{for } m_i \to \infty. \tag{1.7}$$

The result (1.7) does not hold for $p = 1$, but a weaker result, which does hold in this case will be recalled and used in Chapter II.

If a sequence u_m converges in norm to u in $L^p(\Omega)$, $1 < p \leq \infty$ then it also converges weakly to u (the convergence is weak star if $p = \infty$), but the converse statement is not true.

Convolution and regularisation

Convolution provides various constructive processes for approximating non smooth functions by $\mathscr{C}^\infty$ functions. A sequence of regularising functions ρ_n can be constructed as follows: let ρ be a real $\mathscr{C}^\infty$

function with compact support in $\mathbb{R}^n$, so that $\rho \in \mathscr{C}_0^\infty(\mathbb{R}^n) = \mathscr{D}(\mathbb{R}^n)$; we shall suppose, for convenience, that $\rho(x) \geq 0$, that $\rho(x) = 0$ for $|x| \geq 1$, and that $\int_{\mathbb{R}^n} \rho(x)dx = 1$; for a fixed $\eta>0$ we put $\rho_\eta(x) = \frac{1}{\eta^n}\rho\left(\frac{x}{\eta}\right)$, so that we have $\int_{\mathbb{R}^n} \rho_\eta(x)dx = 1$ and $\rho_\eta(x) = 0$ for $|x| \geq \eta$.

For each u in $L^p(\mathbb{R}^n)$, $1\leq p<\infty$, (for example), we define $\rho_\eta * u$ by:

(1.8) $$(\rho_\eta * u)(x) = \int_{\mathbb{R}^n} \rho_\eta(x-y)\, u(y)\, dy\ .$$

It is well known that $\rho_\eta * u$ is a $\mathscr{C}^\infty$ function on $\mathbb{R}^n$ and that, if $1\leq p<\infty$,

(1.9) $$\rho_\eta * u \to u \quad \text{in} \quad L^p(\mathbb{R}^n)\ , \quad \text{when} \quad \eta \to 0\ .$$

This implies that $\mathscr{C}^\infty(\mathbb{R}^n)$ is dense in $L^p(\mathbb{R}^n)$. Similarly it can be shown in the same way that for $1\leq p<\infty$

(1.10) $$\mathscr{C}_0^\infty(\Omega) = \mathscr{D}(\Omega) \quad \text{is dense in} \quad L^p(\mathbb{R}^n)\ .$$

Thus let $\Omega_{1/m}$ be the set of points x of Ω whose distance from Γ, the boundary of Ω, exceeds $1/m$, where m is any sufficiently large integer: then there exists a $\psi_m \in \mathscr{C}_0^\infty(\Omega)$ such that $0 \leq \psi_m \leq 1$, and $\psi_m(x) = 1$ on $\Omega_{1/m}$. If $u \in L^p(\Omega)$, $1\leq p<\infty$, then $\psi_m u \in L^p(\Omega)$ and converges in norm to u, in $L^p(\Omega)$ as m tends to infinity (by Lebesgue's theorem). If $\eta > 0$ is small enough $\eta < \eta(m)$ then $\rho_\eta * (\psi_m u)$ is in $\mathscr{C}_0^\infty(\Omega)$; when η tends to zero, $\rho_\eta * (\psi_m u)$ tends to $\psi_m(u)$ in $L^p(\Omega)$ by (1.9) and thus approximates u.

Weak derivatives

The reader will be assumed to be fairly familiar with the theory of distributions of L. Schwartz [1], but we shall nevertheless recall the definition of a weak derivative resulting from this theory.

If $u \in L^p(\Omega)$, $v \in L^q(\Omega)$, $1\leq p,q\leq\infty$, $q = p$ or $q \neq p$, we say that $v = \partial u/\partial x_i$ in the weak sense, or in the sense of the theory of distributions on Ω, if

$$(1.11)\qquad \int_\Omega u(x)\,\frac{\partial\phi}{\partial x_i}(x)\,dx = -\int_\Omega v(x)\,\phi(x)\,dx\ ,\ \forall\,\phi\in\mathscr{C}_0^\infty(\Omega)\ .$$

More generally, let $\mathscr{A}(D)$ be a differential operator of finite order, whose coefficients a_j are regular functions on Ω,

$$(1.12)\qquad \mathscr{A}(D) = \sum_{[j]\le m} a_j(x)\,D^j\ .$$

If $u\in L^p(\Omega)$ and $v\in L^q(\Omega)$, $1\le p,q\le\infty$, we shall say that

$$(1.13)\qquad \mathscr{A}(D)u = v$$

in the sense of distributions on Ω, if:

$$(1.14)\qquad \sum_{[j]\le m}\int_\Omega (-1)^{[j]}u\,D^j(a_j\phi)\,dx = \int_\Omega v\phi\,dx\ ,\ \forall\,\phi\in\mathscr{C}_0^\infty(\Omega)\ .$$

It follows easily from definition (1.11) that if $v_m = \partial u_m/\partial x_i$ in the distribution sense, $u_m\in L^p(\Omega)$, $v_m\in L^q(\Omega)$, and if $u_m\to u$ in $L^p(\Omega)$ and $v_m\to v$ in $L^q(\Omega)$ in norm or even weakly, then we also have $v = \partial u/\partial x_i$ in the sense of distributions on Ω.

We shall write $\mathscr{D}'(\Omega)$ for the space of scalar distributions on Ω.

1.2 Sobolev spaces

We recall here the main results and definitions regarding Sobolev spaces. For proofs and further results the reader may like to consult R.S. Adams [1], J.L. Lions-E. Magenes [1] and, in the case of $W^{1,1}(\Omega)$, E. Gagliardo [1].

Let Ω be a bounded open set in $\mathbb{R}^n$ with boundary Γ. It will sometimes be necessary to assume certain regularity conditions for the boundary Γ. We shall say that Ω is an open set of class $\mathscr{C}^r$ (where $r\ge 1$ is a natural number to be specified) if

(1.15) Γ is a variety of dimension $n-1$, of class $\mathscr{C}^r$, and Ω is situated locally on one side only of Γ.

We shall say that Ω is an open Lipschitzian set if Γ can be represented locally as the graph of a Lipschitzian function on an open set of $\mathbb{R}^{n-1}$, with Ω being situated locally on one side only of Γ. Finally, we may require that Ω (and also sometimes $\complement\bar{\Omega}$) should satisfy the cone condition. We mean by this (cf. Fig. 1.1) that:

(1.16) there exists a finite covering $(\mathscr{O}_i)_{i \in I}$ of the boundary Γ by a collection of open sets $\mathscr{O}_i$ and for each i, there exists an open cone $\mathscr{C}_i$ of vertex 0, such that $x + \mathscr{C}_i$ does not meet $\mathscr{O}_i \cap \Gamma$ for all $x \in \mathscr{O}_i \cap \Omega$

(or the corresponding condition with $\complement\bar{\Omega}$ in place of Ω).

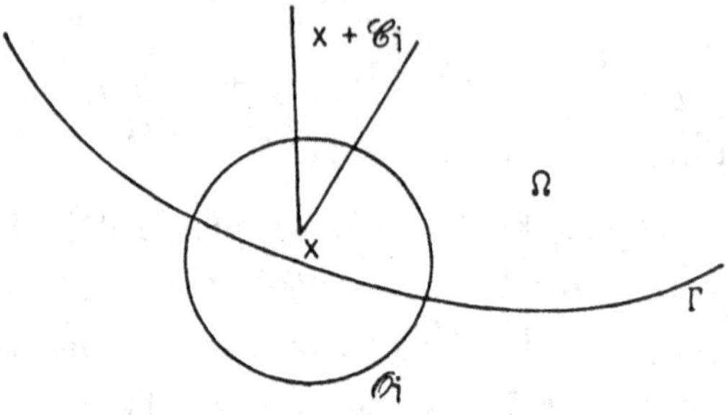

Fig. 1.1

If Ω is any open set of $\mathbb{R}^n$, the Sobolev space $H^1(\Omega)$ is the space of functions u in $L^2(\Omega)$ which have distributional derivatives (i.e. derivatives in the sense of the theory of distributions) $\partial u/\partial x_i$ in $L^2(\Omega)$, $i = 1,\dots,n$ (so that there exist $v_1,\dots,v_n$ in $L^2(\Omega)$ such that (1.11) is satisfied with v replaced by v_i).
$H^1(\Omega)$ is known to be a Hilbert space for the scalar product

(1.17)
$$((u,v))_{H^1(\Omega)} = (u,v) + \sum_{i=1}^{n} (D_i u, D_i v)$$

and the associated norm

(1.18) $$\|u\|_{H^1(\Omega)} = \{((u,u))_{H^1(\Omega)}\}^{1/2} .$$

It can be shown that if Ω satisfies the cone condition, then

(1.19) $\mathscr{C}^\infty(\bar{\Omega})$ is dense in $H^1(\Omega)$

and we have Sobolev's inclusion property:

(1.20) $H^1(\Omega) \subset L^q(\Omega)$ with continuous injection, for $1/q = 1/2 - 1/n$ if $n \geq 2$; and for any finite $q \geq 1$, if $n = 2$

and the compactness result:

(1.21) the injection $H^1(\Omega)$ into $L^q(\Omega)$ is compact for any q such that $1 \leq q < \frac{2n}{n-2}$.

If Ω is an open set of class $\mathscr{C}^1$ at least, we define the space $L^2(\Gamma)$ of (classes of) functions which are real-valued on Γ , and are L^2 for the area measure $d\Gamma$, and we have the trace theorem:

(1.22) there exists a linear continuous mapping γ_0 of $H^1(\Omega)$ into $L^2(\Gamma)$ such that $\gamma_0 u = u_{|\Gamma}$ for all u in $\mathscr{C}^1(\bar{\Omega})$.

The image of $H^1(\Omega)$ is written $H^{1/2}(\Gamma)$; it is a subspace of $L^2(\Gamma)$ which is a Hilbert space for the structure induced by γ_0 . When there is no risk of confusion, we shall simply write u instead of $\gamma_0 u$, the trace of u on Γ .

For an arbitrary bounded open set Ω , we call $H_0^1(\Omega)$ the closure in $H^1(\Omega)$ of the functions $\mathscr{C}^\infty$ with compact support in Ω . By Poincaré's inequality, there exists a finite constant $c_0 = c_0(\Omega) > 0$, depending only on Ω , such that

(1.23) $$\int_\Omega |u(x)|^2 \, dx \leq c_0(\Omega) \sum_{i=1}^n \int_\Omega |D_i u(x)|^2 \, dx ,$$

for all u in $\mathscr{C}_0^\infty(\Omega)$ and, by continuity, for all u in $H_0^1(\Omega)$. Consequently

$$((u,v)) = \sum_{i=1}^{n} (D_i u, D_i v) , \tag{1.24}$$

is an Hilbertian scalar product on $H_0^1(\Omega)$, and the associated norm

$$\|u\| = \{((u,u))\}^{1/2}$$

is equivalent to the norm $\|.\|_{H^1(\Omega)}$ induced by $H^1(\Omega)$.

If Ω is a bounded open set of class $\mathscr{C}^1$, then $H_0^1(\Omega)$ coincides with the kernel of γ_0. We call $H^{-1}(\Omega)$ the dual of $H_0^1(\Omega)$; it is a distribution space and $L^2(\Omega)$ can be identified with a dense subspace of $H^{-1}(\Omega)$. The derivation operators D_i are continuous operators from $L^2(\Omega)$ into $H^{-1}(\Omega)$.

More generally, if p is a finite or infinite number ≥ 1, and m an integer ≥ 1, the Sobolev space $W^{m,p}(\Omega)$ is defined as the space of functions u belonging to $L^p(\Omega)$ which have distribution derivatives $D^j u$ in $L^p(\Omega)$ for any finite integer set j, such that $[j] \leq m$. $W^{m,p}(\Omega)$ is a Banach space when it is endowed with the norm

$$\|u\|_{W^{m,p}(\Omega)} = \sum_{[j]\leq m} \|D^j u\|_{L^p(\Omega)} . \tag{1.25}$$

For $p = 2$ we write $H^m(\Omega)$ in place of $W^{m,p}(\Omega)$ and $H^m(\Omega)$ is endowed with the scalar product and norm defined by

$$\begin{cases} ((u,v))_{H^m(\Omega)} = \sum_{[j]\leq m} (D^j u, D^j v) , \\ \|u\|_{H^m(\Omega)} = \{((u,u))_{H^m(\Omega)}\}^{1/2} \end{cases} \tag{1.26}$$

which make it a Hilbert space.

The properties (1.19) to (1.21) can be generalised as follows: if Ω satisfies the cone condition and $1 \leq p < \infty$, then

(1.27) $\mathscr{C}^\infty(\bar{\Omega})$ is dense in $W^{m,p}(\Omega)$,

(1.28) $W^{m,p}(\Omega) \subset L^q(\Omega)$ with continuous injection, where $1/q = 1/p - m/n$ if $n > mp$; and for any $q \geq 1$ if $n = mp$.

(1.29) $W^{m,p}(\Omega) \subset \mathscr{C}^k(\bar{\Omega})$ with continuous injection, if $\frac{n}{p} + k < m \leq \frac{n}{p} + k + 1$ and k is any non-negative integer.

(1.30) The injection of $W^{m,p}(\Omega)$ into $L^q(\Omega)$ is compact, for any q satisfying $1 \leq q < \frac{np}{n-mp}$ if $n > mp$; and any $q \geq 1$ if $n \leq mp$.

If Ω is an open set of class $\mathscr{C}^{m+1}$, we have trace theorems analogous to (1.22):

(1.31) There exist linear continuous mappings γ_j, $j = 0,\ldots,m-1$, of $W^{m,p}(\Omega)$ into $L^p(\Gamma)$ such that $\gamma_j u = \left.\frac{\partial^j u}{\partial \nu^j}\right|_\Gamma$ for all u in $\mathscr{C}^m(\bar{\Omega})$,

where ν denotes the unit normal on Γ, exterior to Ω.

We shall be more particularly concerned with the space $W^{1,1}(\Omega)$; in this case, we can, following E. Gagliardo [1], complete (1.31) by:

(1.32) The trace mapping γ_0 is a surjective mapping from $W^{1,1}(\Omega)$ on to $L^1(\Gamma)$.

1.3 Vector and tensor function spaces

Let X be a finite-dimensional Euclidean space with scalar product $\xi.\eta$ and norm $|\xi| = \{\xi.\xi\}^{1/2}$; the main examples will be $X = \mathbb{R}^n$, and $X = E_n$ the space of symmetric tensors of order n

carrying the scalar product defined by

$$\xi.\eta = \sum_{i,j=1}^{n} \xi_{ij}\,\eta_{ij} \,. \tag{1.33}$$

When there is no risk of confusion we shall simply write E for E_n.

When using tensor notations we shall adopt the Einstein summation convention in which repeated indices imply summation over the index range, so that for example

$$\begin{cases} |\xi|^2 = \xi_i\,\xi_i = \sum_{i=1}^{n} \xi_i^2 \quad \text{if} \quad \xi \in \mathbb{R}^n \,, \\ \xi.\eta = \xi_{ij}\,\eta_{ij} = \sum_{i,j=1}^{n} \xi_{ij}\,\eta_{ij} \,, \quad \text{if} \quad \xi,\eta \in E_n = E \,, \end{cases} \tag{1.34}$$

and the following notation for partial derivatives will be used:

$$\phi_{,j} = D_j\phi = \frac{\partial\phi}{\partial x_j} \,. \tag{1.35}$$

We write E^D for the subspace of E consisting of the tensors whose trace is zero, $E = E^D \oplus \mathbb{R}\,I$; if $\xi \in E$, ξ^D is its <u>*deviator*</u> whose components are

$$\xi^D_{ij} = \xi_{ij} - \frac{1}{n}\,\xi_{kk}\,\delta_{ij} \qquad (\delta_{ij} \text{ is the Kronecker delta})$$

We write $\mathscr{C}^k(\Omega;X)$ and $\mathscr{C}^k(\bar\Omega;X)$ for the spaces of functions from Ω to X and from $\bar\Omega$ to X respectively which are k times continuously differentiable; the space $\mathscr{C}^k(\bar\Omega;X)$ is, of course, a Banach space for the norm

$$\|u\|_{\mathscr{C}^k(\bar\Omega;X)} = \sum_{[j]\leq k} \ \operatorname{Sup}_{x\in\bar\Omega} |D^j u(x)| \,.$$

We write $L^p(\Omega;X)$ for the space of (classes of) functions from Ω into X which are L^p for the Lebesgue measure $dx = dx_1 \ldots dx_n$. It is a Banach space for the norm

$$\|u\|_{L^p(\Omega;X)} = \left(\int_\Omega |u(x)|^p\,dx\right)^{1/p} , \tag{1.36}$$

if $1 \leq p < \infty$; and, if $p = \infty$, for the norm

(1.37) $$\|u\|_{L^p(\Omega;X)} = \operatorname{ess\,sup}_{\Omega} |u(x)| .$$

For $p = 2$, $L^2(\Omega;X)$ is a Hilbert space for the scalar product and norm ([1])

$$(u,v) = \int_{\Omega} u(x).v(x)\,dx$$

$$|u| = \left(\int_{\Omega} |u(x)|^2\,dx\right)^{1/2}$$

When $X = \mathbb{R}^n$ we shall write $\mathscr{C}^k(\Omega;X) = \mathscr{C}^k(\Omega)^n$, $\mathscr{C}^k(\bar{\Omega};X) = \mathscr{C}^k(\bar{\Omega})^n$, $L^p(\Omega;X) = L^p(\Omega)^n$, etc ...

□

Certain function spaces will play a special role later on. We shall no describe these briefly; they will be studied in more detail as the need arises.

Spaces related to the divergence operator

The space defined below is described and studied in detail in R. Temam [6]

(1.38) $$Y = \{u \in L^2(\Omega)^n \text{ , } \operatorname{div} u \in L^2(\Omega)\} .$$

It is a Hilbert space for the scalar product

(1.39) $$(u,v) + (\operatorname{div} u, \operatorname{div} v) = \int_{\Omega} \big(u(x).v(x) + \operatorname{div} u(x).\operatorname{div} v(x)\big)\,dx .$$

The main properties we shall need are: if $\complement\bar{\Omega}$ satisfies the cone condition (cf. Proposition 1.3 page 23) then

(1.40) $\mathscr{C}^\infty(\bar{\Omega})^n$ is dense in that space.

([1]) It should generally be clear enough from the context which of the various different norms denoted by $|.|$ is meant. For example if u is a function from Ω into X , then $|u|$ is the norm of u in $L^2(\Omega;X)$ and $|u(x)|$ is the norm in X of $u(x)$.

If Ω is of class $\mathscr{C}^1$, we can define the trace of $u.\nu$ on Γ by using the following result.

(1.41) There exists a linear continuous mapping γ_ν of the space Y defined by (1.38), into $H^{-1/2}(\Gamma)$ and $\gamma_\nu u = u.\nu_{|\Gamma}$ for all u in $\mathscr{C}^1(\bar{\Omega})^n$;

$H^{-1/2}(\Gamma)$ is the dual of the space $H^{1/2}(\Gamma)$ (cf. (1.22) and the remark following (1.22)). Furthermore we have the generalised Green's formula

(1.42) $$(u,\text{grad } v) + (\text{div } u,v) = \langle\gamma_\nu u,\gamma_0 v\rangle ,$$

for u in the space (1.38) and $v \in H^1(\Omega)$,

where $\langle .,.\rangle$ denotes the inner product between $H^{1/2}(\Gamma)$ and $H^{-1/2}(\Gamma)$. When there is no risk of confusion we shall write $u.\nu$ instead of $\gamma_\nu u$.

Similarly we can consider the space

(1.43) $$Y = \{\sigma \in L^2(\Omega;E_n) , \text{ div } \sigma \in L^2(\Omega)^n\}$$

where $\text{div } \sigma = \nabla.\sigma$ is the vector whose components are

(1.44) $$\sum_{j=1}^{n} D_j \sigma_{ij} , \quad i = 1,\ldots,n .$$

This is a Hilbert space for the scalar product

(1.45) $$(\sigma,\tau) + (\text{div } \sigma,\text{div } \tau) = \int_\Omega \{\sigma(x).\tau(x) + \text{div } \sigma(x).\text{div } \tau(x)\} \, dx .$$

It can be shown (cf. Proposition 1.3, page 23) that, if $\complement\bar{\Omega}$ satisfies the cone condition, then

(1.46) $\mathscr{C}^\infty(\bar{\Omega};E_n)$ is dense in the space (1.43)

Incidentally we note that if σ is in the space (1.43) each of the row-vectors

$$\sigma_{i.} = \{\sigma_{i1},\ldots,\sigma_{in}\} , \quad i = 1,\ldots,n ,$$

is in the space (1.38). If Ω is of class $\mathscr{C}^1$ it follows from (1.41) and (1.42) that:

(1.47) There exists a continuous linear mapping of the space (1.43) into $H^{-1/2}(\Gamma)^n$ (again denoted by γ_ν) such that $\gamma_\nu\sigma = \sigma.\nu_{|\Gamma}$ for all σ in $\mathscr{C}^1(\bar{\Omega};E_n)$, where $\sigma.\nu$ is the vector whose components are

$$(1.48) \qquad \{\sum_{j=1}^{n} \sigma_{1j}\,\nu_j, \ldots, \sum_{j=1}^{n} \sigma_{nj}\,\nu_j\} .$$

In addition the generalised Green's formula

$$(1.49) \qquad \sum_{i,j=1}^{n} (\sigma_{ij}, D_i u_j) + (\operatorname{div} \sigma, u) = \langle \gamma_\nu \sigma, \gamma_0 u \rangle ,$$

holds for any σ in the space (1.43) and any u in $H^1(\Omega)^n$.

If u is a vector function on Ω, we can associate with it the tensor $\varepsilon(u)$ *(strain tensor)* whose components are defined by

$$(1.50) \qquad \varepsilon_{ij}(u) = \frac{1}{2}\left(\frac{\partial u_j}{\partial x_i} + \frac{\partial u_i}{\partial x_j}\right)$$

If u is in $H^1(\Omega)^n$, $\varepsilon(u)$ is in $L^2(\Omega;E_n)$ and under the same assumptions as in (1.49) the latter relation can be written in the form

$$(1.51) \qquad \big(\sigma,\varepsilon(u)\big) + (\operatorname{div} \sigma, u) = \langle \gamma_\nu \sigma, \gamma_0 u \rangle = \text{"}\langle \sigma.\nu_{|\Gamma}, u_{|\Gamma}\rangle\text{"} .$$

<u>Spaces related to the operator ε</u>

The operator ε having been defined above, it is natural to introduce in linear elasticity theory the space

$$(1.52) \qquad \{u \in L^2(\Omega)^n,\ \varepsilon_{ij}(u) \in L^2(\Omega),\ i,j = 2,\ldots,n\} .$$

This is a Hilbert space for the scalar product

$$(1.53) \quad (u,v) + \big(\varepsilon(u),\varepsilon(v)\big) = \int_\Omega \{u(x).v(x) + \varepsilon(u)(x).\varepsilon(v)(x)\}\, dx$$

and if Ω satisfies the cone condition, then $\mathscr{C}^{\infty}(\bar{\Omega})^n$ is dense in this space.

When Ω is a bounded open Lipschitzian set it follows immediately from the Proposition 1.1 below that

(1.54) The space (1.52) is the same as $H^1(\Omega)^n$.

PROPOSITION 1.1 *(KORN's inequality)*

Let Ω be a bounded open Lipschitzian set. There exists a constant $c_1 = c_1(\Omega)$ depending only on Ω such that

$$(1.55)\quad \sum_{i,j=1}^{n} \int_{\Omega} |D_i u_j(x)|^2\, dx \leq c_1(\Omega) \int_{\Omega} \{|u(x)|^2 + |\varepsilon(u)(x)|^2\}\, dx ,$$

for all u belonging to the space (1.52).

This follows immediately from a closely related result which we shall use elsewhere:

PROPOSITION 1.2

Let Ω be a bounded open Lipschitzian set.

(i) If a distribution u has first order derivatives $D_i u$, $1 \leq i \leq n$, all of which are in $L^2(\Omega)$, then u is in $L^2(\Omega)$ and

$$(1.56)\quad |u|_{L^2(\Omega)/\mathbb{R}} \leq c(\Omega)|\text{grad}\, u| .$$

(ii) If a distribution u has first-order derivatives $D_i u$, $1 \leq i \leq n$ all of which are in $H^{-1}(\Omega)$, then $u \in L^2(\Omega)$ and

$$(1.57)\quad |u|_{L^2(\Omega)/\mathbb{R}} \leq c(\Omega)\|\text{grad}\, u\|_{H^{-1}(\Omega)^n} .$$

The conclusion (i) is established in Deny-Lions [1] under slightly stronger hypotheses for Ω , and in R. Temam [6] (cf. Prop. 1.2 Ch. I) under the hypotheses assumed here. The conclusion (ii) is proved in Magenes-Stampacchia [1] if Ω is of class $\mathscr{C}^1$ and in J. Nečas [1] when Ω is Lipschitzian.

Remark 1.1. Let p be a seminorm, continuous on $L^2(\Omega)$, and which is a norm on the constants. Then (1.56) and (1.57) obviously imply that there exists a constant $c'(\Omega)$, depending only on Ω, such that

$$|u| \leq c'(\Omega)\,\{p(u) + |\text{grad}\, u|\} \tag{1.58}$$

for all u in $L^2(\Omega)$ with grad $u \in L^2(\Omega)^n$ or respectively

$$|u| \leq c'(\Omega)\,\{p(u) + \|\text{grad}\, u\|_{H^{-1}(\Omega)^n}\}, \tag{1.59}$$

for all u in $L^2(\Omega)$ with grad $u \in H^{-1}(\Omega)^n$. For example:

$$p(u) = \left|\int_\Omega u(x)\,dx\right| \quad \text{or} \quad p(u) = \|u\|_{H^{-1}(\Omega)} \tag{1.60}$$

$$\text{or} \quad p(u) = \int_\Gamma |\gamma_0(u)|\,d\Gamma\,, \ \text{with } \Gamma_0 \subset \Gamma\,, \text{ where } \Gamma_0 \text{ has measure } > 0.$$

We deduce from this the

Proof of Proposition 1.1

We have, for all i, j, k,

$$D_{ik}\, u_j = D_i\, \varepsilon_{jk}(u) + D_k\, \varepsilon_{ij}(u) - D_j\, \varepsilon_{ik}(u)\,. \tag{1.61}$$

If u is in the space (1.52) then $D_i u_j$ is in $H^{-1}(\Omega)$ for all i and all j, and by (1.61)

$$D_k D_i u_j \ \text{ is in } H^{-1}(\Omega) \ \text{ for all } k\,. \tag{1.62}$$

It follows from Proposition 1.2 (ii) that $D_i u_j$ is in $L^2(\Omega)$, for all i, j, and therefore that $u \in H^1(\Omega)^n$. The inequality (1.54) is a consequence of (1.59) and (1.61):

$$\begin{aligned} |D_i\, u_j| &\leq c'(\Omega)\,\{\|D_i\, u_j\|_{H^{-1}(\Omega)} + \|\text{grad}\, D_i\, u_j\|_{H^{-1}(\Omega)^n}\} \\ &\leq c''.\{|u| + \sum_{k=1}^{n} (\|D_i\, \varepsilon_{jk}(u)\|_{H^{-1}(\Omega)} + \|D_k\, \varepsilon_{ij}(u)\|_{H^{-1}(\Omega)} + \\ &\qquad + \|D_j \varepsilon_{ik}(u)\|_{H^{-1}(\Omega)})\} \\ &\leq c'''\,\{|u| + |\varepsilon(u)|\}\,. \end{aligned}$$

Remark 1.2 KORN's inequality is proved in a different way in J.Gobert [1] on a less restrictive assumption about Ω (Ω satisfying the cone condition). □

It is interesting to find an explicit expression for the kernel of ε, which is called the *set of rigid displacements* and is denoted by $\mathcal{R}$:

LEMMA 1.1

The kernel of ε[1] *consists of functions* u *of the form*

$$u(x) = a + B.x , \tag{1.63}$$

where $a \in \mathbb{R}^n$, B *is a skew symmetric* $n \times n$ *matrix for general* n , *and if* $n = 3$,

$$u(x) = a + b \times x , \tag{1.64}$$

where $a, b \in \mathbb{R}^3$.

Proof

We set

$$\omega_{ij}(u) = \frac{1}{2} (D_i u_j - D_j u_i) \tag{1.65}$$

and note that

$$\begin{aligned} D_k \omega_{ij}(u) &= \frac{1}{2} (D_{kj} u_i - D_{ik} u_j) \\ &= \frac{1}{2} (D_{jk} u_i + D_{ij} u_k) - \frac{1}{2} (D_{ij} u_k + D_{ik} u_j) \\ &= D_j \varepsilon_{ik}(u) - D_i \varepsilon_{jk}(u) . \end{aligned} \tag{1.66}$$

---

(1) This applies to the kernel in the space of vector distributions $\mathcal{D}'(\Omega)^n$ and a fortiori in any smaller space.

If therefore $\varepsilon(u) = 0$, we have $D_k\omega_{ij}(u) = 0$ for all i, j, k from which it follows also that

$$D_{ik}\, u_j = D_k\, \omega_{ij}(u) = 0\,, \quad \forall\ i, j, k\,.$$

The functions u_j are affine, and $u(x) = a + B.x$ where B is an $n \times n$ matrix; lastly B is skew-symmetric because $\varepsilon(u)$ then reduces to the symmetric part of B. The passage from (1.63) to (1.64) is elementary when $n = 3$.

Remark 1.3 i) It follows from Lemma 1.1 and KORN's inequality that $|\varepsilon(u)|$ is a norm on $H^1(\Omega)^n/\mathscr{R}$, which is equivalent to the natural norm on that space. Likewise if p is a continuous seminorm on $H^1(\Omega)^n$, which is a norm on $\mathscr{R}$, then

$$p(u) + |\varepsilon(u)| \tag{1.67}$$

is a norm on $H^1(\Omega)^n$ equivalent to the initial norm. It is clear that (1.67) is a continuous norm on $H^1(\Omega)^n$ and if we can show that $H^1(\Omega)^n$ is complete for this norm then the equivalence of the two norms will follow from the closed graph theorem. Let us therefore prove the completeness in question: if u_m is a Cauchy sequence for the norm (1.67), then u_m satisfies Cauchy's condition in $H^1(\Omega)^n/\mathscr{R}$ and there exists therefore a sequence $r_m \in \mathscr{R}$ such that $u_m + r_m \to u$ in $H^1(\Omega)^n$. Now

$$p(r_m - r_k) \leqq p(u_m + r_m - u_k - r_k) + p(u_m - u_k)$$

so that r_m is a Cauchy sequence for p, r_m converges to r in $\mathscr{R}$ and u_m converges to $u-r$ in $H^1(\Omega)^n$.

ii) As in Remark 1.1, we can give some examples of continuous seminorms p which are norms on $\mathscr{R}$:

$$p(u) = |u|_{L^2(\Omega)^n} \quad \text{or} \quad p(u) = \|u\|_{H^{-1}(\Omega)^n}$$

$$\text{or} \quad p(u) = \int_{\Gamma_0} |\gamma(u)| d\Gamma ,$$

when Γ_0 is a measurable portion of Γ, of positive measure.

The space $LD(\Omega)$

For any number p satisfying $1 \leq p \leq \infty$, we can define a space, analogous to (1.52), by:

$$\{u \in L^p(\Omega)^n , \varepsilon_{ij}(u) \in L^p(\Omega) , \quad i,j = 1,\ldots,n\} . \tag{1.68}$$

It is a Banach space for the norm

$$\{\int_\Omega (|u(x)|^p + |\varepsilon(u)(x)|^p) dx\}^{1/p} . \tag{1.69}$$

If $1 < p < \infty$, and if Ω satisfies the cone condition, we have a KORN inequality (cf. J. Gobert [1] - [2]):

$$\sum_{i,j=1} \int_\Omega |D_i u_j(x)|^p dx \leq c(\Omega) \int_\Omega (|u(x)|^p + |\varepsilon(u)(x)|^p) dx \tag{1.70}$$

which holds good for all u in the space (1.68), which implies that this space is none other than $W^{1,p}(\Omega)^n$.

On the other hand in the case $p = 1$, which is important in the plasticity context, it follows from a counter-example given by D. Ornstein (1) [1] that KORN's inequality (1.70) is not true for

---

(1) D. Ornstein [1] shows that there is no constant $K < \infty$ such that

$$\int_\Omega |D_{12}u| dx \leq K \{\int_\Omega |D_1^2 u| dx + \int_\Omega |D_2^2 u| dx\} ,$$

for every function u in $\mathscr{C}_0^\infty(\Omega)$, where Ω is the square $(0,1)^2$ in $\mathbb{R}^2$: he proves in fact that there is a sequence u_n of functions of $\mathscr{C}_0^\infty(\Omega)$ such that, for all n:

$$\int_\Omega |D_{12}u_n| dx > n \{\int_\Omega |D_1^2 u_n| dx + \int_\Omega |D_2^2 u_n| dx\} .$$

If KORN's inequality were true in L^1, for the open set Ω, (or for any open superset of Ω) then, by considering the sequence $v_n = (D_1 u_n, D_2 u_n)$ and using the same argument as in Remark 1.1 ($p(v) = |\int_\Omega v\, dx| + \int_{\partial\Omega} |\frac{\partial v}{\partial \tau}| d\Gamma$ where $\frac{\partial v}{\partial \tau}$ is the tangential derivative on $\partial\Omega$) we should arrive at a contradiction.

$p = 1$, and that the space (1.68) is not the same as the space $W^{1,1}(\Omega)^n$; we shall therefore denote it by $LD(\Omega)$,

(1.71) $$LD(\Omega) = \{u \in L^1(\Omega)^n,\ \varepsilon_{ij}(u) \in L^1(\Omega),\ i,j = 1,\dots,n\}.$$

It is a Banach space for the natural norm

(1.72) $$\|u\|_{LD(\Omega)} = \|u\|_{L^1(\Omega)^n} + \sum_{i,j=1}^{n} |\varepsilon_{ij}(u)|_{L^1(\Omega)}.$$

It follows from Proposition 1.3, page 23 that $\mathscr{C}^\infty(\bar{\Omega})^n$ is dense in $LD(\Omega)$. Some other properties of $LD(\Omega)$ will be established later, including in particular a trace theorem for $LD(\Omega)$ and various other properties which can be derived therefrom (cf. Chapter II, Section 1).

1.4 A few technical results

We terminate this section by establishing a few technical results, particular cases of which have already appeared above in several instances.

Once again let Ω be an open subset of $\mathbb{R}^n$; we consider functions from Ω into $\mathbb{R}^m$, and differential operators acting on such functions, of the type

(1.73) $$\mathscr{A}(D) = \sum_{[j]\leq s} a_j D^j,$$

where $a_j \in \mathbb{R}^m$. More precisely let $\mathscr{A}_1,\dots,\mathscr{A}_k$, be k differential operators of the type (1.73) and $\alpha_0,\dots,\alpha_k$ numbers, not necessarily finite, satisfying $1\leq\alpha_j\leq\infty$. We denote by

(1.74) $$\mathscr{H} = \mathscr{H}(\mathscr{A}_1,\dots,\mathscr{A}_k;\ \alpha_0,\dots,\alpha_k;\ \Omega)$$

the space of the u in $L^{\alpha_0}(\Omega)^m$, such that $\mathscr{A}_j u \in L^{\alpha_j}(\Omega)$, $j=1,\dots,k$. We endow this space with the norm

(1.75) $$\|u\|_{L^{\alpha_o}(\Omega)^m} + \sum_{j=1}^{k} \|\mathcal{A}_j u\|_{L^{\alpha_j}(\Omega)}$$

and we have

LEMMA 1.2.

The space $\mathcal{H}$ is a Banach space for the norm (1.75).

Proof

Let u_p be a Cauchy sequence for the norm (1.75). Then the sequences u_p and $\mathcal{A}_j u_p$, $j = 1,\ldots,k$ are Cauchy sequences in $L^{\alpha_o}(\Omega)^m$ and $L^{\alpha_j}(\Omega)$, $j = 1,\ldots,k$ and converge in these spaces to the limits u and g_j, $j = 1,\ldots,k$ respectively. By the properties of continuity of the (weak) derivative (cf. § 1.1), we have $g_j = \mathcal{A}_j u$ for $j = 1,\ldots,k$ so that u belongs to the space $\mathcal{H}$ defined by (1.74) and u_p converges, in norm, in this space to u.

□

We shall say that the space $\mathcal{H}$ is of local type if

(1.76) $$u \in \mathcal{H}\ ,\ \phi \in \mathcal{C}^\infty(\bar{\Omega}) \Rightarrow u\phi \in \mathcal{H}\ .$$

This will be true, for example, if the operators $\mathcal{A}_j$ are all of the first order and $\alpha_o \geq \alpha_j$ for $j = 1,\ldots,k$.

Remark 1.4 The space $LD(\Omega)$ is included, with continuous injection, in $L^{n/(n-1)}(\Omega)^n$ (cf. Chapter II) and we shall be led to consider the space:

$$LD(\Omega) \cap \{u \in L^1(\Omega)^n,\ \text{div}\, u \in L^2(\Omega)\}\ ,$$

In space of dimension 2, this space is of local type, but not when $n \geq 3$ (or to be more accurate it is not known whether the space is of local type when $n \geq 3$).

In the case of a space of local type, we have a density result which can be proved by classical methods:

PROPOSITION 1.3

If the α_j *are all finite,* $1 \leq \alpha_j < \infty$ *, and the space* $\mathcal{H}$ *is of local type and if the open set* $\complement\bar{\Omega}$ *satisfies the cone condition, then the space* $\mathcal{C}^\infty(\bar{\Omega})^m$ *is dense in* $\mathcal{H}$ *.*

<u>Proof</u>

i) Consider the covering of Γ by the open sets $(\mathcal{O}_i)_{i \in I}$ defined by (1.16). These sets together with Ω constitute a covering of $\bar{\Omega}$, and we consider a partition of unity subordinated to this covering:

$$1 = \phi + \sum_{i \in I} \phi_i \ , \tag{1.77}$$

where $\phi \in \mathcal{C}_0^\infty(\Omega)$ and $\phi_i \in \mathcal{C}_0^\infty(\mathcal{O}_i)$.

If $u \in \mathcal{H}$, we write

$$u = \phi u + \sum_{i \in I} \phi_i u \tag{1.78}$$

and as $\mathcal{H}$ is of local type, ϕu and $\phi_i u$ are in $\mathcal{H}$.

We now have to approximate each of these functions by functions in $\mathcal{C}^\infty(\bar{\Omega})^m$.

ii) The function ϕu having a compact support in Ω , the function $v = \widetilde{\phi u}$ which is equal to ϕu in Ω and 0 outside Ω is in $\mathcal{H}(\mathcal{A}_1,\dots,\mathcal{A}_k;\ \alpha_0,\dots,\alpha_k\ ;\ \mathbb{R}^n)$. If ρ_η is a sequence of regularising functions (cf. § 1.1) then $\rho_\eta * v$ is, for all η , a function of $\mathcal{C}^\infty(\mathbb{R}^n)$ with compact support. It follows from (1.9) that when $\eta \to 0$,

$$\rho_\eta * v \to v \quad \text{in} \quad L^{\alpha_0}(\mathbb{R}^n),$$

and

$$\mathcal{A}_j(\rho_\eta * v) = \rho_\eta * \mathcal{A}_j v \to \mathcal{A}_j v \quad \text{in} \quad L^{\alpha_j}(\mathbb{R}^n)$$

for $j = 1,\dots,k$. Under these conditions $\rho_\eta * v$ converges to v in $\mathcal{H}(\mathcal{A}_1,\dots,\mathcal{A}_k\ ;\ \alpha_0,\dots,\alpha_k\ ;\ \mathbb{R}^n)$ and the restriction to Ω of $\rho_\eta * v$ (which is in $\mathcal{C}^\infty(\bar{\Omega})^m$ and even in $\mathcal{C}_0^\infty(\Omega)^m$ for small η) converges to u in $\mathcal{H}(\mathcal{A}_1,\dots,\mathcal{A}_k\ ;\ \alpha_0,\dots,\alpha_k\ ;\ \Omega)$ when $\eta \to 0$.

iii) We now consider a function $v = \phi_i u$ and denote by $\tilde{v}$ the function equal to v in Ω and to 0 outside Ω. We choose a regularising function ρ whose support is contained within the cone $\mathscr{C}_i$ given by (1.16) (cf. Fig. 1.2):

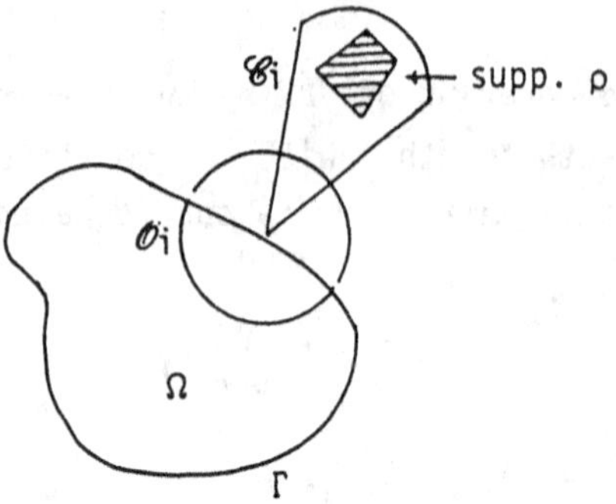

Fig. 1.2

The function $\rho_\eta * \tilde{v}$ is in $\mathscr{C}_0^\infty(\mathbb{R}^n)$ and from (1.9), when $\eta \to 0$, $\rho_\eta * \tilde{v} \to \tilde{v}$ in $L^{\alpha_0}(\mathbb{R}^n)^m$ and on passing to the restrictions of these functions to Ω, we obtain

$$\rho_\eta * \tilde{v}_{|\Omega} \to v = \phi_i u \quad \text{in} \quad L^{\alpha_0}(\mathbb{R}^n)^m . \tag{1.79}$$

On the other hand since $\tilde{v}$ in general exhibits a jump on $\mathscr{O}_i \cap \Gamma$, $\mathscr{A}_j \tilde{v}$ is the sum of the function $\widetilde{\mathscr{A}_j v}$ (equal to $\mathscr{A}_j v$ on Ω and zero outside Ω) and of a distribution μ_j carried by $\mathscr{O}_i \cap \Gamma$. When $\eta \to 0$, $\rho_\eta * \widetilde{\mathscr{A}_j}$ converges to $\widetilde{\mathscr{A}_j v}$ in $L^{\alpha_j}(\mathbb{R}^n)$, and in particular

$$\rho_\eta * \widetilde{\mathscr{A}_j v}_{|\Omega} \to \mathscr{A}_j v \quad \text{in} \quad L^{\alpha_j}(\Omega) . \tag{1.80}$$

Since the support of ρ_η is contained in $\mathscr{C}_i$ it follows from condition (1.16) (for $\complement\bar{\Omega}$), that the support of $\rho_\eta * \mu_j$ (which is contained in supp. ρ_η + supp. μ_j) does not meet Ω, and therefore that $\rho_\eta * \widetilde{\mathscr{A}_j v}_{|\Omega} = \rho_\eta * \mathscr{A}_j \tilde{v}_{|\Omega} = \mathscr{A}_j(\rho_\eta * \tilde{v}_{|\Omega})$ and hence finally, when $\eta \to 0$, that

(1.81) $$\mathscr{A}_j(\rho_\eta * \tilde{v}_{|\Omega}) \to \mathscr{A}_j v \quad \text{in} \quad L^{\alpha_j}(\Omega) .$$

Consequently, we have, when $\eta \to 0$,

(1.82) $$\rho_\eta * \tilde{v}_{|\Omega} \to v = \phi_i u ,$$

in $\mathscr{H}(\mathscr{A}_1,\ldots,\mathscr{A}_k ; \alpha_0,\ldots,\alpha_k ; \Omega)$, and the required result is established.

Remark 1.5 One could obviously take exponents $\alpha_{o1},\ldots,\alpha_{om}$, which are different for the different components $u_1,\ldots,u_m$ of u . One could also, subject to certain conditions allow the coefficients a_j in (1.73) to depend on x .

2. Convex analysis – a review of some basic concepts and results

In this Section we shall recall some basic concepts of convex analysis essentially taken from J.J. Moreau [1], R.T. Rockafellar [1]-[2], I. Ekeland and R. Temam [1]. At the end of this section, and by way of example, we shall apply these to the variational problems arising in linear elasticity theory.

2.1 Convex functions

Proofs of the results recalled in this part can be found in Sections 2 to 5 of Chapter I of Ekeland and Temam [1], or in the other references mentioned above.

We consider functions defined on a vector space V (or on a part of such a space) whose values are in $\mathbb{R}$ or in $\mathbb{R} \cup \{+\infty\}$; such a function is said to be 'proper' if it is not identically $+\infty$. Let $\mathscr{A}$ be a convex subset of V and F a mapping of $\mathscr{A}$ into $\mathbb{R} \cup \{+\infty\}$: the function F is said to be convex (on $\mathscr{A}$) , if

$$(2.1) \qquad F(\lambda u + (1-\lambda)v) \leq \lambda F(u) + (1-\lambda)\, F(v) ,$$

for all u and v belonging to $\mathscr{A}$ and all $\lambda \in]0,1[$. The inequality (2.1) is trivial if $F(u)$ or $F(v) = +\infty$. It is easy to see that if F is convex, then for all $\alpha \in \mathbb{R} \cup \{+\infty\}$, the sets

$$(2.2) \qquad \{u, F(u) \leq \alpha\} \text{ and } \{u, F(u) < \alpha\}$$

are convex, the converse proposition being false.

For any function $F : V \to \mathbb{R} \cup \{\infty\}$, we write dom F for the domain (of finiteness) of F,

$$(2.3) \qquad \text{dom } F = \{u \in V, F(u) < +\infty\} ;$$

this set is convex if F is convex. The epigraph of F is the set

$$(2.4) \qquad \text{epi } F = \{(u,\alpha) \in V \times \mathbb{R} , \quad \alpha \geq F(u)\} .$$

The projection of this set on V is simply dom F and it is easily seen that F is convex if, and only if, epi F is a convex set.

The *indicator function* of a set $\mathscr{A} \subset V$, is the function $\chi_{\mathscr{A}} : V \to \mathbb{R} \cup \{\infty\}$, defined by

$$(2.5) \qquad \chi_{\mathscr{A}}(u) = \begin{cases} 0 & \text{if } u \in \mathscr{A} \\ +\infty & \text{if } u \notin \mathscr{A}. \end{cases}$$

The function $\chi_{\mathscr{A}}$ is convex if and only if $\mathscr{A}$ is convex.

Lastly let us recall that a function $F : \mathscr{A} \to \mathbb{R} \cup \{\infty\}$ is *strictly convex* if the inequality (2.1) holds strictly (i.e. with $<$ in the place of $\leq$) for all $u, v \in \mathscr{A}$, $u \neq v$ and for all $\lambda \in]0,1[$.

Continuity and lower semicontinuity

We shall suppose that the space V is endowed with a topology; generally V will be a normed vector space but in a few rare cases we shall suppose V to be a locally convex topological space of a more general kind (cf. the treatment of conjugate convex functions given below).

A function $F : V \to \mathbb{R} \cup \{\infty\}$ is said to be lower semicontinuous, or more shortly l.s.c. if:

$$F(u) \leq \liminf_{u_j \to u} F(u_j) \tag{2.6}$$

for all u and all $u_j \to u$. A necessary and sufficient condition for lower semicontinuity is that, for all $\alpha \in \mathbb{R}$, the set

$$\{u \in V \text{ , } F(u) \leq \infty\} \tag{2.7}$$

be closed. It can also be shown that F is lower semicontinuous if and only if its epigraph is closed.

If F is a convex l.s.c. function, the sections defined by (2.7) are closed convex sets. The indicator function $\chi_{\mathcal{A}}$ of a set $\mathcal{A} \subset V$ is l.s.c. if and only if $\mathcal{A}$ is closed.

If V is a Banach space, any convex function $F : V \to \mathbb{R} \cup \{+\infty\}$ is continuous on any open set $\mathcal{O} \subset V$ throughout which F is finite; in particular, if the interior of its domain of finiteness, $\overset{\circ}{\text{dom}}\, F$ is non-vacuous, then F is continuous therein. More precisely F is locally Lipschitzian on the interior of its domain (cf. 2.3, Ch. I, Ekeland-Temam [1]).

Conjugate convex functions

It is easy to see that the upper bound of a family of convex functions (resp. l.s.c. functions) is likewise convex (resp. l.s.c.). If F is a function from V into $\mathbb{R} \cup \{+\infty\}$, then the least upper bound of all convex l.s.c. functions which are less than or equal to F itself is a convex l.s.c. function G which is called the 'Γ-regularised function' corresponding to F; G, which may possibly be identically equal to $-\infty$, is the largest convex l.s.c. function less than or equal to F. It can be shown that a necessary and sufficient condition for F to be equal to its corresponding Γ-regularised function is that F should be convex l.s.c. (F, by assumption, not taking the value $-\infty$).

We can instead consider only the continuous affine minorants of F, that is to say, the functions of the form $v \mapsto \ell(v) + \alpha$, where ℓ

is a continuous linear form on V, $\alpha \in \mathbb{R}$, and $F(v) \geq \ell(v) + \alpha$, $\forall\, v \in V$. The Γ-regularised function corresponding to F is likewise the upper envelope of the family of affine continuous minorants of F; we shall see later that it is the envelope of an even more restricted family of functions. If F has at least one continuous affine minorant, then G is not identically $+\infty$ and the epigraph of G is the closed convex hull of the epigraph of F.

Remark 2.1 If F is continuous (or merely l.s.c.) and has a continuous affine minorant, the epigraph of G is the convex hull of the epigraph of F; in this case we say that G is the convexified of F.

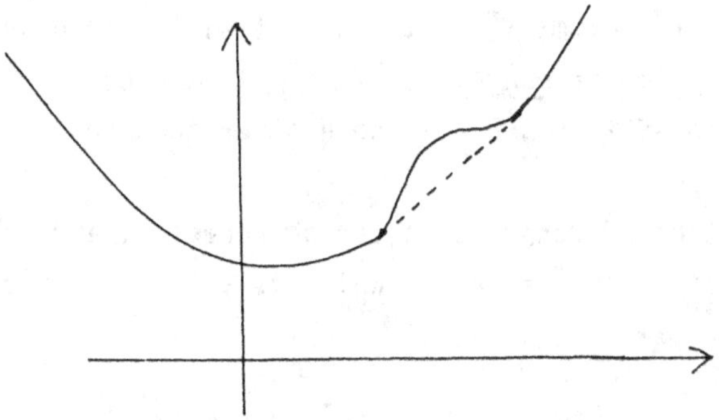

Fig. 2.1.: F and the corresponding 'convexified' curve

This concept, which will find little place in this study, seems destined to play an important part in nonlinear elasticity theory; cf. for example M. Gurtin - R. Temam [1] and the bibliography of that article.

□

Let us now consider two paired topological vector spaces V and V^*, where the scalar product between V and V^* is denoted by $\langle .,. \rangle$. In most cases, V will be a Hilbert space or a reflexive Banach space and V^* will be its dual, both spaces being endowed with the topology defined by the norm. In a few cases V will be a non-reflexive space endowed with the norm topology[1], and V^* will be its dual, endowed with the weak* topology[2].

A continuous affine function on V is a function of the form

$$u(\in V) \mapsto \langle u^*,u\rangle + \alpha \ , \tag{2.8}$$

where $u^* \in V^*$ and $\alpha \in \mathbb{R}$.

If F is a function from V into $\mathbb{R} \cup \{+\infty\}$, and $u^* \in V^*$, the largest of the continuous affine minorants of F of the type (2.8) is that which corresponds to $\alpha = -F^*(u^*)$ where

$$F^*(u^*) = \operatorname*{Sup}_{u\in V} \{\langle u^*,u\rangle - F(u)\} \ . \tag{2.9}$$

The function F^* from V^* into $\mathbb{R} \cup \{+\infty\}$ defined by (2.9) is known as the *polar* or *conjugate* of F. It is a convex l.s.c. function on V^*. When $V = \mathbb{R}$, F^* is none other than the *Legendre transform* of F.

The operation of conjugation can be repeated, and one can therefore define a function $F^{**} = (F^*)^*$ which is a convex l.s.c. function from V into $\mathbb{R} \cup \{+\infty\}$:

$$F^{**}(u) = \operatorname*{Sup}_{u^*\in V^*} \{\langle u^*,u\rangle - F^*(u^*)\} \ . \tag{2.10}$$

It can be shown that F^{**} is simply the Γ-regularised function corresponding to F, which thus appears as the supremum of a particula family of continuous affine minorants of F (viz. those of the form $\phi(u) = \langle u^*,u\rangle - F^*(u^*)$ and which are therefore maximal for a fixed u^*).

As F^* is convex l.s.c. (and also proper), we see that the Γ-regularised function which corresponds to it, namely $F^{***} = (F^*)^{**}$ is identical with F^*,

---

(1) Such as, for example, the space $L^1(\Omega)$ or spaces derived from it.

(2) That is, the topology denoted by $\sigma(V^*,V)$ defined by the family of seminorms $u^*(\in V^*) \mapsto |\langle u^*,u\rangle|$, where $u \in V$; cf. N. Bourbaki [2], F. Trèves [1].

(2.11) $$F^{***} = F^* .$$

Examples

Some examples which will be important later on, derived from the calculus of variations, will be given in Section 2.3. Here we shall confine ourselves to a few simple examples in which the spaces may or may not be of finite dimensions.

(i) If $\mathscr{A}$ is a subset of V and $\chi_{\mathscr{A}}$ is its indicator function, the conjugate $\chi^*_{\mathscr{A}}$ of $\chi_{\mathscr{A}}$, is given by

$$\chi^*_{\mathscr{A}}(u^*) = \operatorname*{Sup}_{u\in V} \langle u^*,u\rangle - \chi_{\mathscr{A}}(u)$$

$$\chi^*_{\mathscr{A}}(u^*) = \operatorname*{Sup}_{u\in \mathscr{A}} \langle u^*,u\rangle$$

and is called *the support function* of the set $\mathscr{A}$; it is a positively homogeneous function. It can be shown that the Γ-regularised function corresponding to $\chi_{\mathscr{A}}$, or in other words the function $\chi^{**}_{\mathscr{A}}$ is simply the indicator function of the closed convex hull of $\mathscr{A}$, $\chi_{\overline{\text{conv}}\,\mathscr{A}}$.

(ii) On $\mathbb{R}$, the conjugate of the function

(2. 3) $$\theta(t) = \frac{1}{\alpha}|t|^\alpha , \quad 1<\alpha<\infty ,$$

is the function $\theta^*(t) = \frac{1}{\alpha'}|t|^{\alpha'}$ where $\alpha' = \frac{\alpha}{\alpha-1}$. If $\alpha = 1$, the conjugate of $\theta(t) = |t|$ is the indicator function of the line segment $[-1,+1]$.

The conjugate of the function

$$\theta(t) = \sqrt{(1 + t^2)}$$

which appears in the theory of minimal surfaces (cf. R. Temam [2]) is the function $\theta^*(t)$ whose value is $-\sqrt{(1-t^2)}$ if $|t| \leq 1$ and $+\infty$ otherwise.

In certain problems serving as plasticity models, we shall meet the function

$$\theta(t) = \frac{t^2}{2} \text{ if } |t| \leq 1 , = +\infty \text{ if } |t| > 1 .$$

An elementary calculation yields

(2.16) $\theta^*(t) = \frac{t^2}{2}$ if $|t| \leq 1$, $= |t| - \frac{1}{2}$ if $t \geq 1$.

Similarly, we may consider in $\mathbb{R}^n$ the functions

$$\xi \mapsto \frac{1}{\alpha}|\xi|^\alpha , \quad \xi \mapsto \sqrt{1+|\xi|^2} ,$$

$$\xi \mapsto \frac{1}{2}|\xi|^2 \text{ if } |\xi| \leq 1 \text{ and } +\infty \text{ if not,}$$

whose respective conjugates are

(2.18) $\eta \mapsto \frac{1}{\alpha'}|\eta|^{\alpha'}$, $\eta \mapsto -\sqrt{1-|\eta|^2}$ if $|\eta| \leq 1$ and $+\infty$ if not

$$\eta \mapsto \frac{1}{2}|\eta|^2 \text{ if } |\eta| \leq 1 \text{ and } |\eta| - \frac{1}{2} \text{ if not.}$$

For example, to determine the conjugate of the third function in (2.17) we have to calculate

$$\sup_{\substack{\xi \in \mathbb{R}^n \\ |\xi| \leq 1}} \{\xi.\eta - \frac{1}{2}|\xi|^2\} .$$

By optimising with respect to the modulus and the direction of ξ separately we find that this supremum is equal to

$$\sup_{0 \leq t \leq 1} \sup_{|\xi| = t} \{\xi.\eta - \frac{t^2}{2}\} = \sup_{0 \leq t \leq 1} \{t|\eta| - \frac{t^2}{2}\}$$

$$= \text{(by (2.16))}$$

$$= \frac{1}{2}|\eta|^2 \text{ if } |\eta| \leq 1$$

$$= |\eta| - \frac{1}{2} \text{ if } |\eta| \geq 1 .$$

(iii) Let V be a Banach space and V^* its dual (norms $\|.\|_V$ and $\|.\|_{V^*}$) , and let θ be an <u>even</u> l.s.c. convex function on $\mathbb{R}$ whose conjugate we denote by θ^* . It is easily shown that the conjugate of the function $\phi(v) = \theta(\|v\|_V)$ is the function

$$\phi^*(v^*) = \theta^*(\|v^*\|_{V^*}) .$$

This applies in particular to the functions $(1/\alpha)\|v\|_V^\alpha$ and $\|v\|_V$ whose respective conjugates are the function $(1/\alpha')\|v^*\|_{V^*}^{\alpha'}$ and the indicator function of the ball $\|v^*\|_{V^*} \leqq 1$.

Finally we give the following example (due to B. Héron) which we shall use on several occasions.

LEMMA 2.1

Let V be a Banach space and V^* its dual. Let v_o, v_o^* be elements of V, V^* respectively, $\mathcal{B}$ a closed subspace of V, and F the convex l.s.c. function on V given by

$$F(v) = \langle v_o^*, v\rangle + \chi_{v_o+\mathcal{B}}(v) .$$

Then for all $v^* \in V^*$,

$$F^*(v^*) = \langle v^* - v_o^*, v_o\rangle + \chi_{\mathcal{B}^o}(v^* - v_o^*) ,$$

where $\mathcal{B}^o$ denotes the polar of $\mathcal{B}$,

$$\mathcal{B}^o = \{v^* \in V^* , \langle v^*, v\rangle = 0 , \forall\ v \in \mathcal{B}\} .$$

<u>Proof</u>

We have

$$F^*(v^*) = \sup_{v-v_0\in\mathcal{B}} \langle v^*-v_0^*, v\rangle = \langle v^*-v_0^*, v_0\rangle + \sup_{w\in\mathcal{B}} \langle v^*-v_0^*, w\rangle .$$

It remains to be shown that the last supremum, which is clearly equal to $\chi^*_{\mathcal{B}}(v^*-v_0^*)$ is also equal to $\chi_{\mathcal{B}^o}(v^*-v_0^*)$, and this is easily

Sub-differentiability

We consider again two paired topological vectors spaces V and V^* .

A function F mapping V into $\mathbb{R} \cup \{+\infty\}$ is said to be *sub-differentiable* at a point $u \in V$ if an exact continuous affine minorant to F can be found at the point u ; or, in other words, if there exists a $u^* \in V^*$ such that

$$(2.20) \qquad F(v) \geq F(u) + \langle u^*, v-u \rangle \ , \ \forall\, v \in V \ .$$

The element u^* is then called the *sub-gradient* of F at u and the set of sub-gradients at u is called the *sub-differential* of F at the point u and is written $\partial F(u)$.

Let F^* be the polar function of F . It is easily seen that $u^* \in \partial F(u)$ if and only if

$$F(u) + F^*(u^*) = \langle u, u^* \rangle$$

or, what amounts to the same thing, having regard to (2.9):

$$(2.22) \qquad F(u) + F^*(u^*) \leq \langle u, u^* \rangle \ .$$

This characterisation of $\partial F(u)$ implies straightaway that if $u^* \in \partial F(u)$ then $u \in \partial F^*(u^*)$. In the most interesting case where F is also convex and l.s.c. then $F = F^{**}$ and we have the equivalence relation

$$(2.23) \qquad u^* \in \partial F(u) \Leftrightarrow \ u \in \ F^*(u^*) \ .$$

Finally we note that the set $\partial F(u)$ (which may be the empty set) is convex and weakly closed ([1]).

([1]) i.e. closed for the topology $\sigma(V^*, V)$, cf. note ([2]) page 30.

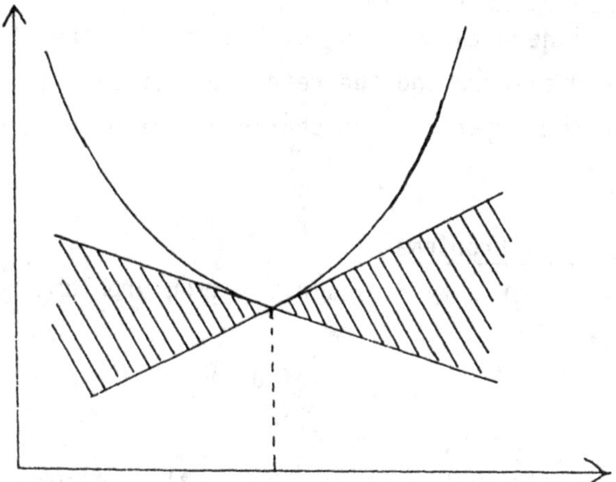

Fig. 2.2: The hatched area represents the directions of the sub-gradients at u

It can be shown (cf. Proposition 5.2, Corollary 2.5, Ch. I, Ekeland-Temam [1]) that if V is a Banach space, F a convex function mapping V into $\mathbb{R} \cup \{+\infty\}$, then F is sub-differentiable at every interior point of its domain ($u \in \overset{\circ}{\widehat{\text{dom } F}}$).

Differentiability and sub-differentiability

A function F from V into $\mathbb{R} \cup \{+\infty\}$ is said to be Gâteaux-differentiable at a point u if there exists a $u^* \in V^*$ such that for all $v \in V$

$$(2.24) \qquad \frac{F(u+\lambda v) - F(u)}{\lambda} \to \langle u^*, v\rangle$$

when λ tends to zero through positive values. The element $u^* \in V^*$ is called the Gâteaux differential of F at u, and is written $F'(u)$. Obviously if $F'(u)$ exists then $u \in \text{dom } F$.

In the case of a convex function F from V into $\mathbb{R} \cup \{+\infty\}$, if F is Gâteaux-differentiable at u, then F is sub-differentiable at the same point and $\partial F(u) = \{F'(u)\}$.

2.2 Variational problems

We recall here the concept of duality in connection with variational problems and also remind the reader of certain standard results concerning the existence and characterisation of the minimum of a convex function.

2.2.1. Dual Problems

If we consider a typical minimisation problem, that of finding

$$\text{(2.25)} \qquad \inf_{u \in V} J(u) \, , \qquad (\mathcal{P})$$

we can associate with this problem $(\mathcal{P})$ another variational problem, known as the dual problem, denoted by $(\mathcal{P}^*)$, which has a number of properties in common with $(\mathcal{P})$.

We use a variant of the duality formalism in W. Fenchel [1]-[2]-[3] and R.T. Rockafellar [1] which has been systematically applied in Ekeland-Temam [1] to problems in the calculus of variations.

This presupposes the following functional framework: let V, V^*, and Y, Y^* be two pairs of topological vector spaces in duality, and let Λ be a continuous linear operator from V into Y, whose transpose Λ^* is a continuous linear operator from Y^* into V^*

$$V^* \xleftarrow{\ \Lambda^*\ } Y^*$$

$$V \xrightarrow{\ \Lambda\ } Y \, .$$

The elements of V and V^* are written $u, v, w, \ldots,$ and $u^*, v^*, w^*, \ldots,$ etc. respectively and those of Y and Y^* are written $p, q, r, \ldots,$ and $p^*, q^*, r^*, \ldots,$. The spaces Y and Y^* will sometimes play very similar roles, and we shall then write, for simplicity, $p, q, r, \ldots,$ for the elements of Y^*.

Lastly we shall suppose that we are given a convex l.s.c. function F defined on V with values in $\mathbb{R} \cup \{+\infty\}$ and similarly a convex l.s.c. function G from Y into $\mathbb{R} \quad \{+\infty\}$, and that the problem $(\mathcal{P})$ can be written in the form

(2.26)
$$\operatorname*{Inf}_{u\in V} \{F(u) + G(\Lambda u)\} \qquad (\mathscr{P})$$

The dual problem $(\mathscr{P}^*)$ can then be written

(2.27)
$$\operatorname*{Sup}_{p^*\in Y^*} \{- F^*(\Lambda^* p^*) - G^*(-p^*)\} \qquad (\mathscr{P}^*)$$

We write inf $\mathscr{P}$ for the infimum of the problem $\mathscr{P}$ and sup $\mathscr{P}^*$ for the supremum of the problem $\mathscr{P}^*$. In general we have

(2.28)
$$-\infty \leq \operatorname{Sup} \mathscr{P}^* \leq \operatorname{Inf} \mathscr{P} \leq +\infty ,$$

but we have a simple criterion at our disposal which allows us to infer that $\operatorname{Sup} \mathscr{P}^* = \operatorname{Inf} \mathscr{P}$ when a certain condition is satisfied.

Suppose that

(2.29) There exists a $u^0 \in V$ such that $F(u^0) < +\infty$ and $G(\Lambda u^0) < +\infty$, G being continuous in Λu^0.

Then

(2.30)
$$\operatorname{Sup} \mathscr{P}^* = \operatorname{Inf} \mathscr{P} ,$$

and if $\operatorname{Inf} \mathscr{P} > -\infty$, this number is finite and $\mathscr{P}^*$ *has at least one solution* (i.e. the maximum in (2.27) is attained by a $\bar{p}^* \in Y^*$).

Extremality relations

Finally suppose that the condition (2.29) holds, and that $\mathscr{P}$ and $\mathscr{P}^*$ have solutions $\bar{u}$ and $\bar{p}^*$ respectively. In these circumstances the following relations, known as extremality relations, are satisfied

(2.31)
$$F(\bar{u}) + F^*(\Lambda^*\bar{p}^*) = \langle \Lambda^*\bar{p}^*, \bar{u}\rangle$$

(2.32)
$$G(\Lambda\bar{u}) + G^*(-\bar{p}^*) = - \langle \bar{p}^*, \Lambda\bar{u}\rangle .$$

All these results are established in the articles mentioned above, or in Chapter III of Ekeland-Temam [1]. Finally we may mention the following special case which will turn out to be of particular importance in the sequel:

An important special case

This is the case where Y is a product of spaces $Y = \prod_{i=1}^{N} Y_i$, while

$$\Lambda = \Lambda_1 \times \dots \times \Lambda_N , \tag{2.33}$$

$$G(p) = \sum_{i=1}^{N} G_i(p_i) , \ \forall\, p = (p_1,\dots,p_N) \in Y , \tag{2.34}$$

where Λ_i is a continuous linear operator of V into Y_i, G_i is a convex l.s.c. function from Y_i into $\mathbb{R} \cup \{+\infty\}$ for $i = 1,2,\dots,N$; where Y_i^* is a dual to Y_i, but where everything else remains unchanged.

The problem (2.26) then becomes

$$\operatorname*{Inf}_{u\in V} \{F(u) + \sum_{i=1}^{N} G_i(\Lambda_i u)\} , \qquad (\mathscr{P}). \tag{2.35}$$

It is easy to see that

$$G^*(p^*) = \sum_{i=1}^{N} G_i^*(p_i^*) , \quad \forall\, p^* = (p_1^*,\dots,p_N^*) \in Y = Y_1 \times\dots\times Y_N ,$$

and that in these circumstances the dual problem (2.27) becomes

$$\operatorname*{Sup}_{p^*\in Y^*} \{-F^*(\Lambda^* p^*) - \sum_{i=1}^{N} G_i^*(-p_i^*)\} . \tag{2.36}$$

The criterion (2.29) can now be written

(2.37) There exists a $u^0 \in V$ such that $F(u^0) < +\infty$ and $G_i(\Lambda_i u^0) < +\infty$, G_i being continuous at $\Lambda_i u^0$, for $i = 1,\dots,N$.

Lastly the extremality relation (2.32) splits up into the N relations

$$G_i(\Lambda_i \bar{u}) + G_i^*(-\bar{p}_i^*) = -\langle \bar{p}_i^*, \Lambda_i \bar{u}\rangle , \quad i = 1,\dots,N . \tag{2.38}$$

2.2.2 Existence theorem and Euler's equation

To demonstrate the existence of a solution in the more usual types of variational problem, appeal is often made to the following results, which we now recall, and a proof of which can be found, for example, in Chapter II of Ekeland-Temam [1].

Let V be a reflexive Banach space, with norm written $\|\cdot\|$, and let $\mathscr{C}$ be a closed non-vacuous convex subset of V. We are given a function J mapping $\mathscr{C}$ into $\mathbb{R}$ and we suppose

(2.39) $\qquad J$ is convex and l.s.c. .

We are interested in the minimisation problem

$$\inf_{v\in\mathscr{C}} J(v) . \tag{2.40}$$

We say that an element u of the set $\mathscr{C}$ is a solution of the problem (2.40) if

$$J(u) = \inf_{v\in\mathscr{C}} J(v)$$

Let us suppose that one or other of the two following conditions is satisfied:

(2.42) $\qquad \mathscr{C}$ is bounded

or

(2.43) $\qquad \lim J(u) = +\infty , \quad$ for $u \in \mathscr{C}$, $\|u\| \to +\infty$;

in this (second) case J is said to be coercive on $\mathscr{C}$.

We then have

PROPOSITION 2.1 *(existence of solution)*

Under the hypotheses described above, and in particular if (2.39), (2.41) and (2.43) hold, then the infimum of problem (2.40) is finite, the problem has at least one solution, and the set of solutions of (2.40) is closed and convex.

PROPOSITION 2.2. *(Uniqueness of solution)*

If the function $J : \mathscr{C} \to \mathbb{R}$ *is strictly convex*

(2.44) $$J(\lambda u+(1-\lambda)v) < \lambda J(u) + (1-\lambda)\, J(v)\ , \ \forall\, u,\ v \in V\ ,\quad u \neq v\ ,\quad \forall\, \lambda \in]0,1[\ ,$$

the problem (2.40) has at most one solution.

PROPOSITION 2.3 *(Euler's inequality)*

Suppose the function $J : \mathscr{C} \to \mathbb{R}$ *to be Gâteaux-differentiable and convex. Then* $u \in \mathscr{C}$ *is a solution of (2.40) if and only if*

(2.45) $$\langle J'(u), v-u\rangle \geq 0\ ,\ \forall\, v \in \mathscr{C}\ .$$

Remark 2.2. The condition (2.45) is also equivalent to

(2.46) $$\langle J'(v), v-u\rangle \geq 0\ ,\quad \forall\, v \in \mathscr{C}\quad ,$$

(always assuming $u \in \mathscr{C}$).

If $J = J_1 + J_2$, where J_1 and J_2 are two convex functions mapping $\mathscr{C}$ into $\mathbb{R}$, and J_1 is Gâteaux-differentiable, then $u \in \mathscr{C}$ is a solution of (2.40) if and only if

(2.47) $$\langle J_1'(u), v-u\rangle + J_2(v) - J_2(u) \geq 0\ ,\quad \forall\, v \in \mathscr{C}$$

or

(2.48) $$\langle J_1'(v), v-u\rangle + J_2(v) - J_2(u) \geq 0\ ,\quad \forall\, v \in \mathscr{C}\ .$$

□

2.3. Applications in the calculus of variations

We shall describe here a few convex functions which arise naturally in the calculus of variations, and we shall also apply the duality techniques mentioned earlier to the variational problems of linear elastic theory.

2.3.1 Convex functions in the calculus of variations

Let Ω be an open subset of $\mathbb{R}^n$ and g a continuous mapping from $\mathbb{R}^m$ into $\mathbb{R}$. To every function u from Ω into $\mathbb{R}^m$, we can associate the function $g \circ u$ from Ω into $\mathbb{R}$ defined by

$$x \mapsto g(u(x))$$

and when this function is integrable over Ω for the Lebesgue measure dx, we can consider the functional

$$\int_\Omega g(u(x))\, dx \ .$$

We recall two interesting results: the first is due to Krasnoselskii [1] (Th.1, p.22). A simpler proof due to J.M. Lasry is given in Ekeland-Temam [2], § 1.2, Chapter IV.

PROPOSITION 2.4

Suppose that g is continuous from $\mathbb{R}^m$ into $\mathbb{R}$ and that for all $u \in V$, where

$$V = L^{\alpha_1}(\Omega) \times \dots \times L^{\alpha_m}(\Omega) \quad (1 \leq \alpha_i < \infty) \ , \tag{2.49}$$

the function $g \circ u$ is Lebesgue-integrable on Ω.

Then the mapping

$$u \mapsto g \circ u \tag{2.50}$$

is a continuous mapping from V into $L^1(\Omega)$.

This result allows us to define the function, from V into $\mathbb{R}$,

$$G : u \mapsto \int_\Omega g(u(x))\, dx$$

which is therefore continuous. It is natural to ask what is the conjugate functional G^*, defined on

$$V^* = L^{\alpha'_1}(\Omega) \times \ldots \times L^{\alpha'_m}(\Omega)$$

where α'_j is the exponent conjugate to α_j (that is $\alpha'_j = \alpha_j/(\alpha_j - 1)$). We have in fact, on more general assumptions about g, the following result:

PROPOSITION 2.5.

Let g be a l.s.c. function mapping $\mathbb{R}^m$ into $\mathbb{R}$. Suppose that there exist numbers $a \geq 0$ and $b \in \mathbb{R}$ such that

$$(2.52) \qquad g(\xi) \geq -a \sum_{i=1}^{m} |\xi_i|^{\alpha_i} + b\,, \quad \forall\, \xi = (\xi_1,\ldots,\xi_m) \in \mathbb{R}^m,$$

and that there exists a $u_0 \in V$ such that $g \circ u_0 \in L^1(\Omega)$.

Then, for all $v \in V^*$,

$$(2.53) \qquad G^*(v) = \int_\Omega g^*(v(x))dx$$

where g^* is the function conjugate to g:

$$(2.54) \qquad g^*(\eta) = \sup_{\xi \in \mathbb{R}^m} \{\xi.\eta - g(\xi)\}\,.$$

Remark 2.3. i) Proposition 2.5 is a particular case of the Propositio 2.1 in Chapter IX of Ekeland-Temam [1].

ii) It may sometimes be useful to consider more generally a function g mapping $\Omega \times \mathbb{R}^m$ into $\overline{\mathbb{R}}$ and the associated functional (when it is defined)

$$G(u) = \int_\Omega g(x,u(x))\,dx\,.$$

Proposition 2.4 then extends to this situation if we suppose g to be a Caratheodory function, that is to say if

(2.55) for almost all $x \in \Omega$, $g(x,.)$ is continuous on $\mathbb{R}^m$, and

(2.56) for all $\xi \in \mathbb{R}^m$, $g(.,\xi)$ is measurable.

Proposition 2.5 extends to the case where g , instead of satisfying the condition (2.52), satisfies

(2.57) $$g(x,\xi) \geq -a \sum_{i=1}^{m} |\xi_i|^{\alpha_i} + b(x) \text{ , p.p. } x \in \Omega, \forall \xi \in \mathbb{R}^m \text{ ,}$$
$$a \geq 0 \text{ , } b \in L^1(\Omega) \text{ ,}$$

where the function g is a normal integrand, which means that

(2.58) for almost all $x \in \Omega$, $g(x,.)$ is l.s.c. on $\mathbb{R}^m$,

and

(2.59) there exists a function $\tilde{g} : \Omega \times \mathbb{R}^m \to \overline{\mathbb{R}}$ which is Borelian and such that $\tilde{g}(x,.) = g(x,.)$ for almost all $x \in \Omega$.

In particular a Caratheodory function is a normal integrand.

If g is a normal integrand satisfying (2.57) and if there is a $u_o \in V$ such that the function $\{x \mapsto g(x,u(x))\}$ is Lebesgue integrable on Ω , then $G^*(v)$ is given by

(2.60) $$G^*(v) = \int_\Omega g^*(x,v(x))\, dx \text{ , } \forall v \in V \text{ ,}$$

where

$$g^*(x,\eta) = \sup_{\xi \in \mathbb{R}^m} \{\xi.\eta - g(x,\xi)\} \text{ , } \forall \eta \in \mathbb{R}^m \text{ , and for almost all } x \in \Omega \text{ .}$$

2.3.2 Examples from linear elasticity theory

We now describe a first application of the principle of duality to linear elasticity, which we shall come back to later. We shall give later on numerous other applications of duality to problems in mechanics.

Let Ω be a bounded open set of $\mathbb{R}^n$, $n = 3$, of class $\mathscr{C}^2$, whose boundary is Γ. We consider an open portion Γ_0 of Γ and we denote by Γ_1 the interior of $\Gamma\backslash\Gamma_0$, so that $\bar{\Gamma}_1 = \Gamma\backslash\Gamma_0$ and $\bar{\Gamma}_0 = \Gamma\backslash\bar{\Gamma}_1$ are the closures with respect to Γ of Γ_1 and Γ_0.

A homogenous isotropic elastic body of form Ω is subjected to specified forces of volume density f in Ω and to tractive forces of surface density g distributed over Γ_1; a specified displacement $u = u(x)$ is imposed on Γ_0 (and even on $\bar{\Gamma}_0$). The forces f and g are assumed to be given by functions satisfying

$$f \in L^2(\Omega)^3 , \quad g \in L^2(\Gamma_1)^3 . \tag{2.62}$$

The displacement on Γ_0 is assumed to be given as the trace (restricted to Γ_0) of a function u_0 in $H^1(\Omega)^3$. Furthermore for technical reasons we shall suppose that $\gamma_0(u_0)$ vanishes outside Γ_0:

$$u_0 \in H^1(\Omega)^3 , \quad \theta_{\Gamma_0} \gamma_0(u_0) = \gamma_0(u_0) \tag{2.63}$$

where θ_{Γ_i} is the characteristic function of Γ_i (in Γ), which has the value 1 on Γ_i and 0 elsewhere $(i = 0,1)$.

Let λ and μ be the Lamé constants of the material, and $\kappa = \lambda + 2\mu/3$ its bulk modulus (or incompressibility), so that $\mu>0$, $\kappa>0$. Then it is well known in linear elasticity theory (cf. for example P. Germain [1]-[2]) that the components u_i of the displacement u are given by the solution of the following variational problem

$$\operatorname*{Inf}_{\substack{u \\ u=u_0 \text{ on } \Gamma_0}} \left\{ \int_\Omega \psi(\varepsilon(u))dx - \int_\Omega f\, u\, dx - \int_\Gamma g\, u\, d\Gamma \right\} \tag{2.64}$$

where, with the Einstein summation convention:

$$\psi(\xi) = \frac{\lambda}{2} \xi_{ii}\, \xi_{jj} + \mu\, \xi_{ij}\, \xi_{ij} , \quad \forall\, \xi \in E , \tag{2.65}$$

or since $\xi_{ij} = \xi^D_{ij} + \frac{1}{3} \xi_{kk}\, \delta_{ij}$,

(2.66) $$\psi(\xi) = \frac{\kappa}{2}\,\xi_{ii}\,\xi_{jj} + \mu\,\xi^D_{ij}\,\xi^D_{ij}\ ,$$

and where ε is the differential operator introduced in (1.50). Incidentally the infimum in (2.64) is to be taken for the set of all $u \in H^1(\Omega)^3$.

We shall write the primal problem (2.64) in the general form considered in (2.26) and to do this we put

$$V = H^1(\Omega)^3\ ,\quad V^* = \text{the dual of } H^1(\Omega)^3 = \{H^1(\Omega)'\}^3$$

$$Y = L^2(\Omega;E) = Y^*\ ,\ \Lambda = \varepsilon$$

$$F(v) = \begin{cases} -L(v) & \text{if } v \in V\ , \text{ and } v - u_0 = 0 \text{ on } \Gamma_0 \\ +\infty & \text{if } v \in V\ , \text{ and } v - u_0 \neq 0 \text{ on } \Gamma_0\ , \end{cases}$$

where

(2.67) $$L(v) = \int_\Omega fv\,dx + \int_{\Gamma_1} gv\,d\Gamma$$

$$G(p) = \int_\Omega \psi(p(x))\,dx\ ,\quad \forall\, p \in Y = L^2(\Omega;E)\ .$$

The problem (2.26) is now identical to problem (2.64) and to find the dual we have to calculate F^* and G^*. To find G^* it is sufficient, by Proposition 2.5, to calculate ψ^* and we shall then have

(2.68) $$G^*(p) = \int_\Omega \psi^*(p(x))dx\ ,\quad \forall\, p \in L^2(\Omega;E)\ .$$

But, for $\eta \in E$

$$\psi^*(\eta) = \sup_{\xi\in E}\,\{\xi.\eta - \psi(\xi)\}$$

$$= \sup_{\xi=\xi^D+sI\in E}\,\{\xi^D.\eta^D + \frac{1}{n}\,\eta_{ii}s - \mu\xi^D_{ij}\,\xi^D_{ij} - \frac{\kappa}{2}\,s^2\}$$

$$= \sup_{s\in\mathbb{R}}\,\{\frac{s}{n}\,\eta_{ii} - \frac{\kappa}{2}\,s^2\} + \sup_{\xi^D\in E^D}\,\{\xi^D.\eta^D - \mu\xi^D.\xi^D\}$$

(2.69) $$\psi^*(\eta) = \frac{(\eta_{ii})^2}{2\kappa\, n^2} + \frac{1}{4\mu}\,\eta^D.\eta^D\ .$$

The calculation of F^* is more delicate and requires an appeal to the generalised Green's formula (1.49). To simplify we restrict ourselves to the calculation of $F(\Lambda^* p)$, for $p \in Y^* = Y$, which is sufficient for our purposes.

LEMMA 2.2

For all $p \in Y = L^2(\Omega;E)$

$$(2.70)\qquad F^*(\Lambda^* p) = \begin{cases} \displaystyle\int_{\Gamma_0} (p.\nu)\, u_0 \, d\Gamma & \text{if } \begin{cases} \operatorname{div} p - f = 0 & \text{in } \Omega \\ p.\nu + g = 0 & \text{on } \Gamma_1 \end{cases} {}^{(1)} \\ +\infty & \text{if not} \end{cases}$$

Proof.

We apply Lemma 2.1 with $V = H^1(\Omega)^3$, $v_0^* = -L$, $v_0 = u_0$, and $\mathscr{B}$ the set of v in $H^1(\Omega)^3$ which vanish on Γ_0. We deduce from this lemma that $F^*(\Lambda^* p)$ is equal to

$$(2.71)\qquad \langle \Lambda^* p + L, u_0 \rangle = \int_\Omega (p.\varepsilon(u_0) + f.u_0)\,dx + \int_{\Gamma_1} g\, u_0 \, d\Gamma ,$$

if $\Lambda^* p + L$ is in $\mathscr{B}^0$, and to $+\infty$ if not.

To say that $\Lambda^* p + L$ is in $\mathscr{B}^0$ is equivalent to saying that

$$(2.72)\qquad \int_\Omega (p.\varepsilon(u) + f.u)\,dx + \int_{\Gamma_1} g.u \, d\Gamma = 0 , \quad \forall\, u \in \mathscr{B} .$$

Writing (2.72) in the first place with u belonging to $\mathscr{C}_0^\infty(\Omega)^3$ we see that

---

(1) If $p \in L^2(\Omega;E)$ and $\operatorname{div} p = -f \in L^2(\Omega)^n$, then by (1.47), the trace $\gamma_\nu(p)$ of $p.\nu$ on Γ has a meaning. In (2.70) $p.\nu|_\Gamma$ and $u_0|_\Gamma$ are 'improper' notations for the traces $\gamma_\nu(p)$ and $\gamma_0(u_0)$. Furthermore the integral is an 'improper' notation for the scalar product $\langle \gamma_\nu(p), \gamma_0(u_0) \rangle$ in the duality between $H^{\frac12}(\Gamma)^3$ and $H^{-\frac12}(\Gamma)^3$ (here we have made use of the condition (2.63)).

$$\sum_{i,j=1}^{3} \langle p_{ij}, \varepsilon_{ij}(u)\rangle - \langle f,u\rangle = 0 \ , \quad \forall\, u \in \mathscr{C}_0^\infty(\Omega)^3 \ ,$$

$$\sum_{i,j=1} \langle \frac{\partial p_{ij}}{\partial x_j} - f_1, u_i\rangle = 0 \ , \quad \forall\, u \in \mathscr{C}_0^\infty(\Omega)^3 \ ,$$

or in other words

(2.73) $$\sum_{j=1}^{3} \frac{\partial p_{ij}}{\partial x_j} - f_i = 0 \quad \text{in the sense of distributions on } \Omega \ ;$$

but if $p \in L^2(\Omega;E)$ satisfies (2.73), then $\operatorname{div} p \in L^2(\Omega)^3$; the trace of the normal component of p on Γ, that is to say $p.\nu$, is defined in $H^{-\frac{1}{2}}(\Gamma)^3$ (cf. (1.47)); and the generalised Green's formula applies, so that we have:

(2.74) $$\int_\Omega p.\varepsilon(u)dx = -\int_\Omega \operatorname{div} p.u \, dx + \int_\Gamma (p.\nu).u \, d\Gamma$$

for all u of $H^1(\Omega)^3$, and taken in conjunction with (2.73) the equation (2.72) becomes

$$\int_\Gamma (p.\nu).u \, d\Gamma + \int_\Gamma g\, u \, d\Gamma = 0 \ , \quad \forall\, u \in \mathscr{B} \ .$$

Since $u = 0$ on Γ_0 for all $u \in \mathscr{B}$ and the trace of u on Γ_1 is arbitrary when u is in $\mathscr{B}$, we obtain

(2.75) $$p.\nu + g = 0 \quad \text{on } \Gamma_1 \ .$$

It can be shown [1] that conversely if $p \in L^2(\Omega;E)$ satisfies (2.73) and (2.75), then $\Lambda^* p + L$ is in $\mathscr{B}^0$.

The last step in the proof is to transform (2.71) with the help of (2.74) (with u replaced by u_0). We then have, for a p satisfying (2.73)-(2.75):

---

[1] A complete justification of this point requires a few (non-essential) additional technical details to be taken into account; these are developed explicitly in the Appendix to Section 2.

$$\langle \Lambda^* p+L, u_0 \rangle = \int_{\Gamma_0} (p.\nu).u_0 \, d\Gamma \quad .$$

□

We can now find an explicit expression for the dual of (2.64). By (2.27), (2.68), (2.69) and Lemma 2.2, it can be written

$$(2.76) \qquad \underset{\substack{\sigma \in L^2(\Omega;E) \\ \text{div } \sigma + f = 0 \text{ in } \Omega \\ \sigma.\nu = g \text{ on } \Gamma_0}}{\text{Sup}} \left\{ \int_{\Gamma_0} (\sigma.\nu).u_0 \, d\Gamma - \frac{1}{2} \mathscr{A}(\sigma,\sigma) \right\}$$

where we have set, for $\sigma, \tau \in L^2(\Omega;E)$,

$$(2.77) \qquad \mathscr{A}(\sigma,\tau) = \frac{1}{9\kappa} \int_\Omega \sigma_{ii} \, \tau_{jj} \, dx + \frac{1}{2\mu} \int_\Omega \sigma^D . \tau^D \, dx \ .$$

We have written $\sigma = -p$ for the generic element of $Y = L^2(\Omega;E)$ in order to conform more closely to the notation usual in mechanics. We recognize in (2.76) the variational problem which furnishes the stress tensor σ of the original problem.

To state the results we distinguish between the case where Γ_0 has positive measure and the case where $\Gamma_0 = \emptyset$.

PROPOSITION 2.6.

Suppose Γ_0 *has positive measure. The problems (2.64) and (2.76) are dual to one another and*

$$(2.78) \qquad \inf\{(2.64)\} = \text{Sup } \{(2.76)\} \in \mathbb{R} \ .$$

The problem (2.76) has an unique solution σ *which represents the stress tensor of the mechanical problem. Problem (2.64) has an unique solution* u *, representing the displacement of the mechanical problem. The two solutions* u *and* σ *are connected by the relations*

$$\sigma^D = 2\mu \, \varepsilon^D(u) \ ,$$

$$tr\ \sigma = 3\kappa\ tr\ \varepsilon(u) = 3\kappa\ div\ u\ .$$

Proof

i) We have established that the problems (2.64) and (2.76) are dual to one another. From first principles, or by using Proposition 2.4 we can see that the function G is continuous on $Y = L^2(\Omega;E)$. Since F is finite at u_0, the criterion (2.29) applies and equality in (2.78) results from (2.30).

ii) The existence and uniqueness of a solution to (2.64) can be deduced from Propositions 2.1 and 2.2 which we use under the following conditions: $V = H^1(\Omega)^3$, J is the functional in (2.64), $\mathscr{C} = \{v \in H^1(\Omega)^3\ ,\ v = 0 \text{ on } \Gamma_0\}$. It is clear that $\mathscr{C}$ is a closed convex non-vacuous set in V and that J is convex and l.s.c. on V.

From (2.66) we have, for all $\xi \in E$,

$$\psi(\xi) \geq c_1 \left\{\xi^D.\xi^D + \frac{(\xi_{ii})^2}{3}\right\}\ ,\ c_1 = \min\left(\frac{3\kappa}{2}, \mu\right),$$

(2.81) $$\psi(\xi) \geq c_1|\xi|^2\ ,$$

so that

(2.82) $$\int_\Omega \psi(\varepsilon(u))dx \geq c_1 \int_\Omega |\varepsilon(u)|^2\ dx\ .$$

Since Γ_0 has positive measure it follows from Remark 1.3 ii) that there exists a $c_2 > 0$ such that

(2.83) $$\int_\Omega \psi(\varepsilon(u))dx \geq c_2\ \|u\|^2_{H^1(\Omega)^3} - \int_{\Gamma_0} |\gamma_0(u)|^2\ d\Gamma\ .$$

From Schwarz's inequality

$$|L(u)| = \left|\int_\Omega f\ u\ dx + \int_\Gamma g\ u\ d\Gamma\right|$$

$$\leq |f|\ \ |u| + \|g\|_{L^2(\Gamma_1)^3}\ \|u\|_{L^2(\Gamma_1)^3}$$

and by the Trace Theorem for $H^1(\Omega)$ there exists a $c_3 > 0$ such that

(2.84) $$|L(u)| \leq \left(|f|+c_3\|g\|_{L^2(\Gamma_1)^3}\right)\|u\|_{H^1(\Omega)^3} = c_4\|u\|_{H^1(\Omega)^3} .$$

We see from (2.83) that the functional to be minimised in (2.64) is minorised by

$$c_2\|u\|^2_{H^1(\Omega)^3} - c_2\int_{\Gamma_0}|\gamma_0(u)|^2\, d\Gamma - c_4\|u\|_{H^1(\Omega)^3} ,$$

and that the condition (2.43) is therefore satisfied on $\mathscr{C}$: by Proposition 2.1 the infimum of (2.65) is finite and (2.64) has at least one solution. Furthermore the function $\int_\Omega \psi(\varepsilon(u))\, dx$ is strictly convex on $\mathscr{C}$ and Proposition 2.2 implies the uniqueness of the solution of (2.64).

iii) Since $\inf\{(2.64)\} \in \mathbb{R} \in (> -\infty)$, the same applies to $\sup\{(2.76)\}$ by (2.30) which implies that

(2.85) the set of σ admissible for (2.76) is non-vacuous.

The existence of a solution to (2.76) can be proved by appealing again to Proposition 2.1 or directly by the result which follows (2.30). Uniqueness is a consequence of Proposition 2.2, the function $\mathscr{A}(\sigma,\sigma)$ being strictly convex on $L^2(\Omega;E)$.

vi) It only remains to prove (2.79) and (2.80). These properties are simple consequences of the extremality relation (2.32) which can be written here in the form

$$\int_\Omega \{\psi(\varepsilon(u)) + \psi^*(\sigma) - \sigma.\varepsilon(u)\}\, dx = 0$$

or, since the integrand is almost everywhere non-negative:

(2.86) $$\psi(\varepsilon(u)) + \psi^*(\sigma) - \sigma.\varepsilon(u) = 0 \ , \quad dx\text{-p.p.}$$

An elementary calculation which we omit enables us to remark that (2.86) is identical to (2.79)-(2.80) (cf. the calculation which led to (2.69)).

□

We now turn to the case where Γ_0 is empty and $\Gamma_1 = \Gamma$.

PROPOSITION 2.7.

Suppose that $\Gamma_0 = \emptyset$.

The problems (2.64) and (2.76) are dual to each other and

$$\text{(2.87)} \qquad inf\{(2.64)\} = sup\ \{(2.76)\}\ .$$

If

$$\text{(2.88)} \qquad L(u) = \int_\Omega f\,u\,dx + \int_{\Gamma_1} g\,u\,d\Gamma = 0\ ,\quad \forall\ u \in \mathcal{R}\ ,$$

problem (2.76) has an unique solution σ *, while problem (2.64) has a solution* u *which is unique to within a rigid displacement, and the relations (2.79) and (2.80) are satisfied.*

If the condition (2.88) does not hold, then the number $inf\ (2.64) = sup\ (2.76)$ *is equal to* $-\infty$ *, the two problems have no solution and there is no* σ *admissible for (2.76).*

<u>Proof.</u>

i) The duality between (2.64) and (2.76) and the equation (2.87) can be established as above.

When condition (2.88) does not hold, there exists a $u_1 \in \mathcal{R}$ such that $L(u_1) \neq 0$, and by considering the elements u of $\mathbb{R}\,u_1$, we see that the infimum of (2.64) is $-\infty$, and that this variational problem has no solution. Since by (2.87), sup (2.76) = $-\infty$, the dual problem (2.76) has no admissible elements $\sigma \in L^2(\Omega;E)$, that is to say there are no elements σ satisfying the constraints $\mathrm{div}\sigma + f = 0$ in Ω , and $\sigma.\nu = g$ on Γ_1 ; <u>*a fortiori*</u> therefore the problem has no solution.

ii) When condition (2.88) is satisfied we apply Propositions 2.1 and 2.2 to a 'quotient problem' derived from (2.64). We write $\dot{u},\dot{v},\ldots,$ for the elements of $H^1(\Omega)^3/\mathscr{R}$ and $u \in \dot{u}$, $v \in \dot{v},\ldots,$ is a representative of $\dot{u}$, $\dot{v},\ldots$. By virtue of (2.88) $\dot{L}(\dot{u}) = L(u)$ defines a linear form which is continuous on $H^1(\Omega)^3/\mathscr{R}$, and we can then consider the variational problem

$$(2.89) \qquad \inf_{\dot{u}\in H^1(\Omega)^3/\mathscr{R}} \{\dot{J}_0(\dot{u})-\dot{L}(\dot{u})\}$$

$$(2.90) \qquad \dot{J}_0(\dot{u}) = \int_\Omega \psi(\varepsilon(u))dx \ , \quad u \in \dot{u} \ .$$

We apply Propositions 2.1 and 2.2 with $V = \mathscr{C} = H^1(\Omega)^3/\mathscr{R}$, $J = \dot{J} - \dot{L}$. Since

$$N(\dot{u}) = |\varepsilon(u)|_{L^2(\Omega;E)} \ , \quad u \in \dot{u} \ , \text{ is a norm on}$$

$H^1(\Omega)^3/\mathscr{R}$ equivalent to the natural norm, the condition (2.43) is satisfied for $\dot{J}$ and hence for $\dot{J} - \dot{L}$. All the other hypotheses in these propositions are easily seen to hold good and the uniqueness and existence of the solution $\dot{u}$ of (2.89) therefore follows. Returning to the problem (2.64) we now see that its infimum is finite $(> -\infty)$, that $u \in \dot{u}$ is a solution of (2.64), and that there are no other solutions (2.64) except those which represent $\dot{u}$. The remaining results stated in Proposition 2.7 can then be proved as in Proposition 2.6.

□

Remark 2.4 i) We have proved the above results on the supposition that Ω is of class $\mathscr{C}^2$. It would have been enough to assume that Ω is an open Lipschitzian set and that $\partial\Omega$ is of class $\mathscr{C}^2$ in the neighbourhood (in $\partial\Omega$) of $\partial\Gamma_0 \cap \partial\Gamma$.

ii) We may also consider the case where $\Omega \subset \mathbb{R}^2$ which would correspond to the plane deformations of an elastic cylindrical bar of rectangular cross-section Ω . We may also consider, though it has no particular mechanical application, the case where $\Omega \subset \mathbb{R}^n$, $n \geq 4$. The modifications needed present no difficulty.

iii) Instead of (2.67) we can take for L any linear form which is continuous on $H^1(\Omega)^3$. Everything which has been said can be restated to apply to this case by making minor modifications where necessary. This enables us to deal with an elastic body subjected to forces more general than f and g, for example to a set of forces distributed with a given surface density on some surface interior to Ω.

iv) We have assumed in Propositions 2.6 and 2.7 that $\Gamma_0 = \emptyset$ or that Γ_0 has positive measure. It is of course possible to consider intermediate cases, the set $\mathscr{R}$ then being replaced by a subset of $\mathscr{R}$ which leaves Γ_0 invariant.

v) It is not necessary to assume that the material is isotropic. For a non-isotropic material, we take in (2.64), (cf. G. Duvaut - J.L. Lions [1]):

(2.91) $$\psi(\xi) = \frac{1}{2} a_{ijkh} \xi_{kh} , \forall \ \xi \in E ,$$

where the a_{ijkh} must satisfy the symmetry conditions

(2.92) $$a_{ijkh} = a_{ijhk} = a_{jikh} = a_{khij} ,$$

and the coercivity condition:

(2.93) there exists an $\alpha_1 > 0$ such that

$$a_{ijkh} \xi_{ij} \xi_{kh} \geq \alpha_1 |\xi|^2 = \alpha_1 \xi_{ij} \xi_{ij} , \quad \forall \ \xi \in E .$$

We recover (2.64) if we take

(2.94) $$a_{ijkh} = \lambda\delta_{ij} \delta_{kh} + \mu(\delta_{ik} \delta_{jh} + \delta_{ih} \delta_{jk}) .$$

In the general case defined by (2.91) to (2.93) the results are essentially the same. The function conjugate to ψ can be written

$$\psi^*(\eta) = \operatorname*{Sup}_{\xi \in E} \{\xi.\eta - \psi(\xi)\} .$$

The above supremum is attained at a point ξ of E which satisfies

$$\eta_{ij} = a_{ijkh}\,\xi_{kh}\,. \tag{2.95}$$

It is easy to see that these relations can be 'inverted' to give

$$\xi_{ij} = A_{ijkh}\,\eta_{kh}\,, \tag{2.96}$$

where the numbers A_{ijkh} satisfy the following conditions:

$$A_{ijkh} = A_{ijhk} = A_{jikh} = A_{khij}\,; \tag{2.97}$$

(2.98) there exists an $\alpha_2 > 0$ such that

$$A_{ijkh}\,\xi_{ij}\,\xi_{kh} \geq \alpha_2|\xi|^2 = \alpha_2\,\xi_{ij}\,\xi_{ij}\,, \quad \forall\, \xi \in E\,;$$

$$a_{ijkh}\,A_{kh\ell m} = \delta_{i\ell}\,\delta_{jm}\,. \tag{2.99}$$

We thus obtain

$$\psi^*(\eta) = \frac{1}{2}\,A_{ijkh}\,\eta_{ij}\,\eta_{kh}\,. \tag{2.100}$$

The only other changes occur in the form of the dual problem (2.76) where one must take

$$\mathscr{A}(\sigma,\tau) = \int_\Omega A_{ijkh}\,\sigma_{ij}\,\sigma_{kh}\,dx \tag{2.101}$$

and in the extremality relations (2.79), (2.80) which have to be replaced by the relations,

$$\sigma_{ij} = a_{ijkh}\,\varepsilon_{kh}(u)\,, \quad i,j = 1,\ldots,n \tag{2.102}$$

which are equivalent to

$$\varepsilon_{ij}(u) = A_{ijkh}\,\sigma_{kh}\,, \quad i,j = 1,\ldots,n\,. \tag{2.103}$$

Appendix of section 2

This appendix is of a purely technical nature and can be omitted on a first reading.

To complete the proof of Lemma 2.2 we still have to show that if $p \in L^2(\Omega;E)$ satisfies (2.73) and (2.75) then (cf. [1] p.47)

$$\int_\Gamma p.\nu\, u\, d\Gamma + \int_{\Gamma_1} g\, u\, d\Gamma = 0\ , \quad \forall\, u \in \mathscr{B}\ . \tag{2.104}$$

The difficulty stems from the fact (cf. [1] p.46) that the first integral, which is in fact a shorthand expression for the inner product of $\gamma_\nu(p)$ and $\gamma_0(u)$ in the duality pairing between $H^{-\frac{1}{2}}(\Gamma)^3$ and $H^{\frac{1}{2}}(\Gamma)^3$ is not necessarily equal to the sum of the products $\langle\gamma_\nu(p),\gamma_0(u)\rangle_{H^{-1/2}(\Gamma_0)^3, H^{1/2}(\Gamma_0)^3}$ and $\langle\gamma_\nu(p),\gamma_0(u)\rangle_{H^{-1/2}(\Gamma_1)^3, H^{1/2}(\Gamma_1)^3}$

To establish the validity of (2.104) we first observe that this relation is true for all u belonging to $\mathscr{B}_0$, the set of elements of $H^1(\Omega)^3$ which are zero in a neighbourhood (in Ω) of Γ_0 ; the result is easily seen to be true in this case because $\gamma_0(u)$ vanishes in a neighbourhood (in Γ) of Γ_0. By continuity, (2.104) remains true for all u in the closure of $\mathscr{B}_0$ in $H^1(\Omega)^3$, and it only remains to show that $\mathscr{B}_0$ is dense in $\mathscr{B}$, the set of elements of $H^1(\Omega)^3$ whose trace on Γ_0 is zero.

This density result is easily proved if Γ_*, the common boundary of Γ_0 and Γ_1 in Γ, has some suitable regularity property. We shall suppose here (cf. a similar situation in R. Kohn - R. Temam [2]):

(2.105) For every x of Γ_*, there is a $\mathscr{C}^2$ diffeomorphism defined in a neighbourhood $\mathscr{O}_x$ of x, which sends $\mathscr{O}_x$ into the cube $(]-1,+1[)^3$ of $\mathbb{R}^3$, $\mathscr{O}_x \cap \Gamma$ into the square $(]-1,+1[)^2$ of $\mathbb{R}^2 \times \{0\}$, and $\mathscr{O}_x \cap \Gamma_*$ into the segment $(]-1,+1[)$ of $\mathbb{R} \times \{0\} \times \{0\}$.

We now consider a covering of Γ_* by a finite number of sets $\mathcal{O}_x$, $x = x^1,\dots,x^N$. It is possible to find a partition of unity subordinated to this covering, with functions $\psi_m^{(i)} \in \mathscr{C}_0^\infty(\mathcal{O}_{x_i})$, vanishing for $x \in \mathcal{O}_{x_i}$ with $d(x, \mathcal{O}_{x_i} \cap \Gamma_*) > 1/m$. This allows us to express u as the sum of a function belonging to $\mathscr{A}_0$ and N functions $\psi_m^{(i)}u$. All that is now required is to show that the norms of $\psi_m^{(i)}u$ in $H^1(\Omega)^3$ can be made arbitrarily small by choosing a large enough m, and this is elementary (note that N remains fixed).

Remark 2.5. i) We have just established that

$$\int_\Gamma p.\nu\, u\, d\Gamma + \int_\Gamma g\, u\, d\Gamma = 0$$

for all p of $L^2(\Omega;E)$ satisfying (2.73)-(2.75) and for all u of $H^1(\Omega)^3$ whose trace on Γ_0 is zero.

ii) We have established that, on hypothesis (2.105), the functions of $H^1(\Omega)^3$ which vanish in the neighbourhood of Γ_0 are dense in the set of functions of $H^1(\Omega)^3$ which vanish on Γ_0. By making use of the techniques of partitioning unity, and of regularisation used in the proof of Proposition 1.3 it follows by standard arguments that:

(2.106) The set of functions of $\mathscr{C}^\infty(\bar{\Omega})^3$ which vanish (i.e. have the value zero) in the neighbourhood of Γ_0, is dense in the set of functions of $H^1(\Omega)^3$ which vanish on Γ.

3. Formulation of the variational problems of plasticity theory

In this section we shall describe the mechanical problem which will constitute our reference problem. We recall the various relations and boundary conditions which have to be satisfied by the displacement field and the stress field. We shall then give the constitutive laws governing the relations between these two fields. Lastly we shall deduce from these (in a formal way) the variational principles whose solutions are respectively the field of displacements and the field of stresses.

3.1 Kinematically admissible sets
Statically admissible sets

The mechanical reference problem is the following: a material body occupies a volume Ω of $\mathbb{R}^3$; we shall suppose that Ω is an open bounded connected set of $\mathbb{R}^3$ with boundary Γ and that Ω is of class $\mathscr{C}^2$ (cf. (1.15)), though some of the developments which follow remain valid on weaker regularity hypotheses. This material body is subjected to internal forces of volume density f in Ω. In addition Γ is the union of a set Γ_0, open in Γ, and its complementary set whose interior we denote by Γ_1 (so that $\Gamma_0 = \Gamma \backslash \bar{\Gamma}_1, \Gamma_1 = \Gamma \backslash \bar{\Gamma}_0$), and the body is subjected on Γ_1 to traction forces of surface density g. We shall suppose f and g to be given functions with

$$f \in L^2(\Omega)^3 \;,\quad g \in L^2(\Gamma_1)^3 \;. \tag{3.1}$$

The object of the problem, from the mechanical point of view, is to determine the new state of equilibrium resulting from the application of the forces f and g. The unknowns are the functions σ and u defined on Ω; σ, a function from Ω into E, representing the stress field ($\sigma(x)$ is the stress tensor at the point x for any point x of Ω) and u, a function from Ω into $\mathbb{R}^3$, representing the displacement vector at x (which means that the material point which was originally situated at x moves to a new equilibrium position at the point $x + u(x)$). The displacement is assumed to be given on Γ_0. As in Section 2 we shall suppose that the displacement on Γ_0 is furnished as the trace, restricted to Γ_0, of a function u_0 in $H^1(\Omega)^3$; for technical reasons we shall furthermore assume that $\gamma_0(u_0)$ vanishes outside Γ_0, so that:

$$(3.2) \qquad u_0 \in H^1(\Omega)^3 , \quad \theta_{\Gamma_0} \gamma_0(u_0) = \gamma_0(u_0) ,$$

where θ_{Γ_i} denotes the characteristic function in Γ of Γ_i, $i = 0,1$.

We must now specify the (minimum) degree of regularity which we shall impose on the functions σ and u. In the elastic case (cf. Section 2.3.2) we sought (and found in the existence results of Propositions 2.6 and 2.7) a function σ in $L^2(\Omega;E)$ and a function u in $H^1(\Omega)^3$. In the case of plasticity we shall still require σ to be in $L^2(\Omega;E)$ but we shall see that there is hardly any hope of obtaining a u in $H^1(\Omega)^3$, *and in fact the whole purpose of the developments in the Part 2 will be aimed at overcoming this very difficulty.*[1] Despite this, our object in Part 1 not being to obtain existence results, we

---

[1] In the physical problem, whose mathematical model we are now studying, we encounter slippage surfaces, that is to say surfaces across which the function u is discontinuous. Now by the Trace Theorem for $H^1(\Omega)$ (or for $W^{1,p}(\Omega)$, cf. (1.22) and (1.31)) no function of $H^1(\Omega)^3$ (nor any function of $W^{1,p}(\Omega)^3$ with $p \geq 1$) can exhibit such a discontinuity.

shall assume almost throughout this first part, that u belongs to $H^1(\Omega)^3$.

We then define the set of kinematically admissible displacements, which we shall denote by $\mathscr{C}_a(u_o)$ or simply by $\mathscr{C}_a$, if there is no risk of confusion

(3.3) $$\mathscr{C}_a = \mathscr{C}_a(u_o) = \{v \in H^1(\Omega)^3 \; , \; v = u_o \;\text{ on }\; \Gamma_o\} \; ,$$

the condition $v = u_o$ on Γ_o signifying that the trace $\gamma_o(v-u_o)$ (cf. (1.22)) vanishes almost everywhere on Γ_o.

Before defining the field of statically admissible stresses let us recall that the field of stresses which provides the solution to the problem satisfies the following equations:

(3.4) $$\nabla.\sigma + f = 0 \quad \left(\text{that is } \sum_{j=1}^{3} \frac{\partial \sigma_{ij}}{\partial x_j} + f_i = 0\right) \quad \text{in } \Omega \; ,$$

(3.5) $$\sigma.\nu = g \quad \left(\text{that is } \sum_{j=1}^{n} \sigma_{ij}\, \nu_j = g_j\right) \quad \text{in } \Gamma_1 \; ,$$

which are derived from the theorems on the conservation of momentum and the continuity properties of σ (cf. for example P. Germain [1]-[2], M. Gurtin [1]. Finally let us recall (cf. (1.47)) that if $\sigma \in L^2(\Omega;E)$ and $\operatorname{div}\sigma = -f \in L^2(\Omega)^3$ (by (3.1) and (3.4)) then we can define the trace of $\sigma.\nu$ on Γ, the latter being in $H^{-\frac{1}{2}}(\Gamma)^3$. The restriction of $\gamma_\nu(\sigma) = \sigma.\nu$ to the open subset Γ_1 of Γ is well-defined and the relations (3.4)-(3.5) are therefore meaningful for such σ, (3.4) being extended in the sense of distributions in Ω, and (3.5) in the sense of the Trace Theorems.

The set of fields of stress which are statically admissible for the problem is denoted by $\mathscr{S}_a(f,g)$ or simply by $\mathscr{S}_a$ if there is no risk of confusion

(3.6) $$\mathscr{S}_a = \mathscr{S}_a(f,g) = \{\sigma \in L^2(\Omega;E) \; , \; \sigma \;\text{ satisfying (3.4)-(3.5)}\} \; .$$

Lastly we write $\mathscr{C}_o$ and $\mathscr{S}_o$ for the sets $\mathscr{C}_a(0)$, $\mathscr{S}_a(0,0)$, respectively so that

(3.7) $\mathscr{C}_0 = \{v \in H^1(\Omega)^3 \;,\; v = 0 \;\text{ on }\; \Gamma_0\}$

(3.8) $\mathscr{S}_0 = \{\sigma \in L^2(\Omega;E) \;,\; \nabla.\sigma = 0 \;\text{ in }\; \Omega \;,\; \sigma.\nu = 0 \;\text{ on }\; \Gamma_1\}$.

Remark 3. . As already mentioned we may later on find it necessary to define sets $\mathscr{C}_a$, $\mathscr{S}_a$, $\mathscr{C}_0$, $\mathscr{S}_0$, for which the states $H^1(\Omega)^3$ and $L^2(\Omega;E)$ are replaced by other spaces.

3.2 The variational problems

The fields u and σ which are the solutions to the problems under consideration must belong to $\mathscr{C}_a$ and $\mathscr{S}_a$ respectively and must also satisfy other relations, which depends on the nature of the material: *the constitutive laws.*

For the class of problems in which we are interested the constitutive law is given by a relation between the stress tensor $\sigma(x)$ and the strain tensor $\varepsilon(u)(x)$, for every point x of Ω . We shall adopt the point of view introduced by J.J. Moreau [2] and consider a material characterised by a given function ψ (the surpotential) which is a proper, l.s.c. convex function from E into $\mathbb{R} \cup \{+\infty\}$, and its conjugate ψ^* from E(1) into $\mathbb{R} \cup \{+\infty\}$, with the same properties (cf. Section 1). Then, following J.J. Moreau [2], for each point x of Ω , we have:

(3.9) $$\sigma(x) \in \partial\psi(\varepsilon(u)(x)) \;,$$

this relation being equivalent, thanks to the properties of the sub-differential (see (2.21) and (2.23)) to

(3.10) $$\varepsilon(u)(x) \in \partial\psi^*(\sigma(x)) \;,$$

or

---

(1) Naturally we identify the Euclidean space E with its dual.

$$(3.11) \qquad \psi(\varepsilon(u)(x)) + \psi^*(\sigma(x)) = \varepsilon(u)(x).\sigma(x) \ .$$

From this starting point we can now show that u and σ are the solutions of variational problems. For u we begin by noting that, by (2.20), the relation (3.9) implies that for $\xi \in E$ and for (almost) all x of Ω

$$(3.12) \qquad \psi(\xi) \geq \psi(\varepsilon(u)(x)) + \sigma(x).(\xi-\varepsilon(u)(x)) \ .$$

In particular if $v \in H^1(\Omega)^3$ we have, for (almost) all x of Ω :

$$(3.13) \qquad \psi(\varepsilon(v)(x)) \geq \psi(\varepsilon(u)(x)) + \sigma(x).(\varepsilon(v)(x)-\varepsilon(u)(x)) \ .$$

and it follows, after integration over Ω, that

$$(3.14) \qquad \Psi(\varepsilon(v)) \geq \Psi(\varepsilon(u)) + \int_\Omega \sigma(x).(\varepsilon(v-u)(x))dx \ ,$$

where we have put

$$(3.15) \qquad \Psi(p) = \int_\Omega \psi(p(x))dx \ , \quad \forall \ p \in L^2(\Omega;E) \ ,$$

the functional $p \mapsto \Psi(p)$ from $L^2(\Omega;E)$ into $\mathbb{R} \cup \{+\infty\}$ being clearly convex, l.s.c. and proper.

We can transform the integral on the right-hand side of (3.15) by using the generalised Green's formula (1.51):

$$(3.16) \qquad \int_\Omega \sigma.\varepsilon(v-u)dx = - \int_\Omega (\text{div } \sigma).(v-u)dx + \int_\Gamma (\sigma.\nu).(v-u)d\Gamma \ .$$

We restrict ourselves at present to an element v of $\mathscr{C}_a$ and utilise the fact that $\sigma \in \mathscr{S}_a$; making use of the Remark 2.5 we obtain

$$\int_\Omega \sigma.\varepsilon(v-u)dx = \int_\Omega f.(v-u)dx + \int_{\Gamma_1} g.(v-u)d\Gamma$$
$$= (\text{using } (2.67)) = L(v-u) \ .$$

Reverting to (3.14) we see that

$$\Psi(\varepsilon(v)) - L(v) \geq \Psi(\varepsilon(u)) - L(u) \ , \ \forall\ v \in \mathscr{C}_a \ , \tag{3.17}$$

and we conclude that if σ , u is the solution to the problem under consideration,

(3.18) u attains the minimum in $\mathscr{C}_a$ of $\Psi(\varepsilon(v)) - L(v)$.

Similarly, to write down the variational problem whose solution is σ , we begin by observing that, by virtue of (2.20), the relation (3.10) implies

$$\psi^*(\xi) \geq \psi^*(\sigma(x)) + \varepsilon(u)(x).(\xi-\sigma(x)) \ , \tag{3.19}$$

for all $\xi \in E$ and for (almost) all x of Ω . If $\tau \in L^2(\Omega;E)$ we have, for (almost) all x of Ω

$$\psi^*(\tau(x)) \geq \psi^*(\sigma(x)) + \varepsilon(u)(x).(\tau(x)-\sigma(x)) \tag{3.20}$$

and hence by integration over Ω

$$\Psi^*(\tau) \geq \Psi^*(\sigma) + \int_\Omega \varepsilon(u)(x).(\tau(x)-\sigma(x))dx \ , \tag{3.21}$$

where we have set in (3.21),

$$\Psi^*(\tau) = \int_\Omega \psi^*(\tau(x))dx \ , \ \forall\ \tau \in L^2(\Omega;E) \ ; \tag{3.22}$$

It follows from Proposition 2.5 that Ψ^* is indeed the conjugate of Ψ on $L^2(\Omega;E)$ ([1]) .

([1]) Since $\psi \geq 0$, it suffices to verify that there exists a $\tau_0 \in L^2(\Omega;E)$ such that $\Psi(\tau_0) < \infty$. But ψ being proper there is an $s \in E$ such that $\psi(s) < \infty$, and the constant function $\tau_0(x) = s$ meets the required conditions. Note that Ψ considered as a function on $L^\alpha(\Omega;E)$, $1 \leq \alpha < \infty$ also has Ψ^* , as its conjugate, on $L^{\alpha'}(\Omega;E)$ where $\alpha' = \alpha/(\alpha-1)$.

With the help of the generalised Green's formula (1.51) we write

(3.23) $$\int_\Omega (\tau-\sigma).\varepsilon(u)dx = -\int_\Omega (\mathrm{div}(\tau-\sigma)).u\, dx + \int_\Gamma ((\tau-\sigma).\nu)u\, d\Gamma .$$

We now restrict ourselves to an element τ of $\mathscr{S}_a$ and make use of Remark 2.5 and the fact that $\sigma \in \mathscr{S}_a$ and $u \in \mathscr{C}_a$. We then find (cf, Note ([1]) on p.46)

(3.24) $$\int_\Omega (\tau-\sigma).\varepsilon(u)dx = \int_{\Gamma_0} ((\tau-\sigma).\nu).u_0\, d\Gamma$$

and (3.21) then implies:

(3.25) σ attains the minimum in $\mathscr{S}_a$ of $\Psi^*(\tau) - \int_{\Gamma_0} (\tau.\nu).u_0\, d\Gamma$.

Remark 3.2. The case presented above corresponds to that of a homogeneous material. In the case of a non-homogeneous material, we need to introduce a non-negative Caratheodory function $\psi = \psi(x,\xi)$ from $\Omega \times E$ into $\mathbb{R} \cup \{+\infty\}$ (cf. Remark 2.3) which is convex, l.s.c. and proper in ξ, for almost all x of Ω. The constitutive law (3.9)-(3.10) takes the form

(3.26) $$\sigma(x) \in \partial\psi(x,\varepsilon(u)(x)) , \ \varepsilon(u)(x) \in \partial\psi^*(x,\sigma(x)) ;$$

where, in (3.26) ψ^* is the conjugate of ψ with respect to the second argument for almost all x, and the sub-differentials $\partial\psi$, $\partial\psi^*$ are similarly taken with respect to the second argument. The problems (3.18) and (3.25) remain unchanged provided that we define

(3.27) $$\Psi(p) = \int_\Omega \psi(x,p(x))dx, \quad \forall\, p \in L^2(\Omega;E)$$

(3.28) $$\Psi^*(\tau) = \int_\Omega \psi^*(x,\tau(x))dx , \quad \forall\, \tau \in L^2(\Omega;E) .$$

The functions Ψ and Ψ^* are proper l.s.c. convex functions conjugate to each other on $L^2(\ ;E)$. The further results can be extended without difficulty to this case. □

3.3 Examples of constitutive laws

In conclusion, we shall describe here the most familiar examples of functions ψ, and in particular the function which corresponds to the Hencky model of plastic behaviour, which will be at the centre of the study which follows.

i) Linear elasticity

This is the simplest case, already implicitly described in Section 2.3.2. The function ψ, for an isotropic material, is that given in (2.65)-(2.66):

$$(3.29) \qquad \psi_e(\xi) = \frac{\lambda}{2}\, \xi_{ii}\, \xi_{jj} + \mu \xi_{ij}\, \xi_{ij}$$
$$= \frac{\kappa}{2}\, \xi_{ii}\, \xi_{jj} + \mu \xi^D_{ij}\, \xi^D_{ij} \ , \ \forall\, \xi \in E \ ,$$

where we have used the Einstein summation convention and

$\xi_{ij} = \xi^D_{ij} + \frac{1}{3}\xi_{kk}\,\delta_{ij}$, $\kappa = \lambda + 2\mu/3$, $\kappa>0$, $\mu>0$.

The calculation of ψ^*_e has already been carried out in Section 2.3.2. :

$$(3.30) \qquad \psi^*_e(\eta) = \frac{1}{18\kappa}\, \eta_{ii}\, \eta_{jj} + \frac{1}{4\mu}\, \eta^D_{ij}\, \eta^D_{ij} \ .$$

Note that the functions ψ_e and ψ^*_e are both positive definite quadratic forms over E. The constitutive law can be written

$$(3.31) \qquad \sigma^D = 2\mu\, \varepsilon^D(u) \ , \ \sigma_{ii} = 3\kappa\, \varepsilon_{ii}(u) \ ,$$

at every point x of Ω (compare this with (2.79)-(2.80)), or more succinctly

$$(3.32) \qquad \varepsilon(u) = A\sigma \ ,$$

where $A\sigma$ is the tensor whose components are

$$(3.33) \qquad (A\sigma)_{ij} = \frac{1}{2\mu}\, \sigma_{ij} - \frac{\lambda}{2\mu(3\lambda+2\mu)}\, \sigma_{kk}\, \delta_{ij} \ .$$

In the case of a non-isotropic material the functions ψ_e and ψ^*_e

are those given in (2.91) and (2.100), with the corresponding properties of the coefficient a_{ijkh}, A_{ijkh}, and the constitutive law is again of the form (3.32), the tensor $A\sigma$ having as its components (compare with 2.103)) :

$$(A\sigma)_{ij} = A_{ijkh}\,\sigma_{kh} .$$

In all these cases we can write

$$\psi_e^*(\eta) = \frac{1}{2} A(\eta,\eta) \tag{3.35}$$

where

$$A(\xi,\eta) = A\xi.\eta \quad , \tag{3.36}$$

the tensor $A\xi$ being given by (3.33) in the isotropic case and by (3.34) in the general case.

□

For all the examples which follow K is a closed convex non-vacuous subset of E, and contains 0 in its interior. In a certain number of cases, K is of the form $K^D \oplus \mathbb{R}\, I$, where K^D is a non-vacuous closed convex subset of E^D which contains 0 in its interior and $\mathbb{R}\, I$ represents the set of spherical tensors. The convex set K is given in general under the form

$$K = \{\sigma \in E \ , \ \mathcal{F}(\sigma) \leq 0\} \ , \tag{3.37}$$

where $\mathcal{F}$ is a convex function, continuous on E, with $\mathcal{F}(0) < 0$. The most familiar examples are the von Mises convex for which

$$\mathcal{F}(\sigma) = \frac{1}{2}\,|\sigma^D|^2 - k^2 \ , \ |\sigma^D|^2 = \sigma^D_{ij}\,\sigma^D_{ij} \quad , \tag{3.38}$$

(cf. (1.34)), and the Tresca convex, where

(3.39) $$\mathcal{F}(\sigma) = \operatorname*{Max}_{i,j} |\sigma_i - \sigma_j| - k' ,$$

k and k' being numbers >0 , and $\sigma_1, \sigma_2, \sigma_3$ denoting the eigenvalues of σ . We may also mention the convex set considered by W.H. Yang for which

(3.40) $$\mathcal{F}(\sigma) = \operatorname{Max}\{|2\sigma_1-\sigma_2-\sigma_3|,|2\sigma_2-\sigma_1-\sigma_3|, 2\sigma_3-\sigma_1-\sigma_2|\} - k'' \quad (k''>0)$$ (1)

ii) Model of the perfectly plastic elastic type

This is the Hencky model in which we shall be especially interested (cf. H. Hencky [1], R. Hill [1], W.T. Koiter [1], W. Prager and P. Hodge [1]). Here the function ψ^* is simpler to define than ψ ; it can be written

(3.41) $$\psi^*(\eta) = \begin{cases} \psi_e^*(\eta) = \frac{1}{18\kappa}\eta_{ii}\eta_{jj} + \frac{1}{4\mu}\eta_{ij}^D\eta_{ij}^D & \text{if } \eta \in K \\ +\infty \text{ if not;} \end{cases}$$

λ, μ, κ have the same meaning as in the elastic model and (3.41) has to be compared with (3.30). The function ψ which is the conjugate of of ψ^* cannot in general be written down explicitly. However it may be noted that if $K = K^D \mathbb{R} I$, then

(3.42) $$\psi(\xi) = \frac{\kappa}{2}\xi_{ii}\xi_{jj} + \psi^D(\xi^D) ,$$

where ψ^D is the conjugate of $(\psi^D)^*$ and $(\psi^D)^*: E^D \to \mathbb{R}$ is the restriction of $(\psi^D)^*$ to E^D .

In the case of von Mises' convex set,

(3.43) $$\psi^D(\xi^D) = \begin{cases} \mu\,\xi_{ij}^D\xi_{ij}^D & \text{if } \xi_{ij}^D\xi_{ij}^D \leq k^2/2\mu^2 \\ k((2\xi_{ij}^D\xi_{ij}^D)^{1/2} - k/2\mu) & \text{if } \xi_{ij}^D\xi_{ij}^D \geq k^2/2\mu^2 . \end{cases}$$

---

(1) The convexity of this particular function $\mathcal{F}$ is proved by W.H. Yang [1] for $n=3$ and in J.J. Moreau [6] where more general results valid in any space dimensions can be found.

In this case the constitutive law can be written in the following form

(3.44)
$$\begin{cases} \varepsilon(u) = A\sigma + \boldsymbol{\lambda} \\ \lambda_{ij}(\xi_{ij}-\sigma_{ij}) \leqslant 0 \ , \ \forall\, \xi \in K \ . \end{cases}$$

The second relation (2.43) states that $\boldsymbol{\lambda}$ is in the sub-differential with respect to σ of the indicator function of K : $\boldsymbol{\lambda} = 0$ if σ is in the interior of K ; if $\sigma \in \partial K$ and K has a tangent plane at σ then $\boldsymbol{\lambda}$ is directed along the outward normal to K . Lastly if $\sigma \in \partial K$ and K is arbitrary, $\boldsymbol{\lambda}$ lies within the cone conjugate to the tangent cone to K at σ .

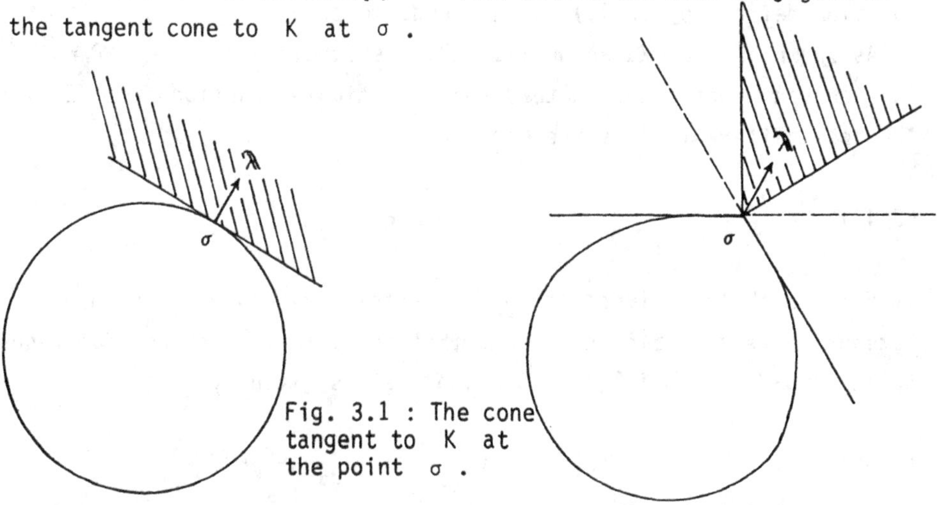

Fig. 3.1 : The cone tangent to K at the point σ .

In the non-isotropic case, we can also put (see (3.35)-(3.36))

(3.45)
$$\psi^*(\eta) = \begin{cases} \psi_e^*(\eta) = \frac{1}{2} A\eta.\eta & \text{if } \eta \in K \\ +\infty & \text{if not,} \end{cases}$$

and (3.44) remains valid, but we know of no explicit expression for ψ in this case.

□

We now describe various models of which it could be said that the perfectly elastic plastic model appears, in a certain sense, to be a limiting case (this will be defined more precisely in a few cases in Chapter III).

iii) Elasto-visco-plastic type model [1]

Given a parameter $\alpha>0$, the function ψ^* , which we denote by ψ^*_α, can be written (cf. G. Duvaut - J.L. Lions [1])

$$\psi^*_\alpha(\eta) = \frac{1}{2} A\eta.\eta + \frac{1}{4\alpha} [\eta - P_K\eta]^2 , \tag{3.46}$$

where P_K is the projection operator on K in E . Comparing this expression with (3.41)-(3.45) we see that ψ^*_α represents a *penalised* [2] form of ψ^* the function corresponding to the perfectly elastic plastic model, and it is easy to see that ψ^*_α converges to the function defined by (3.45) as α tends to zero.

As shown in E.H. Zarantonello [1], the function $\eta \mapsto \frac{1}{2}[\eta-P_K\eta]^2$ is a convex, continuous, Gâteaux-differentiable function from E into $\mathbb{R}$ whose differential is the mapping

$$\eta \mapsto \eta - P_K\eta . \tag{3.47}$$

It follows that the function ψ^*_α is (strictly) convex, continuous and Gâteaux-differentiable on E, and that its Gâteaux differential (equal to its sub-differential, cf. Section 2.1) is given by

$$(\psi^*_\alpha)'(\eta) = \partial\psi_\alpha(\eta) = A\eta + \frac{1}{2\alpha}[\eta - P_K\eta] \tag{3.48}$$

which enables us to express the constitutive law (3.10) of the material in the form

$$\varepsilon(u) = A\sigma + \frac{1}{2\alpha}[\sigma - P_K\sigma] . \tag{3.49}$$

It follows that $\varepsilon(u) = A\sigma$ if $\sigma \in K$ (since in this case $\sigma = P_K\sigma$) and that $\varepsilon(u)$ is given by (3.49) otherwise. For the von Mises convex set (3.37) (3.39) we have

---

[1] See Remark 3.3 as regards the interest, from the point of view of mechanics, of the models considered in ii), iii), iv), v) and vi).

[2] For the 'penalty' concept in the calculus of variations, cf. R. Courant [1], J.L. Lions [2].

(3.50) $$\varepsilon(u) = \begin{cases} A\sigma & \text{if } |\sigma^D| \leq \sqrt{2}\,k \\ A\sigma + \dfrac{1}{2\alpha}\dfrac{|\sigma^D| - \sqrt{2}\,k}{|\sigma^D|}\sigma^D & \text{if } |\sigma^D| \geq \sqrt{2}\,k \end{cases}$$

or, putting $s_+ = \max(s,0)$:

(3.51) $$\varepsilon(u) = A\sigma + \frac{1}{2\alpha}\frac{(|\sigma^D| - \sqrt{2}\,k)_+}{|\sigma^D|}\sigma^D$$

We know of no explicit expression for $\psi(\xi)$.

iv) Rigid-visco-plastic type model

This is closely related to the previous model. We replace A by 0 in (3.46) :

(3.52) $$\psi_\alpha^*(\eta) = \frac{1}{4\alpha}[\eta - P_K\eta]^2 ,$$

which leads to the constitutive law (compare with (3.49)):

(3.53) $$\varepsilon(u) = \frac{1}{2\alpha}[\sigma - P_K\sigma]$$

and in this case $\varepsilon(u) = 0$ if $\sigma \in K$ (rigid model). For the von Mises convex set we have

(3.54) $$\varepsilon(u) = \begin{cases} 0 & \text{if } |\sigma^D| \leq \sqrt{2}\,k \\ \dfrac{1}{2\alpha}\dfrac{|\sigma^D| - \sqrt{2}\,k}{|\sigma^D|}\sigma^D & \text{if } |\sigma^D| \geq \sqrt{2}\,k \end{cases}$$

It follows that $\varepsilon_{ii}(u) = \operatorname{div} u = 0$, and the model is therefore *incompressible*.

v) Norton-Hoff type model

We recall the definition of the gauge-function of a convex set K ; this is the function defined by

(3.55) $$\theta(\xi) = \inf\{s,\ \xi \in sK,\ s>0\} .$$

Given a parameter $q > 1$, the Norton-Hoff type model is obtained, following A. Friaâ [1] (cf. also R. Temam [12]), by choosing $\psi^* = \psi_q^*$ to be of the form

$$(3.56) \qquad \psi^*(\eta) = \frac{1}{2} A\eta.\eta + \frac{\kappa_1}{q} (\theta(\eta))^q \quad (\kappa_1 > 0) .$$

Of course when K is of the form $K^D + \mathbb{R} I$, then

$$(3.57) \qquad \theta(\eta) = \theta(\eta^D) = \inf\{s, \xi^D \in sK^D, s>0\} .$$

As $\theta(\xi) \leq 1$ if and only if $\xi \in K$, we see that, when $q \to \infty$, the function ψ_q^* tends to the function ψ^* of the elastic perfectly plastic model, or in other words to the function defined by (3.41) and the Norton-Hoff model thus appears as a formal perturbation of the elastic perfectly plastic model.

The function ψ_q^* is convex and continuous on E. We do not know any explicit expression for $\psi_q = (\psi_q^*)^*$ and we shall not state the constitutive law in an explicit form in the general case. However, for the von Mises convex set, it is easily seen that

$$(3.58) \qquad \theta(\xi) = \theta(\xi^D) = \frac{|\xi^D|}{\sqrt{2}\,k} = \frac{1}{\sqrt{2}\,k} (\xi_{ij}^D \, \xi_{ij}^D)^{1/2}$$

$$(3.59) \qquad \theta'(\xi) = \partial\theta(\xi) = \frac{1}{\sqrt{2}\,k|\xi^D|} \xi^D$$

and that the constitutive law can be written

$$(3.60) \qquad \varepsilon(u) = A\sigma + \frac{\kappa_1}{(\sqrt{2}\,k)^q} |\sigma^D|^{q-2} \sigma^D$$

which constitutes the usual form of the Norton-Hoff law (cf. H.J. Hoff [1], F.H. Norton [1]).

vi) Models with hardening

We introduce an auxiliary finite-dimensional Euclidean space and a function θ from Ω into E_1 which represents the hardening parameter.

We take, for each $\alpha \geq 0$, a family of non-vacuous closed convex sets $\mathcal{K}_\alpha$ in $E \times E_1$, whose intersection is $E \times \{0\}$, and such that $\mathcal{K}_0 = K \times E_1$. The constitutive relations can be written (instead of (3.9)-(3.10))

(3.61) $$(\sigma(x),\theta(x)) \in \partial\psi_\alpha(\varepsilon(u)(x)(x),0)$$

(3.62) $$(\varepsilon(u)(x),0) \in \partial\psi_\alpha^\star(\sigma(x),\theta(x)) ,$$

where ψ_α and $\psi_\alpha^\star$ are conjugate proper l.s.c. convex functions from $E \times E_1$ into $\mathbb{R} \times \{+\infty\}$, $\psi_\alpha^\star$ being defined as follows :

$$\psi_\alpha^\star(\xi,\theta) = \begin{cases} \frac{1}{2} A\xi.\xi + \frac{1}{2}|\theta|^2 & \text{if } (\xi,\theta) \in \mathcal{K}_\alpha \\ +\infty & \text{if not .} \end{cases}$$

We shall return to Chapter III to the question of the modification needed to problems (3.18)-(3.25) and the relationship with the elastic perfectly plastic model when α tends to zero. We end by giving two familiar examples of spaces E_1 and convex families $\mathcal{K}_\alpha$:

(3.63) K is the von Mises convex set, $E_1 = \mathbb{R}$ and for $\alpha \geq 0$

$$\mathcal{K}_\alpha = \{(\xi,\theta) \in E \times \mathbb{R} \ , \ |\xi^D| \leq \sqrt{2}\, k + \alpha\theta\}$$

(3.64) K is the von Mises convex set, $E_1 = E^D$ and for $\alpha \geq 0$

$$\mathcal{K}_\alpha = \{(\xi,\theta) \in E \times E^D \ , \ \xi - \alpha\theta \in K \ , \text{ i.e. } |\xi^D - \alpha\theta| \leq \sqrt{2}\, k\}.$$

We refer the reader to B. Halphen and N.Q. Son [1] for a complete study of models with hardening; (3.63) corresponds to an isotropic hardening model and (3.64) a linear hardening model (known as Prager's model).

<u>Remark 3.3</u>. It is well known that the mathematical models which we have described above are not, from the mechanics standpoint, the most satisfactory models for taking account of the corresponding physical phenomena such as plasticity, visco-plasticity, etc... They are all in

fact non-linear elastic models. The most realistic models are those which take account of the 'history' of the material and this includes a time variable. Nevertheless certain of the models described above can be of interest from the mechanical standpoint under certain special circumstances, and from the mathematical standpoint the study of static models is probably an indispensable first step towards an understanding of the far more complex dynamic problems.

As regards the perfectly-plastic elastic model (Hencky model) in which we shall more especially be concerned, one knows (cf. B.Nayroles [1]) that it is used for calculating maximum permissible loads in structures (cf. P. Hodge [1], J. Salencon [1]) and that it provides a suitable method of solving the problems arising in material forming. From the mathematical standpoint this problem has been studied by several authors, in particular by G. Anzelotti [1], G. Anzelotti and M. Giaquinta [2]. E. Christiansen [1]-[2], O. Debordes [1]-[2], O. Debordes and B. Nayroles [1], G. Duvaut and J.L. Lions [1], H. Hencky [1], R. Hill [1], R. Kohn and G. Strang [1]-[2]-[3], R. Kohn and R. Temam [1]-[2], H. Matthies, G. Strang and E. Christiansen [1], B. Mercier [1], J.J. Moreau [1]-[3]-[4]-[5], J. Naumann [1]-[2], B. Nayroles [1]-[2], W. Prager and P. Hodge [1], G. Strang [1], G. Strang and R. Temam [1]-[2]-[3], P. Suquet [1], [2]-[3]-[4]-[5], R. Temam [6]-[7]-[8]-[9]-[10].

4. Duality of the variational problems

We shall begin here the study of the variational problems relating to the Hencky plasticity model, which were described in Section 3. We shall first recall the data of the problem and make a few remarks on the displacement (or strain) problem (Section 4.1). We then go on to determine the dual problem, which turns out to be the stress problem, and study the relations between primal and dual (Section 4.2).

4.1 The displacement (or strain) problem

4.1.1. Definition of the problem

Let us recall the data : an elastic perfectly plastic body occupies a domain Ω of $\mathbb{R}^3$; Ω is a bounded open connected set of $\mathbb{R}^3$ of class $\mathscr{C}^2$, whose boundary is Γ . This boundary is split into three subsets Γ_0, Γ_1, Γ_*; Γ_0 and Γ_1 are open in Γ , and Γ_* is the common boundary in Γ of Γ_0 and Γ_1 . We shall suppose that the regularity condition (2.105) is satisfied on Γ_*[1] .

This body is subjected to forces in Ω of volume-density f and to traction forces on Γ_1 , of surface-density g ; the functions f and g are assumed to be given with

---

(1) It is not essential to assume that Γ is of class $\mathscr{C}^2$. It is sufficient to suppose that Γ is of class $\mathscr{C}^2$ in a neighbourhood (in Γ) of $\Gamma_* \quad \bar{\Gamma}_1$ and of course that (2.105) holds.

(4.1) $$f \in L^2(\Omega)^3 \ , \quad g \in L^2(\Gamma_1)^3 \ .$$

Apart from the field of stresses σ (which will appear at the dual problem level), the unknown is the field of displacements $u : \Omega \mapsto \mathbb{R}^3$. The displacement is assumed to be imposed on Γ_0, and is the solution of the variational problem (3.18), which we shall now try to express in a rather more explicit form.

As in Sections 2 and 3 the displacement on Γ_0 (or on $\bar{\Gamma}_0$) is assumed to be specified as the trace, restricted to Γ_0, of a function u_0 in $H^1(\Omega)^3$, and to avoid technical difficulties it is assumed that $\gamma_0(u_0)$ vanishes outside Γ_0 :

(4.2) $$u_0 \in H^1(\Omega)^3 \ , \quad \theta_{\Gamma_0} \gamma_0(u_0) = \gamma_0(u_0) \ ,$$

where θ_{Γ_i} denotes the characteristic function in Γ of Γ_i for $i = 0,1$.

We now introduce the sur-potential ψ. As before let E be the space of symmetric tensors of order 2, E^D the subspace of E of tensors with zero trace, and I the identity tensor, so that $E = E^D \oplus \mathbb{R} I$. We take a set $K \subset E$ such that

(4.3) $$K \subset E \quad \text{is closed and convex}$$

(4.4) $K = K^D \oplus \mathbb{R} I$, $K^D = K \cap E^D$ is bounded and contains 0 in its interior.

The sur-potential is defined through its conjugate ψ^*, as in (3.4),

(4.5) $$\psi^*(\eta) = \begin{cases} \frac{1}{2} A\eta.\eta & \text{if } \eta \in K \\ +\infty & \text{if not,} \end{cases}$$

where $A\eta$ is the vector given by (3.33) [1] ,

---

[1] We could, more generally, take $A\eta$ given by (3.34) for non-isotropic materials but, to simplify a little, we shall not do this.

$$(4.6)\qquad (A\eta)_{ij} \begin{cases} = \dfrac{1}{2\mu}\,\eta_{ij} - \dfrac{\lambda}{2\mu(3\lambda+2\mu)}\,\eta_{kk}\,\delta_{ij} \\[2ex] = \dfrac{1}{2\mu}\,\eta^D_{ij} + \dfrac{1}{9\kappa}\,\eta_{kk}\,\delta_{ij} \end{cases}$$

$$(4.7)\qquad A\eta.\eta = \frac{1}{2\mu}\,\eta^D_{ij}\,\eta^D_{ij} + \frac{1}{9\kappa}\,\eta_{ii}\,\eta_{jj}\,,$$

where λ,μ are the Lamé constants, $\kappa = \lambda + 2\mu/3$ is the bulk modulus (also known as the incompressibility, or coefficient of incompressibility) and $\mu>0$, $\kappa>0$.

The function ψ is the conjugate of ψ^* $(= \psi^{**})$ and is given by

$$(4.8)\qquad \psi(\xi) = \operatorname*{Sup}_{\eta\in E}\,\{\xi.\eta - \psi^*(\eta)\} = \operatorname*{Sup}_{\eta\in K}\,\{\xi.\eta - \frac{1}{2}\,A\eta.\eta\}\,.$$

Remembering that $K = K^D \oplus \mathbb{R}\,I$ and that $\xi.\eta = \xi^D_{ij}\,\eta^D_{ij} + \frac{1}{3}\,\xi_{ii}\,\xi_{jj}$, we obtain :

$$\psi(\xi) = \operatorname*{Sup}_{s=\eta_{jj}\in\mathbb{R}}\,\{\frac{1}{3}\,\xi_{ii}s - \frac{1}{18\kappa}\,s^2\} + \operatorname*{Sup}_{\eta^D\in K^D}\,\{\xi^D.\eta^D - \frac{1}{4\mu}\,\eta^D.\eta^D\}$$

$$(4.9)\qquad \psi(\xi) = \frac{\kappa}{2}\,\xi_{ii}\,\xi_{jj} + \psi(\xi^D)$$

$$(4.10)\qquad \psi(\xi^D) = \operatorname*{Sup}_{\eta^D\in K^D}\,\{\xi^D.\eta^D - \frac{1}{4\mu}\,\eta^D\,\eta^D\}\,.$$

Remark 4.1. We cannot in general give a more explicit expression for ψ . However if K is the von Mises convex set, $|\sigma^D| \leq \sqrt{2}\,k$ (see (3.37)-(3.38)), then

$$\psi(\xi^D) = \operatorname*{Sup}_{|\eta^D|\leq\sqrt{2}\,k}\,\{\xi^D.\eta^D - \frac{1}{4\mu}\,\eta^D.\eta^D\} = \operatorname*{Sup}_{0\leq s\leq\sqrt{2}\,k}\,\{|\xi^D|s - \frac{s^2}{4\mu}\}$$

$$(4.11)\qquad \psi(\xi^D) = \phi(|\xi^D|)$$

where $\phi : \mathbb{R} \mapsto \mathbb{R}$ is the function

$$\phi(s) = \begin{cases} \mu\, s^2 & \text{if} \quad |s| \leq \dfrac{k}{\sqrt{2}\,\mu} \\[2ex] \sqrt{2}\, k|s| - \dfrac{k^2}{2\mu} & \text{if} \quad |s| \geq \dfrac{k}{\sqrt{2}\,\mu} \end{cases} \tag{4.12}$$

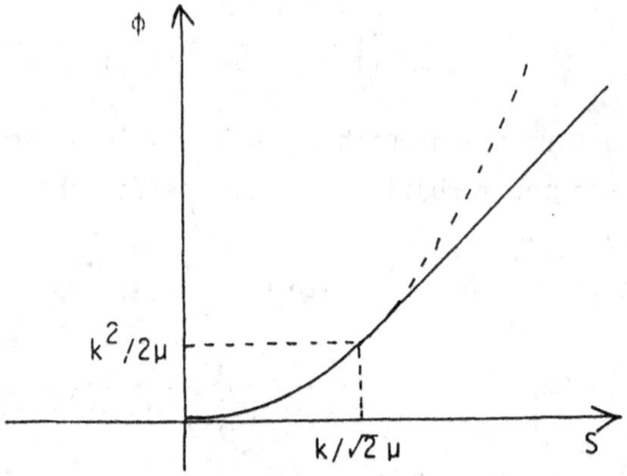

Fig. 4.1 : The function $\phi(s)$, $s \geq 0$

□

If we impose on u the H^1-regularity ([1]), then u will belong to $\mathscr{C}_a = \mathscr{C}_a(u_o)$ (cf. (3.3))

$$\mathscr{C}_a = \mathscr{C}_a(u_o) = \{v \in H^1(\Omega)^3 \, , \, v = u_o \text{ on } \Gamma_o\} \, . \tag{4.13}$$

We put for $v \in H^1(\Omega)^3$,

$$L(v) = \int_\Omega f\, v\, dx + \int_{\Gamma_1} g\, v\, d\Gamma \tag{4.14}$$

and for $\tau \in L^2(\Omega;E)$

$$\Psi(\tau) = \int_\Omega \psi(\tau(x))dx = \frac{\kappa}{2}\int_\Omega |\text{tr}\ \tau(x)|^2 dx + \int_\Omega \psi(\tau^D(x))dx \, . \tag{4.15}$$

The displacement u is the solution to the problem $\mathscr{P}$ defined hereafter (cf. Hencky's principle (3.18)):

([1]) As already mentioned this rather too strong regularity condition will be weakened later on.

(4.16)
$$\operatorname*{Inf}_{v\in\mathscr{C}_a} \{\Psi(\varepsilon(v)) - L(v)\}$$

or

(4.17)
$$\operatorname*{Inf}_{v\in\mathscr{C}_a} \{\frac{\kappa}{2}\int_\Omega (\operatorname{div} v)^2 dx + \int_\Omega \psi(\varepsilon^D(v))dx - L(v)\} .$$

Remark 4.2 As has already been pointed out in Remark 3.3 the Hencky model does not give a satisfactory description of the phenomena of plasticity, since it takes no account of the history of the material. The Hencky model is in fact more closely related to a non-linear elasticity model with threshold (cf. P. Germain [3]), and a model providing a more realistic representation of plasticity is that of Prandtl-Reuss. None the less the Hencky model is interesting to study for the various reasons mentioned in the Introduction.

4.1.2. A few remarks

Before going on to study duality we shall establish certain properties of the functions ψ and Ψ and make a few remarks.

LEMMA 4.1.

i) There are two numbers k_o, k_1 *with* $0<k_o\leq k_1<+\infty$ *such that*

(4.18)
$$k_o(|\xi^D|-1) \leq \psi(\xi^D) \leq k_1|\xi^D| , \quad \forall\ \xi \in E .$$

ii) ψ *is a finite convex continuous function from* E^D *into* $\mathbb{R}$

iii) The mapping $p \mapsto \psi(p)$ *is well-defined and continuous from* $L^1(\Omega;E^D)$ *into* $L^1(\Omega)$.

Proof

i) By hypothesis K^D contains a ball of E^D, $B(0,k_o')$ of centre 0 and radius k_o' and is contained in a ball $B(0, k_1')$ of centre 0 and radius k_1', with $0<k_o'\leq k_1'<\infty$. It follows therefore that

$$\psi(\xi^D) = \operatorname*{Sup}_{\eta\in K^D} \{\xi^D.\eta - \frac{1}{4\mu}\eta.\eta\}$$

$$\psi(\xi^D) \geq \sup_{\substack{\eta\in E^D \\ |\eta|\leq k_0'}} \{\xi^D\eta - \frac{1}{4\mu}\eta.\eta\}.$$

This latter supremum is equal to

$$(4.19)\qquad \begin{cases} \mu|\xi^D|^2 & \text{if } |\xi^D| \leq \dfrac{k_0'}{\sqrt{2}\,\mu} \\[2ex] k_0'(\sqrt{2}|\xi^D| - \dfrac{k_0'}{2\mu}) & \text{if } |\xi^D| \geq \dfrac{k_0'}{\sqrt{2}\,\mu} \end{cases}$$

and this expression is bounded below by $k_0(|\xi^D|-1)$ provided k_0 is small enough. Similarly

$$\psi(\xi^D) \leq \sup_{\eta\in K^D} \{\xi^D.\eta\}$$

$$\leq \sup_{\substack{\eta\in E^D \\ |\eta|\leq k_1'}} \{\xi^D.\eta\} = k_1'|\xi^D| ,$$

whence we obtain the inequality on the right of (4.18) with $k_1 = k_1'$.

ii) By (4.18) the function ψ is of course convex and finite at every point of E^D . Its domain of finiteness in E^D is thus the whole of E^D and therefore by a result recalled in Section 2.1, ψ is continuous in E^D .

iii) By (4.18), $\psi(p)$ is an L-integrable function for all $p \in L(\Omega;E^D)$. We thus obtain an everywhere-defined mapping from $L^1(\Omega;E^D)$ into $L^1(\Omega)$. It follows from Proposition 2.4 that this mapping is continuous. Naturally Ψ , defined by

$$\Psi : p \mapsto \Psi(p) = \int_\Omega \psi(p(x))dx ,$$

which is a mapping from $L^1(\Omega;E^D)$ into $\mathbb{R}$, is convex and everywhere continuous. □

We end with a few remarks on Problem (4.17).

A first question to be asked in connection with this variational problem is whether the infimum is finite or $-\infty$. This is not certain *a priori* since ψ has only *linear* growth at infinity. An answer to this question will be provided by Limit Analysis, a subject which is tackled in the next section. The idea of Limit Analysis is to introduce into (4.16) a parameter $\lambda \geq 0$, i.e. to consider the problem $\mathcal{P}_\lambda$

$$\text{(4.20)} \qquad \underset{v\in\mathcal{C}_a}{\text{Inf}} \; \{\Psi(\varepsilon(v)) - \lambda L(v)\} \; ,$$

to find for what non-negative values of λ this infimum is finite, and to see whether the value $\lambda = 1$ occurs amongst these values.

We have the following :

PROPOSITION 4.1.

i) The set of values of $\lambda \geq 0$ *for which the infimum in (4.20) is finite,* $(> \infty)$ *forms a semiclosed interval* $[0,\bar{\lambda})$

ii) If $0 \leq \lambda < \bar{\lambda}$ *then any minimising sequence* $\{u_m\}$ *of the problem* $(\mathcal{P}_\lambda)$ *is bounded in the following sense*

$$\text{(4.21)} \qquad \varepsilon(u_m) \text{ is bounded in } L^1(\Omega;E)$$

$$\text{(4.22)} \qquad \text{div } u_m \text{ is bounded in } L^2(\Omega) \; .$$

Proof.

i) Since $\Psi \geq 0$, the infimum of $\mathcal{P}_\lambda$ is finite (≥ 0) for $\lambda = 0$. To establish the first part of Proposition 4.1 it will therefore be sufficient to show that if $0 < \lambda < \lambda'$ and $\inf \mathcal{P}_{\lambda'} > -\infty$, then $\inf \mathcal{P}_\lambda > -\infty$. This however follows at once from the fact that

$$\text{(4.23)} \quad \Psi(\varepsilon(v)) - \lambda L(v) = \frac{\lambda}{\lambda'} \{\Psi(\varepsilon(v)) - \lambda' L(v)\} + \left(1 - \frac{\lambda}{\lambda'}\right) \Psi(\varepsilon(v))$$

$$\geq \frac{\lambda}{\lambda'} \{\Psi(\varepsilon(v)) - \lambda' L(v)\} \geq \frac{\lambda}{\lambda'} \text{Inf } \mathcal{P}_{\lambda'} > -\infty \; .$$

The point $\bar{\lambda}$ is the supremum of the non-negative values of λ for which $\inf \mathcal{P}_\lambda > -\infty$. We do not know for the moment (but this point will be cleared up in Section 5) whether $\inf \mathcal{P}_{\bar\lambda} > -\infty$.

ii) If $0 \leq \lambda < \bar\lambda$, there is a λ' with $\lambda < \lambda' < \bar\lambda$ such that $\inf \mathcal{P}_{\lambda'} > -\infty$. We now use (4.23) with v replaced by u_m :

$$(4.24)\quad \Psi(\varepsilon(u_m)) - \lambda L(u_m) \geq \frac{\lambda}{\lambda'} \{\Psi(\varepsilon(u_m)) - \lambda' L(u_m)\} + (1-\frac{\lambda}{\lambda'})\ \Psi(\varepsilon(u_m))$$

$$\geq \frac{\lambda}{\lambda'} \inf \mathcal{P}_{\lambda'} + (1-\frac{\lambda}{\lambda'})\ \Psi(\varepsilon(u_m)) .$$

Since by hypothesis $\Psi(\varepsilon(u_m)) - \lambda L(u_m)$ converges to $\inf \mathcal{P}_\lambda$ this sequence is bounded above and $\Psi(\varepsilon(u_m))$ thus remains bounded above. The second part of the Proposition now follows from the fact that, by (4.15) and (4.18)

$$\Psi(\varepsilon(u_m)) = \frac{\kappa}{2}\int_\Omega |\operatorname{div} u_m|^2 dx + \int_\Omega \psi(\varepsilon^D(u_m))dx$$

$$\geq \frac{\kappa}{2}\int_\Omega |\operatorname{div} u_m|^2 dx + k_0 \int_\Omega (|\varepsilon^D(u_m)|-1)dx .$$

Remark 4.3 i) In Section 5 we shall determine the number $\bar\lambda$ as the value of the infimum in an auxiliary variational problem.

ii) To determine the set of values of $\lambda \leq 0$ for which $\inf \mathcal{P}_\lambda > -\infty$, it suffices to replace f and g by $-f$ and $-g$ in the preceding expressions. The results are similar.

iii) If $\{u_m\}$ is a minimising sequence for $\mathcal{P}_\lambda$ and $\lambda < \bar\lambda$, then $u_m = u_0$ on Γ_0 for all m. It follows from an inequality of the Poincaré type, to be established in Chapter II, that if $\Gamma_0 \neq \emptyset$ then u_m is bounded in $L^1(\Omega)^3$. If $\Gamma_0 = \emptyset$, u_m remains bounded in $L^1(\Omega)^3/\mathcal{R}$.

iv) As $L^1(\Omega)$ *is not a reflexive space* (see Section 1.1 and (1.7)), we cannot extract from a sequence $\{u_m\}$ satisfying (4.21) a subsequence such that $\varepsilon(u_m)$ converges weakly in $L^1(\Omega;E)$.

It is here that we encounter the essential mathematical difficulty regarding the existence of solutions of $\mathcal{P}$. *The object of Chapter II will be to establish that the minimising sequences for* $\mathcal{P} = \mathcal{P}_1$

have (when $1<\bar{\lambda}$) cluster-points (points of accumulation) which are, in a sense to be precisely defined later, 'weak' solutions of $\mathscr{P}$.

4.2 The dual problem : the problem in terms of stress

Our object now is to determine the dual of the problem $\mathscr{P}$ by using the general methods described in Section 2.2.1. : we shall then see appear the problem for the stresses (i.e. (3.25)). The duality of problem $\mathscr{P}$ has been studied by several authors using similar methods, G. Duvaut - J.L. Lions [1], B. Mercier [1], B. Nayrolles [1]-[2]. The treatment given here (taken from G. Strang - R. Temam [1]) follows the approach of W. Fenchel [1]-[2]-[3], R.T. Rockafellar [1], I. Ekeland - R. Temam [1] and is better adapted to the developments which we have in view.

□

We shall write the problem $\mathscr{P}$ in the form (2.35). To recover the corresponding framework, we put

$$V = H^1(\Omega)^3 \ , \ Y = Y_1 \times Y_2 = L^1(\Omega) \times L^2(\Omega;E^D) \ ;$$

the space Y is none other than $L^2(\Omega;E)$, the above factorisation corresponding to the factorisation $\mathbb{R}\,I \oplus E^D$ of the space E. An element p of Y can be written indifferently in the form of a pair (p_1,p_2) or in the form of a tensor $p_1 I + p_2$ of E. Let us also recall that the inner product of Y (or Y_2) is

$$\int_\Omega p_{ij}\, q_{ij}\, dx \ .$$

The operator Λ is the operator ε which likewise we factorise into $\Lambda_1 \times \Lambda_2$, $\Lambda_1 = \text{div}$, $\Lambda_2 = \varepsilon^D$. We again define, for $v \in V$,

$$F(v) = \begin{cases} -\,L(v) & \text{if} \quad v = u_0 \quad \text{on} \quad \Gamma_0 \\ +\infty & \text{if not} \end{cases}$$

and for all $p = (p_1,p_2)$ of Y ,

$$G(p) = G_1(p_1) + G_2(p_2)$$

$$G_1(p_1) = \frac{\kappa}{2}\int_\Omega (p_1)^2 dx \ , \quad \forall\ p_1 \in Y_1$$

$$G_2(p_2) = \int_\Omega \psi(p_2)dx \ , \quad \forall\ p_2 \in Y_2 \ .$$

Clearly F is a convex l.s.c. function on V and by Lemma 4.1 the function G is everywhere finite, convex and continuous on Y .

The space V^* will be the dual of V and Y^* will be taken equal to Y .

With these notations the problem $\mathscr{P}$-(4.17) is identical to the problem $\mathscr{P}$ in (2.35). The dual is given by (2.36) and to express it in an explicit form we have to determine the functions $F^* : V^* \mapsto \mathbb{R} \cup \{+\infty\}$, and $G^* : Y \mapsto \mathbb{R} \cup \{+\infty\}$ conjugate to F and G .

The calculation of F^* was done in Lemma 2.2 for elements of the form $\Lambda^* p$ which is sufficient; for any p of $Y^* = Y$:

$$(4.25)\quad F^*(\Lambda p) = \begin{cases} \displaystyle\int_{\Gamma_0} (p.\nu)u_0 \, d\Gamma & \text{if } \begin{cases} \operatorname{div} p - f = 0 & \text{in } \Omega \\ p.\nu + g = 0 & \text{on } \Gamma \end{cases} , \text{ [1]} \\ +\infty & \text{if not.} \end{cases}$$

G_1^* and G_2^* are found by applying Proposition 2.5 whose hypothesis are clearly satisfied ((2.52) holds with $a = b = 0$). We obtain

$$G_1^*(p_1) = \frac{1}{2\kappa}\int_\Omega (p_1)^2 dx \ , \ \forall\ p_1 \in Y_1 = L^2(\Omega)$$

$$G_2^*(p_2) = \int_\Omega \psi^*(p_2)dx \quad , \ \forall\ p_2 \in Y_2 = L^2(\Omega;E^D) \ ;$$

or, by (4.5)

---

[1] See note [1] p.44.

$$(4.26)\quad G_2^*(p_2) = \begin{cases} \frac{1}{2}\int_\Omega Ap_2.p_2\,dx & \text{if } p_2(x) \in K \text{ (or } K^D) \text{ for almost all } x, \\ +\infty & \text{if not.} \end{cases}$$

We now introduce the tensor $\sigma = -p = -p_1 I - p_2$ and obtain

$$(4.27)\quad G^*(\sigma) = \begin{cases} \frac{1}{2}\mathscr{A}(\sigma,\sigma) & \text{if } \sigma^D(x) \in K^D \text{ for almost all } x, \\ +\infty & \text{if not.} \end{cases}$$

where $\mathscr{A}$ is the bilinear form written down in (2.77)

$$(4.28)\quad \mathscr{A}(\sigma,\tau) = \frac{1}{9\kappa}\int_\Omega \operatorname{tr}\sigma.\operatorname{tr}\tau\,dx + \frac{1}{2\mu}\int_\Omega \sigma^D.\tau^D\,dx,$$

$\forall\ \sigma,\tau \in L^2(\Omega;E)$.

We make explicit the dual given by (2.36), writing $\sigma = -p$ the generic variable of Y. We obtain

$$(4.29)\quad \sup_\sigma \{-\frac{1}{2}\mathscr{A}(\sigma,\sigma) + \int_{\Gamma_0}(\sigma.\nu)u_0\,d\Gamma\},$$

where the supremum is taken over the set of σ, called *the set of* σ *admissible for* $\mathscr{P}^*$, which satisfy all the following conditions :

$$(4.30)\quad \begin{cases} \sigma \in L^2(\Omega;E) \\ \operatorname{div}\sigma + f = 0 \text{ in } \Omega \\ \sigma.\nu = g \text{ on } \Gamma_1 \\ \sigma^D(x) \in K^D \text{ for almost all } x \in \Omega . \end{cases}$$

We thus recover, apart from a change of notation and the substitution of a maximisation for a minimisation, the problem (3.25).

The relations between $\mathscr{P}$ and $\mathscr{P}^*$ are given by

THEOREM 4.1.

The following conditions are equivalent

(i) $\text{Inf}\, \mathcal{P} > -\infty$

(ii) *There exists at least one* σ *admissible for* $\mathcal{P}^*$

(iii) $\text{Inf}\, \mathcal{P} = \text{Sup}\, \mathcal{P}^* \in \mathbb{R}$

If any (and hence all) of these conditions are satisfied, the problem $\mathcal{P}^*$ *has an unique solution* σ .

Proof

Condition (ii) is necessary and sufficient for $\text{Sup}\, \mathcal{P}^* > -\infty$. Thus (iii) implies (i) and (ii), while (ii) implies (i) since by (2.28) $\text{Sup}\, \mathcal{P}^* \leq \text{Inf}\, \mathcal{P}$. The first part of the Theorem will therefore have been proved if we can show that (i) implies (iii). To do this we have only to apply (2.30). Now, by hypothesis $\text{Inf}\, \mathcal{P} > -\infty$ and so we have to show that there is a $u^0 \in V$ such that $F(u^0) < +\infty$, $G(\Lambda u^0) < +\infty$, G being continuous at Λu^0 (cf. (2.29)). Since G is everywhere finite and continuous on Y the element $u^0 = u_0$ possesses all the required properties and the desired result follows.

The problem $\mathcal{P}^*$ could not have any solution if the set of admissible σ were empty. When the conditions (i)(ii)(iii) are all satisfied, it is once again (2.30) which implies the existence of a solution to the problem $\mathcal{P}^*$. Note that the existence of a solution to $\mathcal{P}^*$ could also be deduced from Proposition 2.1. Finally the uniqueness of the solution to $\mathcal{P}^*$ is ensured by Proposition 2.2, after noting that the quadratic function $\sigma \mapsto \mathcal{A}(\sigma,\sigma)$ is strictly convex (since it is ≥ 0 and vanishes only if $\sigma = 0$).

Remark 4.4. We have in all cases

$$\text{Inf}\, \mathcal{P} = \text{Sup}\, \mathcal{P}^* \in \mathbb{R} \cup \{-\infty\} .$$

Extremality relations

Let us suppose that, conditions (i) (ii) (iii) of Theorem 4.1 being satisfied, the problem $\mathcal{P}$ has a solution u ; the problem $\mathcal{P}^*$ then by Theorem 4.1, has a solution and the extremality conditions (2.31)

(2.32) and (2.38) can be written down. The relation (2.31) is trivial, while the two relations (2.38) can be written (for i = 1,2) :

(4.31) $$\frac{\kappa}{2}\int_\Omega (\text{div } u)^2 dx + \frac{1}{18\kappa}\int_\Omega (\text{tr } \sigma)^2 dx = \frac{1}{3}\int_\Omega (\text{div } u)(\text{tr } \sigma) dx$$

(4.32) $$\int_\Omega \psi(\varepsilon^D(u))dx + \int_\Omega \psi^*(\sigma^D)dx = \int_\Omega \varepsilon^D(u).\sigma^D\, dx .$$

that is

$$\int_\Omega \left\{\frac{\kappa}{2}(\text{div } u)^2 + \frac{1}{18\kappa}(\text{tr } \sigma)^2 - \frac{1}{3}(\text{div } u)(\text{tr } \sigma)\right\}dx = 0$$

$$\int_\Omega \left\{\psi(\varepsilon^D(u)) + \psi^*(\sigma^D) - \varepsilon^D(u).\sigma^D\right\}dx = 0 .$$

The two integrands being non-negative we obtain the relations

(4.33) $$\frac{\kappa}{2} \text{ div } u + \frac{1}{18\kappa}(\text{tr } \sigma)^2 dx = \frac{1}{3} \text{ div } u.\text{tr } \sigma$$

(4.34) $$\psi(\varepsilon^D(u)) + \psi^*(\sigma^D) = \varepsilon^D(u).\sigma^D$$

valid for almost all x of Ω . The first reduces to

(4.35) $$\text{tr } \sigma = 3\kappa \text{ div } u$$

which appeared in the linear elasticity case (cf. (2.80)). The equation (4.33) gives a less simple relation between σ^D and $\varepsilon^D(u)$: at a point x where $\sigma^D(x)$ belongs to *the interior* of K^D , we have as in the linear elasticity case (cf. (2.79) :

(4.36) $$\sigma^D = 2\mu\, \varepsilon^D(u) .$$

At a point x where $\sigma^D(x)$ belongs to the boundary of K^D , there is no relationship between σ^D and $\varepsilon^D(u)$.

Remark 4.5. To find the dual of the problem $\mathcal{P}_\lambda$ introduced in (4.20), it is sufficient to replace f and g by λf and λg in (4.29) and (4.30). We find the same functional to maximise as in (4.29)

$$(4.37)\qquad \operatorname*{Sup}_{\sigma} \; -\frac{1}{2}\mathscr{A}(\sigma,\sigma) + \int_{\Gamma_0} (\sigma.\nu)u_0 \, d\Gamma\} \quad ,$$

the set of σ being now those such that

$$(4.38)\qquad \begin{cases} \sigma \in L^2(\Omega;E) \\ \operatorname{div} \sigma + \lambda f = 0 \quad \text{in} \quad \Omega \\ \sigma.\nu = \lambda g \quad \text{on} \quad \Gamma_1 \\ \sigma^D(x) \in K^D \quad \text{for almost all} \quad x \in \Omega \end{cases}$$

As $0 \in K^D$, if σ is admissible for $\mathcal{P}^*_\lambda$ (i.e. if σ satisfies all the conditions in (4.38)) and if $0<\lambda'<\lambda$, then $\lambda'\sigma/\lambda$ is admissible for $\mathcal{P}^*_{\lambda'}$. Thus the set of $\lambda \geq 0$ for which $\mathcal{P}^*_\lambda$ has at least one admissible σ (and for which $\operatorname{Sup} \mathcal{P}^*_\lambda > -\infty$), is a semiclosed interval, say $[0,\tilde{\lambda})$. Using Theorem 4.1 applied to $\mathcal{P}_\lambda$, we arrive at the conclusions of Proposition 4.1.i), and deduce of course that $\tilde{\lambda} = \bar{\lambda}$.

5. Limit Analysis

We have seen that the infimum of the problem $\mathscr{P}$ is finite ($> -\infty$) if a certain number $\bar{\lambda}$ appearing in Lemma 4.2 exceeds 1. The determination of this number $\bar{\lambda}$ is therefore important, from the mechanical standpoint, because by knowing this number we can tell whether or not the material can support the forces f and g imposed upon it (it can support them if $\operatorname{Inf} \mathscr{P} > -\infty$ and breaks down if $\operatorname{Inf} \mathscr{P} = -\infty$) . By using ideas suggested by the method of calculating the breaking stress normally used in structural mechanics, we shall show (*completely rigorously*) that $\bar{\lambda}$ can be determined as the infimum of an auxiliary variational problem, called the *Limit Analysis* problem.

In Section 5.1 we shall introduce and study the limit analysis problem $\mathscr{P}_{AL}$ (cf. (5.12)-(5.13)) and its relationship with the strain analysis (the initial problem $\mathscr{P}$).

In Section 5.2 we determine the dual of $\mathscr{P}_{AL}$ and find once more, though in a somewhat different way, the results previously obtained.

5.1 The Limit Analysis problem $\mathscr{P}_{AL}$

We first describe a heuristic method of arriving at the limit analysis problem.

Since the question to be decided is whether $\operatorname{Inf} \mathscr{P} > -\infty$, or in other words, after Theorem 4.1, whether there exist σ which are admissible for $\mathscr{P}^*$, we can see that u_0 and the form $\mathscr{A}$ in (4.29) do not play any important role in this regard. We can therefore eliminate them by making $u_0 = 0$ and $\mathscr{A}(\sigma,\sigma) = 0$, which can be done *formally* by making μ tend to $+\infty$ (cf. (4.28), $\kappa = \lambda + 2\mu/3$) . In the limit $\psi_\infty^*(\xi)$ has the value 0 if $\xi^D \in K^D$ and $+\infty$ if not, which means, by conjugation, that $\psi_\infty(\xi)$ has the value

$$\operatorname*{Sup}_{\eta^D \in K^D} \xi.\eta = \operatorname*{Sup}_{\substack{t\in\mathbb{R}\\ \eta^D\in K^D}} \{\tfrac{1}{3}(\operatorname{tr}\xi)t + \xi^D.\eta^D\}$$

(5.1)
$$\psi_\infty(\xi) = \begin{cases} \psi_\infty(\xi^D) & \text{if } \operatorname{tr}\xi = 0 \\ +\infty & \text{if not ,} \end{cases}$$

$\psi_\infty(\xi^D)$ being the support function of K^D in E^D:

(5.2)
$$\psi_\infty(\xi^D) = \operatorname*{Sup}_{\eta^D\in K^D} \xi^D.\eta^D .$$

This suggests that we should study the problem $\mathcal{P}$ when ψ is replaced by ψ_∞ and u_0 by 0 ; and this is what we shall now do.

□

We begin by a few remarks on ψ_∞ . Firstly ψ_∞ is positively homogeneous on E^D (and on E),

(5.3)
$$\psi_\infty(r\xi^D) = r\,\psi_\infty(\xi^D) \ , \ \forall\ r\geq 0 \ ;$$

and secondly, by the assumptions about K^D (cf. (4.4)), there exist numbers k_0' and k_1' , with $0<k_0'\leq k_1'<\infty$, such that

(5.4)
$$k_0'|\xi^D| \leq \psi_\infty(\xi^D) \leq k_1'|\xi^D| \ , \ \forall\ \xi^D \in E^D .$$

To prove this it is sufficient to note that K^D contains a ball $B(0,k_0')$ of E^D of centre 0 and radius $k_0'>0$ and is contained in a ball $B(0,k_1')$ of centre 0 and radius k_1' , and to observe that

$$k_0'|\xi^D| = \operatorname*{Sup}_{\eta^D\in B(0,k_0')} \xi^D.\eta^D \leq \operatorname*{Sup}_{\eta^D\in K^D} \xi^D.\eta^D \leq \operatorname*{Sup}_{\eta^D\in B(0,k_1')} \xi^D.\eta^D = k_1'|\xi^D| .$$

In the case of von Mises' convex set (3.38):

(5.5) $$\psi_\infty(\xi^D) = \sqrt{2}\, k|\xi^D| .$$

It is clear from (5.1) that, for $v \in H^1(\Omega)$,

(5.6) $$\Psi_\infty(\varepsilon(v)) = \int_\Omega \psi_\infty(\varepsilon(v))dx$$

is equivalent to

(5.7) $$\Psi_\infty(\varepsilon(v)) = \begin{cases} \int_\Omega \psi_\infty(\varepsilon^D(v))dx & \text{if } \operatorname{div} v = 0 \\ +\infty & \text{if not.} \end{cases}$$

The simplified problem, corresponding to $\psi = \psi_\infty$ and $u_0 = 0$, can therefore be written

(5.8) $$\operatorname*{Inf}_v \{\Psi_\infty(\varepsilon^D(v)) - L(v)\} ,$$

where the infimum is taken over all v simultaneously satisfying the following three conditions

(5.9) $$\begin{cases} v \in H^1(\Omega)^3 \\ \operatorname{div} v = 0 \\ v = 0 \text{ on } \Gamma_0 : \end{cases}$$

The determination of the problem dual to $\mathscr{P}$, which was carried out in the general case in Section 4, applies here without any change. The function G_1 becomes the indicator function of 0 in $L^2(\Omega)$ and $G_2(p_2) = \Psi_\infty(p_2)$, $\forall\, p_2 \in Y_2 = L^2(\Omega;E^D)$.

Thus G_1^* is 0 and G_2^* is given by

$$G_2^*(p_2) = \Psi_\infty(p_2) = \begin{cases} 0 & \text{if } p_2(x) \in K^D \text{ for almost all } x \text{ of } \Omega \\ +\infty & \text{if not .} \end{cases}$$

The dual problem simplifies (as expected) and becomes

$$\text{(5.10)} \qquad \sup_{\sigma} \{0\} \ ,$$

the supremum being over the same set of σ as before that is to say the set defined by (4.30). This means that $\text{Sup}\ \mathcal{P}^*$ has the value 0 or $-\infty$ according as to whether or not there exist σ satisfying (4.30).

Theorem 4.1 which specifies the relationship between $\mathcal{P}$ and $\mathcal{P}^*$ (and in particular which gives the equation $\text{Inf}\ \mathcal{P} = \text{Sup}\ \mathcal{P}^*$) does not apply as such here. In fact the functions G_1 and (hence) G are neither of them continuous at any point of Y so that (2.29) does not hold and (2.30) cannot be applied. Nevertheless the conclusions of the theorem are valid and we shall for the moment assume the truth of the following lemma which will be proved later on.

LEMMA 5.1.

The conclusions of Theorem 4.1 apply to the problem (5.8) and its dual (5.10). In particular, in all cases

$$\text{Inf}\{(5.8)\} = \text{Sup}\{(5.10)\} \quad (= 0 \ \text{ or } \ -\infty) \ .$$

□

We shall now introduce the limit analysis problem $\mathcal{P}_{AL}$ which is closely related to (5.8). It can be written as

$$\text{(5.11)} \qquad \inf_{v} \frac{\psi_\infty(\varepsilon^D(v))}{L(v)} \ ,$$

where the infimum is over all v satisfying the conditions (5.9) and $L(v) > 0$. This particular optimisation problem is not convex but because of the homogeneity property (5.3) of ψ_∞, it comes to the same thing to consider instead the following problem, which <u>*is convex:*</u>

$$\text{(5.12)} \qquad \inf_{v} \Psi_\infty(\varepsilon^D(v))$$

where this time the infimum is over all v satisfying the conditions:

$$(5.13)\qquad \begin{cases} v \in H^1(\Omega)^3 \ , \ \text{div } v = 0 \ , \ v = 0 \ \text{ on } \ \Gamma_0 \\ L(v) = 1 \end{cases}$$

We have the

THEOREM 5.1

The infimum of problem (4.16) or (5.8) is finite $(> -\infty)$ *and there exist* σ *satisfying (4.30) if and only if the infimum of the problem* $\mathcal{P}_{AL}$ *defined by (5.12) and (5.13) is* ≥ 1 .

Proof.

The infima of problems (4.16) and (5.8) are either both finite or else equal to $-\infty$. Thanks to Theorem 4.1 and to Lemma 5.1 the two infima are both finite if there exist σ satisfying (4.30); otherwise they both have the value $-\infty$.

We now make use of the homogeneity property (5.3) for ψ_∞ (and Ψ_∞). If the infimum of the problem (5.12)-(5.13) is ≥ 1 this means that

$$\Psi_\infty(\varepsilon(v\ )) \geq 1$$

for all v satisfying (5.13); and then by homogeneity

$$(5.14)\qquad \Psi_\infty(\varepsilon(v)) - L(v) \geq 0 \ ,$$

for all v satisfying (5.9) with $L(v) > 0$. The relation (5.14) being trivial for those v satisfying (5.9) with $L(v) \leq 0$, we conclude that the infimum of problem (5.8) is ≥ 0 (and in fact by Lemma 5.1 the infimum is actually zero in this case).

If now the infimum of the problem (5.12)-(5.13) is <1 , then there exists a v_0 satisfying (5.13) such that

$$\Psi_\infty(\varepsilon^D(v_0)) < 1 \ .$$

Hence for any $s>0$

$$\Psi_\infty(\varepsilon^D(s\ v_0)) - L(s\ v_0) = s(\Psi_\infty(\varepsilon^D(v_0)) - 1),$$

and letting s tend to $+\infty$, we conclude from this that the infimum in (5.8) is $-\infty$.

□

Lemma 5.1 remains to be proved.

Proof of Lemma 5.1

We apply to problem (5.8) the duality framework of Section 2 but under slightly different conditions which will allow us to obtain the same dual (5.10) as before but nevertheless still permit us to satisfy the condition (2.29) and to apply (2.30)

We take for V the space

$$W = \{v \in H^1(\Omega)^3\ ,\ \text{div}\ v = 0\ ,\ v = 0 \ \text{ on } \ \Gamma_0\} \tag{5.15}$$

V^* will be the dual of W and $Y^* = Y = L^2(\Omega;E^D)$. The operator Λ is the operator ε^D; we put $F(v) = -L(v)$, $\forall v \in V$, and $G(p) = \Psi_\infty(p)$, $\forall p \in L^2(\Omega;E^D)$. The problem (5.8) is now identical to the problem (2.26). To determine the dual (2.27) we have to find F^* and G^*; G^* is simply the indicator function of the set

$$\{\sigma \in L^2(\Omega;E^D),\ \sigma^D(x) \in K^D \ \text{ p.p.}\}\ .$$

The calculation of F^* is more delicate; we shall prove that for $p \in Y$,

$$F^*(\Lambda^* p) = \begin{cases} 0 & \text{if there is a } p_0 \in L^2(\Omega) \text{ such that} \\ & \qquad \sigma = -p_0 I - p \in \mathscr{S}_a \\ +\infty & \text{otherwise} \end{cases} \tag{5.16}$$

($\mathscr{S}_a$ is the set defined in (3.8)). Assuming this for the moment we can express (2.27) in an explicit form and we find as the dual of (5.8) the problem (5.10) ([1]). The essential difference is that, in the present instance, the convex function G is everywhere finite and continuous on Y and the condition (2.29) is satisfied with, for example, $u^0 = 0$; we can therefore repeat without change the argument used to prove Theorem 4.1, to obtain the conclusions of Lemma 5.1.

Proof of (5.16)

To establish (5.16) we write

$$F^*(\Lambda^* p) = \sup_{v \in W} \{<\Lambda^* p, v> + L(v)\} .$$

This supremum is 0 if $\Lambda p + L = 0$ (in W') and $+\infty$ if not. Suppose $\Lambda^* p + L = 0$, then

$$(5.17) \qquad \int_\Omega p.\varepsilon^D(v)dx + \int_\Omega f\, v\, dx + \int_{\Gamma_1} g\, v\, d\Gamma = 0 , \ \forall v \in W .$$

For $v \in W$, $\varepsilon^D(v) = \varepsilon(v)$ since $\operatorname{div} v = 0$. Replacing $\varepsilon^D(v)$ by $\varepsilon(v)$ in the first integral and writing down (5.17) for v in $W \cap \mathscr{C}_0^\infty(\Omega)^3$, we arrive at

$$(5.18) \qquad <- \operatorname{div} p + f, v> = 0 , \ \forall \ v \in W \cap \mathscr{C}_0^\infty(\Omega)^3 .$$

By a well-known result in the theory of Navier-Stokes equations (cf. for example O.A. Ladyzhenskaya [1] or R. Temam [6] Proposition 1.1, Ch.I) this implies the existence of a distribution p_0 such that

$$(5.19) \qquad - \operatorname{div} p + f = \operatorname{grad} p_0 .$$

By Proposition 1.2 ii), p_0 is necessarily in $L^2(\Omega)$. If we now put $\sigma = - p - p_0 I$, then $\sigma \in L^2(\Omega;E)$ and the condition (5.19) can be written

---

(1) This was not obvious *a priori* because a given variational problem may well have several different duals.

(5.20) $$\operatorname{div} \sigma + f = 0 \text{ in } \Omega .$$

We now revert to (5.17) where we replace $p.\varepsilon^D(v)$ by $-\sigma.\varepsilon(v)$. The use of the generalised Green's formula (1.51) is legitimate, and having regard to (5.20), allows us to write (5.17) in the form

(5.21) $$\int_\Gamma (\sigma.\nu)v \, d\Gamma + \int_{\Gamma_1} g \, v \, d\Gamma + 0 , \quad \forall v \in W .$$

As $v = 0$ on Γ_0, this is equivalent to

(5.22) $$\int_{\Gamma_1} (\sigma.\nu + g)v \, d\Gamma = 0 , \quad \forall v \in W .$$

We are able to characterise the trace of the elements of $H^1(\Omega)^3$ which have zero divergence. This was done in C. Foias - R. Temam [1], and this set of trace is

(5.23) $$\{w \in H^{1/2}(\Gamma)^3 , \int_\Gamma w.\nu \, d\Gamma = 0\} .$$

The set of traces of the elements of W is therefore

(5.23) $$\{w \in H^{1/2}(\Gamma)^3 , w = 0 \text{ on } \Gamma_0 , \int_{\Gamma_1} w.\nu \, d\Gamma = 0\} .$$

We must therefore have $\int_{\Gamma_1} (\sigma.\nu+g)w \, d\Gamma = 0$ for any w belonging to the space defined by (5.23), and it follows that there exists an $\alpha \in \mathbb{R}$ such that

$$\sigma.\nu + g = \alpha\nu \text{ on } \Gamma_1 .$$

As the distribution p_0 in (5.19) is defined only to within an additive constant, we can, on replacing p_0 by $p_0 + \alpha$, assume that $\alpha = 0$ and we then obtain as the final condition

(5.24) $$\sigma.\nu + g = 0 \text{ on } \Gamma_1 .$$

This concludes the proof of (5.16) and with it that of Lemma 5.1. □

Remark 5.1 Theorem 5.1 can easily be extended to cover the problems $\mathscr{P}_\lambda$ considered in (4.20) and in Proposition 4.1: by repeating without change the proof of Theorem 5.1 we see that the infimum of the problem $\mathscr{P}_\lambda$ is finite $(> -\infty)$ if and only if the infimum of the Limit Analysis problem is $\geq \lambda$.

We can therefore give some additional precision to Proposition 4.1. Firstly the number $\bar{\lambda}$ is the infimum of $\mathscr{P}_{AL}$:

$$(5.25) \qquad \bar{\lambda} = \text{infimum of } \mathscr{P}_{AL} \quad \text{(i.e. (5.12)-(5.13))};$$

secondly the infimum of $\mathscr{P}_\lambda$ is still finite when $\lambda = \bar{\lambda}$, a point which had not been specifically covered in the statement and proof of the Proposition 4.1.

The introduction of problems $\mathscr{P}_\lambda$ with a parameter λ varying between 0 and a limiting value $\bar{\lambda}$ is a standard technique in the calculation of load limits in structural mechanics; cf. for example P.G. Hodge [1] , J. Salençon [1].

Remark 5.2. We can re-interpret the second part of Proposition 4.1:

(5.2) if $\text{Inf } \mathscr{P}_{AL} > 1$, then every minimising sequence of the problem $\mathscr{P}$-(4.16) is bounded in the sense of (4.21)-(4.22).

Remark 5.3. Although the theory of load limits is a completely standard subject in structural mechanis (cf. for example the references mentioned under Remark 5.1) we have not found problem (5.12)-(5.13) explicitly stated in the mechanical literature. The problem *seems* to have been introduced for the first time in G. Strang - R. Temam [2].

5.2 Dual of the Limit Analysis problem - Applications

We shall now find an explicit form of the dual of the Limit Analysis problem (5.12)-(5.13), which will enable us to give another proof of Theorem 5.1.

□

We have to apply the general principles given in Section 2.2 and write the problem $\mathscr{P}_{AL}$ in the form (2.26). To do this we take as the space V the space W already introduced in (5.15).

(5.26) $$W = \{v \in H^1(\Omega)^3, \operatorname{div} v = 0 , v = 0 \text{ on } \Gamma_0\} ;$$

V^* will be the dual of W and we put $Y^* = Y = L^2(\Omega;E^D)$, $\Lambda = \varepsilon^D$. We choose the functions F and G as follows: for all v of V,

$$F(v) = 0 \text{ if } L(v) = 1 \text{ , and } = +\infty \text{ otherwise,}$$

and for all p of Y,

$$G(p) = \Psi_\infty(p) = \int_\Omega \psi_\infty(p)dx .$$

The problem $\mathscr{P}_{AL}$ ((5.12)-(5.13)) is then identical to the problem (2.26). To determine the dual (2.27) we need to calculate F^* and G^* As regards G^*, we have by Proposition 2.5,

$$G^*(p) = \int_\Omega \psi_\infty^*(p)\,dx \text{ , } \forall\, p \in Y^* = Y$$

that is to say

$$G^*(p) = \begin{cases} 0 \text{ if } p(x) \in K^D \text{ for almost all } x \text{ in } \Omega \\ \\ +\infty \text{ otherwise} \end{cases}$$

The calculation of F^* is more difficult. We shall establish

LEMMA 5.2.

For all p of $Y^* = Y = L^2(\Omega;E^D)$

$$F^*(\Lambda^* p) = \begin{cases} \text{if there is a } p_0 \in L^2(\Omega) \text{ and } \lambda \in \mathbb{R} \\ \text{such that} \begin{cases} \operatorname{div} \sigma + \lambda f = 0 \text{ in } \Omega \\ \sigma.\nu = \lambda g \text{ on } \Gamma_1 \\ \sigma = -p_0 I - p \end{cases} \\ \\ +\infty \text{ otherwise.} \end{cases}$$

Assuming this result for the moment we can now express, with the help of (2.27), the problem $\mathscr{P}_{AL}^*$ in an explicit form, namely

(5.27)
$$\operatorname*{Sup}_{\sigma}\ \{\lambda\}$$

where the supremum is over all $\sigma = -p_0 I - p \in L^2(\Omega;E)$ for which there exists a $\lambda \in \mathbb{R}$ such that

(5.28)
$$\begin{cases} \operatorname{div} \sigma + f = 0 & \text{in } \Omega \\ \sigma.\nu = \lambda g & \text{on } \Gamma_1 \\ \sigma^D(x) \in K^D & \text{for almost all } x \in \Omega \,. \end{cases}$$

To say that σ satisfies (5.28) amounts to saying that σ is admissible for the problem $\mathscr{P}_\lambda^*$ (cf. (4.38) and Remark 4.5); $\mathscr{P}_{AL}^*$ thus furnishes the supremum of those λ for which $\mathscr{P}_\lambda^*$ possesses admissible σ . □

The relations between $\mathscr{P}_{AL}$ and $\mathscr{P}_{AL}^*$ are easy to study: $\inf \mathscr{P}_{AL} \in \mathbb{R}$ since this number is ≥ 0 ; moreover as the function G is continuous at every point of Y , condition (2.29) is satisfied (by every point u^0 of W such that $L(u^0) = 1$) and we can apply (2.30). Thus

(5.29)
$$\operatorname{Inf} \mathscr{P}_{AL} = \operatorname{Sup} \mathscr{P}_{AL}^* \qquad (\geq 0) \,,$$

and there is a solution to $\mathscr{P}_{AL}^*$, that is to say there is a σ satisfying (5.28) when λ is the number (5.29) denoted by $\tilde{\lambda}$.

Let us restate this last point a little more precisely. There exists a σ admissible for $\mathscr{P}_\lambda$ when $\lambda = \tilde{\lambda}$ but there is no σ admissible for $\mathscr{P}_\lambda^*$ when $\lambda > \tilde{\lambda}$. On comparing this with Remark 4.5 we see that $\tilde{\lambda} = \bar{\lambda}$ and we can therefore complete the result given in Remark 4.5 by the statement that there are σ which are admissible for $\mathscr{P}_\lambda^*$ for all λ in the closed interval $[0, \bar{\lambda}]$.

Bearing in mind Theorem 4.1 (applied to $\mathscr{P}_\lambda$) *we have thus proved, by means of a stress analysis, all the conclusions of Theorem 5.1 which had already been established by a strain analysis.*

□

To conclude this section we prove Lemma 5.2.

Proof of Lemma 5.2

Let v_0 be any element of W such that $L(v_0) = 1$. If we call $\mathcal{A}$ the kernel of L in W, F is simply the indicator function of $v_0 + \mathcal{A}$. To calculate F^* we can now apply Lemma 2.1 (with $v_0^* = 0$). We find that

$$F^*(v^*) = \langle v^*, v_0\rangle + \chi_{\mathcal{A}^0}(v^*)$$

that is to say $F^*(v^*) = \langle v^*, v_0\rangle$ if $v^* \in \mathcal{A}^0$, and $= +\infty$ otherwise. We confine ourselves to $v^* = \Lambda^* p$. To say that $\Lambda^* p \in \mathcal{A}^0$ is equivalent here to saying that

$$\langle \Lambda^* p, v\rangle = 0\ , \text{ for all } v \in W \text{ such that } Lv = 0.$$

By an elementary theorem of linear algebra this is true if and only if the two linear forms on W, $\Lambda^* p$ and L, are linearly dependent, or in other words, there exists a $\lambda \in \mathbb{R}$ such that $\Lambda^* p + \lambda L = 0$ (in W').

Thus $F^*(\Lambda^* p) = \langle \Lambda^* p, v_0\rangle$ if there exists a $\lambda \in \mathbb{R}$ such that $\Lambda^* p + \lambda L = 0$ and if not, then $F^*(\Lambda^* p) = 0$.

When proving (5.16) we characterised the condition $\Lambda p + L = 0$ in W'. Replacing L by λL, we find that this condition means that

$$(5.30)\quad \begin{cases} \text{There is a } p_0 \in L^2(\Omega) \text{ such that} \\ \sigma = - p_0 1 - p \text{ satisfies } \operatorname{div}\sigma + \lambda f = 0 \text{ in } \Omega \\ \qquad\qquad \text{and } \sigma.\nu = \lambda g \text{ on } \Gamma_1 . \end{cases}$$

When $\Lambda p + \lambda L = 0$, we have $\langle \Lambda^* p, v_0\rangle = -\lambda L(v_0) = -\lambda$ (by definition of v_0). In conclusion therefore $F^{*0}(\Lambda^* p)$ has the value $-\lambda$ if (5.30) holds and the value $+\infty$ if not; which proves Lemma 5.2.

6. Relaxation of the boundary condition

As has already been said, it is not possible for us to prove in general the existence of a solution to the strain problem $\mathscr{P}$ (cf. (4.16)). Anyway the existence of a solution (of the type postulated) seems to be ruled out on mechanical grounds because the displacement fields observed in practice can exhibit surface discontinuities and even discontinuities at the boundary if plastification occurs there. In this latter case the boundary condition $u = u_0$ would not be satisfied at some or all the points of Γ_0. It is clear that in this case *the variational principle (4.16) corresponds to the search for u in a too restricted class of functions and that it cannot take proper account of the mechanical phenomenon involved.*

In this Section we describe a first stage towards an extension of the variational principle (4.16): we relax partially the boundary conditions. We introduce a new variational problem $\mathscr{PR}$ which generalises the initial problem $\mathscr{P}$, and in which the boundary condition on Γ_0 is partially abandoned. This new problem seems to be able to take account of the phenomena involved in plastification over the whole or part of Γ_0[1]. After having defined the problem $\mathscr{PR}$ (in Section 6.1) and giving its dual (in Section 6.2) we go on to study its relationship with the initial problem (in Section 6.3).

In Chapter II we shall deal with the next stage in the extension of the problem $\mathscr{P}$, when the problem $\mathscr{PR}$ will itself be generalised.

---

[1] See Remark 6.1.

6.1 The relaxed problem

The relaxed problem involves the function ψ_∞ introduced in (5.2) and which reduces on E^D to the support function of K^D (in E^D),

$$(6.1) \qquad \psi_\infty(\xi^D) = \sup_{\eta^D \in K^D} \xi^D.\eta^D \ , \quad \forall \xi^D \in E^D .$$

On the other hand, for $p \in L^1(\Gamma)^3$ (or $p \in L^1(\Gamma_0)^3$) we define on Γ (or on Γ_0) the tensor $\mathcal{T}(p)$ whose components are given by

$$(6.2) \qquad \mathcal{T}_{ij}(p) = \frac{1}{2}(p_i \nu_j + p_j \nu_i) \ , \quad i,j = 1, 2, 3,$$

where the ν_i are the components of the vector ν, the outward normal unit-vector on Γ. We similarly write $\mathcal{T}^D(p)$ for the deviator of $\mathcal{T}(p)$, whose components are therefore given by

$$(6.3) \qquad \mathcal{T}^D_{ij}(p) = \mathcal{T}_{ij}(p) - \frac{1}{3}\, p.\nu\, \delta_{ij} .$$

The relaxed problem $\mathcal{PR}$ can then be written as

$$(6.4) \qquad \operatorname*{Inf}_{v} \{\Psi(\varepsilon(v)) + \int_{\Gamma_0} \psi_\infty(\mathcal{T}^D(\gamma_\tau(u_0-v)))d\Gamma - L(v)\} ,$$

where the infimum is taken over all v such that

$$(6.5) \qquad \{v \in H^1(\Omega)^3 \ , \ v.\nu = u_0.\nu \ \text{ on } \Gamma_0\} .$$

([1]) The operator $\mathcal{T}$ may be likened to the operator ε, the derivations D_i being replaced by multiplication by ν_i. The operator $\mathcal{T}$ also appears in works on the mechanics of rupture: cf. N.Q. Son [1].

We denote by $\gamma_\tau(w)$ (or w_τ) the tangential component on Γ of a vector w

$$\gamma_\tau(w) = w_\tau = w - (w.\nu)\nu \ .$$

In comparison with $\mathscr{P}$, the problem $\mathscr{PR}$ has been given a weaker boundary condition: $v.\nu = u_0.\nu$ on Γ_0 instead of $v = u_0$ on Γ_0; on the other hand the functional to be minimised has the additional term

$$\int_{\Gamma_0} \psi_\infty(\mathscr{T}^D(u_0 - v))d\Gamma \ . \tag{6.6}$$

If we note that this term vanishes if $v = u_0$ on Γ_0($\psi_\infty(0) = 0$ by (5.3)), and that the set (6.5) is larger than $\mathscr{C}_a$, we conclude that $\mathscr{PR}$ is an extension of $\mathscr{P}$: we are minimising the same functional for $\mathscr{PR}$ as for $\mathscr{P}$ but over a larger set. In particular

$$\text{Inf}\,\mathscr{PR} \leq \text{Inf}\ \mathscr{P} \ . \tag{6.7}$$

We shall see that in actual fact the two sides in (6.7) are equal.

6.2 Dual of the relaxed problem

To determine the dual of $\mathscr{PR}$ we have, as in the preceding Sections, to use the general framework of Section 2.2. To write the problem in the form (2.35) we put $V = H^1(\Omega)^3$, $Y = Y_1 \times Y_2 \times Y_3$ where $Y_1 = L^2(\Omega)$ and $Y_2 = L^2(\Omega;E^D)$ as in Section 4, while

$$Y_3 = \{q \in L^2(\Gamma_0)^3 \ , \ q.\nu = 0 \ \text{ on } \ \Gamma_0\} \ . \tag{6.8}$$

The operator Λ is of the form $\Lambda_1 \times \Lambda_2 \times \Lambda_3$ where $\Lambda_1 = \text{div}$ and $\Lambda_2 = \varepsilon^D$ as in problem $\mathscr{P}$, while, for $v \in H^1(\Omega)^3$, $\Lambda_3 v$ is the tangential component on Γ_0 of the trace of v on Γ; Λ_3 is a continuous function from V into Y_3 by the Trace Theorem (1.22).

The functions F and G are chosen as follows : for all v of V,

$$F(v) = \begin{cases} -L(v) & \text{if } v.\nu = u_o.\nu \text{ on } \Gamma_o \\ +\infty & \text{if not ,} \end{cases}$$

and for $p = (p_1, p_2, p_3) \in Y$,

$$G(p) = G_1(p_1) + G_2(p_2) + G_3(p_3)$$

$$G_1(p_1) = \frac{\kappa}{2}\int_\Omega (p_1)^2 dx \quad , \forall\, p_1 \in Y_1 ,$$

$$G_2(p_2) = \int_\Omega \psi(p_2)\,dx \quad , \forall\, p_2 \in Y_2 ,$$

$$G_3(p_3) = \int_{\Gamma_o} \psi_\infty(\mathscr{T}^D((u_o)_\tau - p_3))\,d\Gamma \quad , \forall\, p_3 \in Y_3 .$$

The space V^* will be the dual of V and Y^* is taken equal to Y. All the required conditions in Section 2.2 are satisfied so that the problem (6.4)-(6.5) is identical to the problem (2.35). To determine the dual we calculate F^* and G^*. For G^* we have

$$G^*(p) = G_1^*(p_1) + G_2^*(p_2) + G_3^*(p_3) \quad , \forall\ p = (p_1, p_2, p_3) \in Y ,$$

the calculation of G_1^* and G_2^* having already been done in Section 4. The calculation of G_3^* is complicated and we shall leave it for the moment; lastly for F^* we have the following result which will be proved in Section 6.3.

LEMMA 6.1.

For all p *of* Y

$$F^*(\Lambda^* p) = \begin{cases} \displaystyle\int_{\Gamma_o} (p_3.u_{o\tau} - (\sigma.\nu)u_o)\,d\Gamma & \text{if } \begin{cases} \operatorname{div} \sigma + f = 0 & \text{in } \Omega \\ \sigma.\nu = g & \text{on } \Gamma_1 \\ p_3 = (\sigma.\nu)_\tau & \text{on } \Gamma_o \end{cases} \\ +\infty & \text{otherwise,} \end{cases}$$

where we have put $\sigma = -p_1 I - p_2$.

We can now formulate the problem $\mathcal{PR}^*$ which corresponds to (2.36):

$$\text{(6.9)} \qquad \sup_p \{-\tfrac{1}{2}\mathcal{A}(\sigma,\sigma) - G_3^*(-p_3) - \int_{\Gamma_0} (p_3 u_0 - (\sigma.\nu)u_0)\, d\Gamma\}$$

where $\sigma = -p_1 I - p_2$, and the supremum is taken over the set of p satisfying the following conditions:

$$\text{(6.10)} \qquad \begin{cases} p = (p_1, p_2, p_3) \in Y \\ \operatorname{div}\sigma + f = 0 \quad \text{in } \Omega \\ \sigma.\nu = g \quad \text{on } \Gamma_1 \\ \sigma^D(x) \in K^D \quad \text{for almost all } x \text{ of } \Omega \\ p_3 = (\sigma.\nu)_\tau \quad \text{on } \Gamma_0 . \end{cases}$$

This formulation of $\mathcal{PR}^*$ is not completely explicit since we have not not yet calculated G_3^* ; $G_3^*(-p_3)$ is in general complicated but we shall establish in Section 6.3 (cf. Lemma 6.3) that if $p_3 = (\sigma.\nu)_\tau$ where p and $\sigma = -p_1 I - p_2$ satisfy the conditions (6.10) then

$$\text{(6.11)} \qquad G_3^*(-p_3) = -\int_{\Gamma_0} p_3\, u_{0\tau}\, d\Gamma .$$

Under these conditions p_3 *disappears completely from the problem (6.9)-(6.10) which becomes identical to the problem (4.29)-(4.30).* This observation will enable us to compare $\mathcal{P}$ and $\mathcal{PR}$.

6.3 Relationship with the initial problem

We have the following

THEOREM 6.1

i) $\mathcal{P}^*$ *and* $\mathcal{PR}^*$ *are identical,* $\mathcal{P}$ *and* $\mathcal{PR}$ *admit of the same dual;*

ii) *Inf* $\mathcal{P}$ = *Inf* $\mathcal{PR}$ = *Sup* $\mathcal{P}^*$ = *Sup* $\mathcal{PR}^*$ $(\in \mathbb{R} \cup \{\infty\})$;

iii) *Every minimising sequence of* $\mathcal{P}$ *is also a minimising sequence of* $\mathcal{PR}$;

iv) *If the number in ii) is* $> -\infty$, $\mathcal{P}^*$ *and* $\mathcal{PR}^*$ *have a unique solution* σ *which is the stress tensor of the problem.*

Proof

Assuming for the moment that (6.11) is true, we have already observed that $\mathcal{P}^*$ and $\mathcal{PR}^*$ are identical and, by Theorem 4.1

$$\text{(6.12)} \qquad \operatorname{Inf} \mathcal{P} = \operatorname{Sup} \mathcal{P}^* = \operatorname{Sup} \mathcal{PR}^* .$$

To prove (ii) we still need to establish that

$$\text{(6.13)} \qquad \operatorname{Inf} \mathcal{PR} = \operatorname{Sup} \mathcal{PR}^* .$$

This is obvious, from (6.7) and (6.12), if $\operatorname{Sup} \mathcal{PR}^* = -\infty$. If $\operatorname{Sup} \mathcal{PR}^* > -\infty$ then $\operatorname{Inf} \mathcal{PR} > -\infty$ by (2.28); and (6.13) will now follow from (2.30) if we can show that (2.29) is satisfied. To verify (2.29) we note that G is everywhere finite and continuous on Y : this has already been observed for G_1 and G_2 ; for G_3 it follows from the inequalities (5.4) and Proposition 2.4 which together imply that the (convex) function

$$q \mapsto \int_{\Gamma_0} \psi_\infty(\mathcal{T}^D q)\, d\Gamma ,$$

is everywhere finite and continuous on $L^1(\Gamma_0)^3$ and hence on Y_3 . It is therefore possible to satisfy (2.29) with $u^0 = u_0$.

This proves (ii); statements (iii) and (iv) then become evident.

Remark 6.1. We have not found the problem $\mathcal{PR}$ ((6.4)-(6.5)) explicitly described in the mechanical literature. It should be possible to obtain it from the virtual work theorem (cf. for example P. Germain [1]-[2]) by supposing that there is plastification over the whole or part of Γ_0 .

Nevertheless (cf. G. Strang - R. Temam [2]), Theorem 6.1 shows that $\mathcal{PR}$ is an acceptable extension, from the mechanical standpoint, of the initial Hencky problem:

- the infimum (giving the energy) remains the same,
- the dual (giving the stress tensor) is unchanged.

□

To end this section we shall prove Lemma 6.1 and (6.11).

Proof of Lemma 6.1.

We use Lemma 2.1; the function F has the required form with $v_0^* = -L$, $v_0 = u_0$ and

(6.14) $$\mathscr{B} = \{v \in H^1(\Omega)^3 \,,\ v.\nu = 0 \ \text{ on } \ \Gamma_0\} \,.$$

We deduce from Lemma 2.1 that

(6.15) $$F^*(\Lambda^* p) = \langle \Lambda^* p + L, u_0 \rangle$$

if

(6.16) $$\langle \Lambda^* p + L, v \rangle = 0 \,, \quad \forall\ v \in \mathscr{B} \,,$$

and that $F^*(\Lambda^* p) = +\infty$ if not.

Condition (6.16) can be written

(6.17) $$\int_\Omega (p_1 \operatorname{div} v + p_2 . \varepsilon^D(v) + fv)\,dx + \int_\Gamma p_3 v_\tau \, d\Gamma + \int_{\Gamma_1} g\, v\, d\Gamma = 0, \ \forall\ v \in \mathscr{B}.$$

To interpret (6.16)-(6.17) we first write down this condition for the v belonging to $\mathscr{C}_0^\infty(\Omega)^3$, after having noted that $\sigma = -p_1 I - p_2$:

$$-\int_\Omega \sigma_{ij}\, v_{i,j}\, dx + \int_\Omega fv\, dx = 0 \,, \quad \forall\ v \in \mathscr{C}_0^\infty(\Omega)^3$$

which is equivalent to

(6.18) $$\operatorname{div} \sigma + f = 0 \ \text{ in } \ \Omega \,.$$

When p (and σ) satisfy (6.18) then $\sigma \in L^2(\Omega;E)$, $\operatorname{div} \sigma \in L^2(\Omega)$ and we can apply the generalised Green's formula (1.51) enabling us to rewrite (6.17) in the form

(6.19) $$-\int_\Gamma (\sigma.\nu) v \, d\Gamma + \int_{\Gamma_0} p_3\, v_\tau\, d\Gamma + \int_{\Gamma_1} g\, v\, d\Gamma = 0 \,, \quad v \in \mathscr{B} \,.$$

This implies that

(6.20) $$p_3 = (\sigma.\nu) \quad \text{on} \quad \Gamma_0$$

(6.21) $$g = \sigma.\nu \quad \text{on} \quad \Gamma_1 .$$

Conversely it is easy to see that if p (and σ) satisfy (6.20)-(6.21) then (6.19) holds for all $v \in \mathscr{C}^1(\bar{\Omega})^3$ whose normal component vanishes in a neighbourhood of Γ_0. This set of functions is, by Lemma 6.2 below, dense in $\mathscr{B}$ and (6.19) therefore holds by continuity ([1]). Thus the conditions (6.18) (6.20) and (6.21) are together equivalent to (6.16).

When these conditions are satisfied we can again use the generalised Green's formula (1.51) and transform the expression (6.15) for $F^*(\Lambda^* p)$ into:

$$F^*(\Lambda^* p) = \langle \Lambda^* p + L, u_0 \rangle$$
$$= -\int_\Omega \sigma_{ij}\, \varepsilon_{ij}(u_0)dx + \int_\Omega f\, u_0\, dx + \int_{\Gamma_0} p_3\, u_{0\tau}\, d\Gamma + \int_{\Gamma_1} g\, u_0\, d\Gamma$$
$$= \int_{\Gamma_0} (p_3\, u_{0\tau} - (\sigma.\nu)u_0)d\Gamma \quad .$$

This proves Lemma 6.1.

LEMMA 6.2.

The set of v *belonging to* $\mathscr{C}^2(\bar{\Omega})^3$ *whose normal component vanishes in a neighbourhood of* Γ_0 *is dense in the set (6.14).*

<u>Proof</u>

At each point x of Γ we define an orthonormal trihedron τ_1, τ_2, ν consisting of two tangent vectors τ_1, τ_2, and the outward normal unit-vector. Since Ω is of class $\mathscr{C}^2$ it is possible to extend

---

([1]) The expression $\int_\Gamma (\sigma.\nu)v\, d\Gamma$ is a convenient 'abuse' of notation used here to represent the scalar product of $\sigma.\nu_{|\Gamma}$ and $v_{|\Gamma}$ defined in terms of the duality between the spaces $H^{\frac{1}{2}}(\Gamma)^3$ and $H^{-\frac{1}{2}}(\Gamma)^3$. The formula

$$\int_\Gamma (\sigma.\nu)v\, d\Gamma = \int_{\Gamma_0} (\sigma.\nu)v\, d\Gamma + \int_{\Gamma} (\sigma.\nu)v\, d\Gamma$$

is not automatic; cf. ([1]) p.46 and ([1]) p.47.

these vector fields τ_1, τ_2, ν defined on Γ into orthonormal vector fields of class $\mathscr{C}^2$ defined in an open neighbourhood $\mathscr{O}$ of Γ.

By making use of the partition of unity subordinated to the covering of $\bar{\Omega}$ by $\mathscr{O}$ and Ω, we can restrict ourselves to functions in $\mathscr{B}$ with compact support in $\mathscr{O}$. Such a function can be written as

$$v = a\,\tau_1 + b\,\tau_2 + c\,\nu ,$$

where a, b, c are in $H^1(\Omega)$ and $c = 0$ on Γ_0. By (1.19) the functions a and b are limits in $H^1(\Omega)$ of functions in $\mathscr{C}^\infty(\bar{\Omega})$, a_m and b_m. By (2.106), c is the limit in $H^1(\Omega)$ of a sequence of functions c_m of class $\mathscr{C}^\infty(\bar{\Omega})^3$, which have the value zero in a neighbourhood of Γ_0. The sequence $v_m = a_m\,\tau_1 + b_m\,\tau_2 + c_m\nu$ has the required properties and converges to v in $H^1(\Omega)^3$.

□

Finally we prove (6.11).

LEMMA 6.3

If $\sigma \in L^2(\Omega;E)$, $div\ \sigma \in L^2(\Omega)$ *and* $\sigma^D(x) \in K^D$ *for almost all* x *of* Ω, *then*

$$(6.22)\qquad G_3(-(\sigma.\nu)_\tau) = \int_{\Gamma_0} (\sigma.\nu)\ u_{0\tau}\ d\Gamma .$$

Proof

i) By definition of G_3:

$$G_3^*(-p_3) = \sup_{\tilde{p}_3 \in Y_3} \{-\int_\Gamma p_3\,\tilde{p}_3\,d\Gamma - \int_\Gamma \psi_\infty(\mathscr{T}^D(u_0 - \tilde{p}_3))d\Gamma\} ,$$

or, putting $\tilde{p}_3 = u_{0\tau} - q$, $q \in Y_3$:

$$G^*(-p_3) = -\int_{\Gamma_0} p_3\,u_0\ d\Gamma + \sup_{q\in Y_3}\ \{\int_{\Gamma_0} p_3\,q\ d\Gamma - \int_{\Gamma_0} \psi_\infty(\mathscr{T}^D q)\ d\Gamma\}.$$

The relation (6.22) will be proved $(p_3 = (\sigma.\nu)_\tau)$ if we can show that this last supremum, which is ≤ 0, is in fact equal to 0, or in other words if we can show that

$$(6.23) \qquad \int_{\Gamma_0} p_3\, q\, d\Gamma - \int_{\Gamma_0} \psi_\infty(\mathscr{T}^D q)\, d\Gamma \leq 0$$

for all $q \in L^2(\Gamma_0)^3$, such that $q.\nu = 0$ on Γ_0. To prove (6.23) it would be sufficient to show that

$$(6.24) \qquad p_3\, q - \psi_\infty({}^D q) \leq 0$$

almost everywhere on Γ_0 whenever $q \in Y_3$ and $p_3 = (\sigma.\nu)_\tau$ σ satisfying the conditions specified in the Lemma. We shall first establish (6.24)by supposing σ to be a little more regular, and then on the present hypotheses.

ii) Let us first prove (6.24) when σ satisfies the conditions of the Lemma and in addition belongs to $\mathscr{C}(\bar{\Omega};E)$. We then have, by continuity, $\sigma^D(x) \in K^D$, $\forall\, x \in \Gamma_0$. On the other hand by utilising at each point x of Γ_0 an orthonormal base e_1, e_2 e_3 with $e_3 = \nu$, we can see that, at the point x, $p_3 = ((p_3)_1,(p_3)_2,0)$, $q = (q_1,q_2,0)$

$$(6.25) \qquad \mathscr{T}^D(q) = \begin{pmatrix} 0 & 0 & \frac{1}{2}q_1 \\ 0 & 0 & \frac{1}{2}q_2 \\ \frac{1}{2}q_1 & \frac{1}{2}q_2 & 0 \end{pmatrix}$$

and

$$(6.26\,) \qquad p_3.q = (\sigma.\nu)_\tau.q = \sigma^D.\mathscr{T}^D(q) \;.$$

The left-hand side of (6.24) is simply

$$\sigma^D \mathscr{T}^D(q) - \psi_\infty(\mathscr{T}^D q) \;,$$

and this is ≤ 0 by definition of ψ_∞ and because $\sigma^D(x) \in K^D$.

iii) In the case where σ does not satisfy the additional condition we shall obtain the desired result by approximation.

([1]) Cf. Lemma 7.1 and Remark 7.1.ii), Ch. II; we do not of course use the result of Lemma 6.3 in the proof of Lemma 7.1.

By an approximation result established in Chapter II [1] every $x \in \Gamma$ has an open neighbourhood $\mathcal{O}$ on which one can approximate σ by a sequence of functions $\sigma_m \in \mathscr{C}^\infty(\mathcal{O} \cap \bar{\Omega};E)$, with $\sigma_m^D(x) \in K^D$, $\forall\, x \in \mathcal{O} \cap \bar{\Omega}$ and such that for $m \to \infty$

$$\sigma_m \to \sigma \;,\quad \operatorname{div} \sigma_m \to \operatorname{div} \sigma$$

in $L^2(\mathcal{O} \cap \bar{\Omega};E)$ and $L^2(\mathcal{O} \cap \bar{\Omega})^3$ respectively. The inequality (6.24) holds (for $p_3 = (\sigma_m.\nu)_\tau$) on $\Gamma_0 \cap \mathcal{O}$ and to obtain the result for σ (on $\Gamma_0 \cap \mathcal{O}$), it suffices to see that $(\sigma_m.\nu)_\tau$ converges in some appropriate sense. By an algebraic calculation similar to that carried out in paragraph ii) of this proof it can be seen that at every point $x \in \Gamma$,

$$(\sigma_m.\nu)_\tau = (\sigma_m^D.\nu)_\tau \;,\quad (\sigma.\nu)_\tau = (\sigma^D.\nu)_\tau \;.$$

The functions $(\sigma_m.\nu)_\tau$ are therefore bounded in $L^\infty(\Gamma_0 \cap \mathcal{O})^3$ (since K^D is bounded) and since, by the Trace Theorem (1.47), the converge to $(\sigma.\nu)_\tau$ in $H^{-\frac{1}{2}}(\Gamma_0 \cap \mathcal{O})^3$ we conclude that the inequality (6.24) which holds on $\Gamma_0 \cap \mathcal{O}$ for the σ_m thus also holds on $\Gamma_0 \cap \mathcal{O}$ for σ.

The Lemma is therefore proved.

Remark 6.2 By (6.25) if q is a tangent vector on Γ

$$|\mathscr{T}^D(q(x))| = \frac{1}{\sqrt{2}}\,|q(x)| \;, \tag{6.27}$$

for almost all x of Γ.

Chapter II

Solution of the variational problems in the finite-energy spaces

Introduction

We saw in Chapter I that there was little hope of finding a solution to the strain problem I(4.16) in the Sobolev space $H^1(\Omega)^3$ or even in $W^{1,1}(\Omega)^3$: one aspect of the mathematical difficulty is that the *a priori* estimates which one obtains for a minimising sequence (see (4.21) (4.22)) are not good enough to ensure adequate compactness properties for these sequences.

The aim of this Chapter is to prove the existence of a solution to the strain problem in a larger function space and, using the analogy of other problems in the calculus of variations, we shall concern ourselves with *the largest possible class of functions, the class of all finite-energy functions*. This mathematical concept needs to be more precisely defined but we can already get a general idea of it by considering the following question which was already mentioned in the Introduction to this book:

(☆) We consider a sequence of sufficiently 'smooth' functions (say $u_m \in \mathscr{C}^1(\bar{\Omega})^3$) which converges uniformly to a u (belonging to $\mathscr{C}(\bar{\Omega})^3$), such that the strain energy for each displacement field

$$\Psi(\varepsilon(u_m)) = \int_\Omega \psi(\varepsilon(u_m))dx ,$$

remains bounded. What further information do we have about u ?

The answer to this question leads us to consider the space BD(Ω) which has been introduced in H. Matthies, G. Strang and E. Christiansen [1] and P. Suquet [1] (cf. also J.J. Moreau [3]) then studied by G. Strang and R. Temam [1][3], R. Temam [9][10][11]. A considerable part of this Chapter is devoted to the study of this space (Sections 2 and 3). This study is preceded, in Section 1, by various supplementary theorems and other additional matter on the space LD(Ω) (already introduced in Chapter I); this space is somewhat like the space BD(Ω) though it does not share some of its most important properties; and in particular its bounded closed sets do not have a certain weak compactness property. Nevertheless we have thought it more convenient to take this opportunity of demonstrating, for the space LD(Ω) , certain properties common to both LD(Ω) and BD(Ω) ; this is the object of Section 1. Sections 2 and 3 contain the main results on BD(Ω) , and in particular the most important of these which are Theorems 2.1 to 2.4 and which require techniques recalling those used in the theory of Sobolev spaces, but also differing from them in several essential points. We would also draw attention to the approximation results in Theorems 3.2 to 3.4 which are indispensable in what follows. Finally the weak compactness of the bounded closed sets of BD(Ω) enables us to give a simple answer to the question (☆) posed earlier.

Sections 4 and 5 are devoted to the study of convex functions of a measure: when f is a convex function of no greater than linear growth at infinity and satisfying certain other technical conditions, and if μ is a bounded measure, it is possible to define a bounded measure, which we denote by f(μ) , which coincides with f∘h dx when μ is a function (μ = h or μ = h dx). These functions of measure studied in Section 4 enables us, in Section 5, to approach from a new angle convex functionals of measure (which are convex mappings of the space of bounded measures into the set $\mathbb{R}$ or $\mathbb{R} \cup \{+\infty\}$). Section 5 also contains various useful results on the approximation of convex functions of a measure.

The aim of Sections 1 to 5 is to introduce all the indispensable tools needed to study the generalised strain problem. From Section 6 onwards we resume the study of the problem outlined in Sections 4 and 6 of Chapter I. In Section 6 we give a generalised form of the strain problem making use of the concepts introduced in the preceding sections: this problem $\mathcal{Q}$ is an extension of the problem $\mathcal{PR}$ of Section I.6 which is itself in turn an extension of the problem $\mathcal{P}$ of Section I.4. Section 7 is devoted to studying the duality between the generalised strain problem and the stress problem as it appeared in Sections I.4 and I.6: this raises fresh difficulties because the product $\sigma.\varepsilon(u)$ is not automatically defined when σ is an admissible stress tensor $(\sigma \in L^2(\Omega;E))$ and u belongs to the field of admissible displacements $(u \in BD(\Omega))$.

Section 8 is the cumulation of Chapter II: it gives the existence theorem for the solution of the generalised strain problem and the relations between this solution and the stress field.

1. Further results on the space LD (Ω)

We continue our study of the space $LD(\Omega)$ introduced in Chapter I (cf. 1.71). We shall establish the following four fundamental results:

(i) a trace theorem: if $u \in LD(\Omega)$, its trace on Γ is defined and belongs to $L^1(\Gamma)^n$;

(ii) an inclusion result in a space L^p (of the same type as the Sobolev inclusions);

(iii) a regularity result;

(iv) a compactness result (compactness of the injection of $LD(\Omega)$ in certain spaces L^p).

The results (i), (ii), (iv) above are related to the corresponding results in the theory of Sobolev spaces, without however being immediately deducible from them: $LD(\Omega)$ is related to the space $W^{1,1}(\Omega)^n$ which is a subspace of the former but $LD(\Omega) \neq W^{1,1}(\Omega)^n$ because, as already pointed out in Chapter I, Korn's inequality does not hold good in L^1. Nevertheless (i) and (ii) are the results which would follow if $LD(\Omega) = W^{1,1}(\Omega)$ were true.

After a few remarks on the deformation operator given in Section 1.1 we state the main results in Section 1.2 and give various applications of these. The results (i), (ii), (iii), (iv) are then proved in Sections 1.3 to 1.6. It is not necessary to read these proofs for an understanding of what follows thereafter. The treatment is in an n-dimensional space (with arbitrary n), and this involves no additional complications.

1.1. Some properties of the strain (or deformation) operator

We shall suppose that Ω is a bounded open connected set in $\mathbb{R}^n$ with $n \geq 2$.

We recall that if $u = (u_1,\ldots,u_n)$ is a function mapping $\mathbb{R}^n$ into $\mathbb{R}^n$, then the strain tensor $\varepsilon(u)$ is the tensor of order 2 whose components, for the canonical basis of $\mathbb{R}^n$, can be written

$$(1.1) \qquad \varepsilon_{ij}(u) = \frac{1}{2}\left(\frac{\partial u_i}{\partial x_j} + \frac{\partial u_j}{\partial x_i}\right) .$$

We write $\mathcal{R}$ for the kernel of the operator ε in any space comprised between $\mathcal{C}_0^\infty(\Omega)^n$ and $\mathcal{D}'(\Omega)^n$, the space of vector-distributions on Ω (cf. Section I.1) ; $\mathcal{R}$ is the set of functions u of the form

$$(1.2) \qquad u(x) = a + B.x ,$$

where $a \in \mathbb{R}^n$, B is an $n \times n$ skew symmetric matrix, and when $n = 3$,

$$(1.3) \qquad u(x) = a + b \wedge x ,$$

$a, b \in \mathbb{R}^3$ (cf. Lemma I.1.1).

The following result allows us to characterise the image of ε.

PROPOSITION 1.1.

Let $\varepsilon_{ij} = \varepsilon_{ji}$ be a family of distributions on Ω, $\varepsilon_{ij} \in \mathcal{D}'(\Omega)$, $i,j = 1,\ldots,n$. Then the following conditions are equivalent:

i) there exists a distribution $u \in \mathcal{D}'(\Omega)^n$ such that

$$(1.4) \qquad \varepsilon_{ij} = \varepsilon_{ij}(u) = \frac{1}{2}\left(\frac{\partial u_i}{\partial x_j} + \frac{\partial u_j}{\partial x_i}\right) , \quad \forall\, i,j .$$

ii) For every family of functions $\phi_{ij} \in \mathcal{C}_0^\infty(\Omega)^n$, $i,j = 1,\ldots,n$, such that

$$(1.5)\qquad \sum_{i=1}^{n} \frac{\partial \phi_{ij}}{\partial x_j} = 0 \ , \ i = 1,\ldots,n \ ,$$

we have

$$\sum_{i,j=1}^{n} \langle \varepsilon_{ij}, \Phi_{ij} \rangle = 0 \ .$$

This proposition is a corollary of more general topological theorems (cf. G. De Rham [1]) (1). A direct proof is given in J.J. Moreau [5].

The result implies that if a sequence u_m is such that the $\varepsilon_{ij}(u_m)$ converge in some very weak sense (for example in the sense of distributions) to a limit $\varepsilon_{ij} \in \mathcal{D}'(\Omega)$, then there exists a distribution $u \in \mathcal{D}'(\Omega)^n$ such that $\varepsilon_{ij} = \varepsilon_{ij}(u)$, $i,j = 1,\ldots,n$.

1.2. Statement of the main results and applications

We recall the definition given in I.(1.71) of the space $LD(\Omega)$

$$(1.6)\qquad LD(\Omega) = \{u \in L^1(\Omega)^n \ , \ \varepsilon_{ij}(u) \in L^1(\Omega) \ , \ i,j = 1,\ldots,n\} \ .$$

It is easily seen that this is a Banach space for the natural norm

$$(1.7)\qquad \|u\|_{LD(\Omega)} = \|u\|_{L^1(\Omega)^n} + \sum_{i,j=1}^{n} \|\varepsilon_{ij}(u)\|_{L^1(\Omega)} \ .$$

We have already observed that this space is distinct from the space $W^{1,1}(\Omega)^n$ which it contains, since Korn's inequality does not hold in L^1 : by a counter-example given in D. Ornstein [1] (for $n = 2$), there are u of $LD(\Omega)$ whose first derivatives are not all integrable. Furthermore, as a particular case of Proposition I.1.3, we had

$$(1.8)\qquad \mathscr{C}^\infty(\bar{\Omega})^n \text{ is dense in } LD(\Omega) \ .$$

---

(1) A similar result used in the theory of the Navier-Stokes equations enables us to characterise those distributions which are gradients; cf. G. De Rham [1] or for example R. Temam [5], Prop. 1.1 Ch.I.

We suppose that

(1.9) The open set Ω is of class $\mathscr{C}^1$

and we have:

THEOREM 1.1 *(Trace theorem)*

There exists a surjective continuous linear operator γ_o *from* $LD(\Omega)$ *onto* $L^1(\Gamma)^n$ *such that*

(1.10) $$\gamma_o(u) = u_{|\Gamma}\ , \quad \text{for all } u \in LD(\Omega) \cap \mathscr{C}(\bar{\Omega})^n$$

Furthermore for all i,j, *and all* $\phi \in \mathscr{C}^1(\bar{\Omega})$, *we have the generalised Green's formula*

(1.11) $$\frac{1}{2}\int_\Omega (u_i \frac{\partial\phi}{\partial x_j} + u_j \frac{\partial\phi}{\partial x_i})dx + \int_\Omega \phi\, \varepsilon_{ij}(u)\, dx = \int_\Gamma \phi\, \mathscr{T}_{ij}(\gamma_o(u))d\Gamma$$

where $\mathscr{T}_{ij}$ *is the operator defined in I.(6.2).*

THEOREM 1.2 *(Sobolev inclusion)*

The space $LD(\Omega)$ *is contained into* $L^{n^*}(\Omega)^n$, $n^* = n/(n-1)$, *and the injection is continuous.*

THEOREM 1.3. *(Regularity)*

If $u \in \mathscr{D}'(\Omega)^n$ *and if* $\varepsilon_{ij}(u) \in L^1(\Omega)$ *for all* $i,j = 1,\dots,n$, *then* u *is a function of* $L^1(\Omega)^n$.

THEOREM 1.4. *(Compactness theorem)*

The injection of $LD(\Omega)$ *into* $L^p(\Omega)^n$ *is compact, for all* p *such that* $1 \leq p < n^* = n/(n-1)$.

These theorems will be proved in Sections 1.3 to 1.6. For the moment we shall draw a few conclusions from them.

PROPOSITION 1.2

On the quotient space $LD(\Omega)/\mathscr{R}$, *the expression*

(1.12) $$\sum_{i,j=1}^{n} \|\varepsilon_{ij}(u)\|_{L^1(\Omega)^n}$$

induces a norm equivalent to that induced by $LD(\Omega)$.

Proof

It is clear that (1.12) induces on $LD(\Omega)/\mathscr{R}$ a norm dominated by the norm induced by that of $LD(\Omega)$. By the closed graph theorem, the proposition will have been established, if we can show that $LD(\Omega)/\mathscr{R}$ is complete for the norm associated with (1.12).

Accordingly, let $\dot{u}_m$ be a sequence of $LD(\Omega)/\mathscr{R}$ which is a Cauchy sequence for (1.12) and for each m , let u_m be a representative of $\dot{u}_m$ in $LD(\Omega)$, $(u_m \in \dot{u}_m)$. We then have

$$\sum_{i,j=1}^{n} \|\varepsilon_{ij}(u_m - u_k)\|_{L^1(\Omega)} = \operatorname*{Inf}_{\substack{v_m \in \dot{u}_m \\ v_k \in \dot{u}_k}} \sum_{i,j=1}^{n} \|\varepsilon_{ij}(v_m - v_k)\|_{L^1(\Omega)}$$

and, for all i and j , $\varepsilon_{ij}(u_m)$ is a Cauchy sequence in $L^1(\Omega)$ which converges in that space to a limit ξ_{ij} . By Proposition 1.1, there exists a distribution $u \in \mathscr{D}'(\Omega)^n$ such that $\xi_{ij} = \varepsilon_{ij}(u), \forall i,j$. Lastly it follows from the Regularity Theorem 1.3 that $u \in L^1(\Omega)^n$, i.e. $u \in LD(\Omega)$, and it is easily verified that $\dot{u}_m$ converges to $\dot{u}$ in $LD(\Omega)/\mathscr{R}$ for the norm induced by (1.12).

PROPOSITION 1.3.

For all $u \in LD(\Omega)$, *there exists an* $r = r(u) \in \mathscr{R}$ *such that*

$$(1.13) \qquad \|u-r\|_{LD(\Omega)} \leq c_1(\Omega) \sum_{i,j=1}^{n} \|\varepsilon_{ij}(u)\|_{L^1(\Omega)} ,$$

where the constant c_1 *depends only on* Ω .

Proof

By definition of the norm in $LD(\Omega)/\mathscr{R}$:

$$(1.14) \qquad \operatorname*{Inf}_{r \in \mathscr{R}} \|u-r\|_{LD(\Omega)} = \|\dot{u}\|_{LD(\Omega)/\mathscr{R}} .$$

By Proposition 1.2 there exists a constant c_1 , depending only on Ω

such that

$$(1.15)\qquad \operatorname*{Inf}_{r\in\mathscr{R}} \|u-r\|_{LD(\Omega)} \leq c_1(\Omega) \sum_{i,j=1}^{n} \|\varepsilon_{ij}(u)\|_{L^1(\Omega)} ,$$

for all u of $LD(\Omega)$. As $\mathscr{R}$ is a finite-dimensional space, the infimum in (1.15) is attained: there is an $r(u)\in\mathscr{R}$ such that

$$\|u-r(u)\|_{LD(\Omega)} = \operatorname*{Inf}_{r\in\mathscr{R}} \|u-r\|_{LD(\Omega)} .$$

PROPOSITION 1.4.

If p *is a continuous seminorm on* $LD(\Omega)$ *which is a norm on* $\mathscr{R}$ *(so that* $p(u) = 0$, $u\in\mathscr{R}$ *implies* $u = 0$) , *then*

$$(1.16)\qquad p(u) + \sum_{i,j=1}^{n} \|\varepsilon_{ij}(u)\|_{L^1(\Omega)}$$

is a norm on $LD(\Omega)$ *equivalent to the initial norm.*

Proof

It follows from the assumptions that (1.16) defines a norm on $LD(\Omega)$, which is dominated by the initial norm, i.e. that $p(u) \leq c\,\|u\|_{LD(\Omega)}$ for all u . The proposition will therefore follow from the closed-graph theorem if we can show that $LD(\Omega)$ is complete for the norm (1.16).

Accordingly let u_m be a Cauchy sequence for (1.16). The sequence $\varepsilon_{ij}(u_m)$ is, for all i and j , a Cauchy sequence in $L^1(\Omega)$; it converges in that space to ξ_{ij} and by Proposition 1.1 there is a $u\in\mathscr{D}'(\Omega)^n$ such that $\xi_{ij} = \varepsilon_{ij}(u)$, $\forall\, i,j$. By the Regularity Theorem (Theorem 1.3), $u\in L^1(\Omega)^n$, and thus $u\in LD(\Omega)$ and $\varepsilon_{ij}(u_m-u)$ converges to 0 in $L^1(\Omega)$ for all i,j .

Thanks to Proposition 1.3 we see that, for all m , there exists a $r_m = r_m(u_m-u)\in\mathscr{R}$, such that $u_m - u - r_m$ converges to 0 in $LD(\Omega)$. By writing

$$p(r_m-r_k) \leq p(r_m-u_m+u) + p(u_m-u_k) + p(u_k-u-r_k) ,$$

we see that $p(r_m-r_k) \to 0$ when m and k tend to infinity (since $p(u_m-u_k) \to 0$ and u_m is a Cauchy sequence for (1.16)). This means that r_m is a Cauchy sequence in $\mathscr{R}$ and converges to a limit r and that u_m converges to $u+r$ in $LD(\Omega)$.

Remark 1.1. i) By combining Proposition 1.3 with Theorem 1.2, we find of course that, for all $u \in LD(\Omega)$, there exists an $r = r(u)$ such that

$$(1.17) \qquad \|u-r(u)\|_{L^{n^*}(\Omega)^n} \leq c_2(\Omega) \sum_{i,j=1}^{n} \|\varepsilon_{ij}(u)\|_{L^1(\Omega)} .$$

ii) We can in fact arrange for the mapping $u \mapsto r(u)$ to be linear (and therefore continuous); R. Kohn gives in [1] and explicit construction for $r(u)$.

More generally we can obtain such an operator r by proceeding as follows: we first take a basis $w_1,\ldots,w_M$, of the space $\mathscr{R}$, which we suppose to have dimension M , and continuous linear forms $\theta_1,\ldots,\theta_M$ on $LD(\Omega)$, such that $\langle\theta_i,w_j\rangle = \delta_{ij}$, $\forall\, i,j$. We can then set

$$(1.18) \qquad r(u) = \sum_{j=1}^{M} \langle\theta_j,u\rangle w_j .$$

For example in 3-dimensional (physical) space (corresponding to M=6), if we suppose for simplicity that the origin is at the centre of gravity of Ω , (i.e. $\int_\Omega x\,dx = 0$) we can take $w_i = e_i$, $w_{i+3} = e_i \wedge x$, for $i = 1,2,3$, where e_1, e_2, e_3 is the canonical basis of $\mathbb{R}^3$; then $\theta_i(u) = \frac{1}{|\Omega|}\int_\Omega u_i(x)dx$, $i = 1,2,3$ $(|\Omega| = \int_\Omega dx)$, and $\theta_4(u)$, $\theta_5(u)$, $\theta_6(u)$ are the components of $I_0^{-1}.\int_\Omega x\wedge u(x)\,dx$, where I_0 is the linear operator (the inertia tensor of Ω at the origin), defined by

$$b \in \mathbb{R}^3 \mapsto \int_\Omega x \wedge (b\wedge x)dx \in \mathbb{R}^3 .$$

Remark 1.2. i) In Proposition 1.4 it is possible to take for $p(u)$, any norm (in $\mathscr{R}$) of $r(u)$. Another and more interesting example, is to define $p(u)$, when Γ_0 is a non-vacuous open subset of Γ by

(1.19) $$p(u) = \int_{\Gamma_0} |u(x)|dx \ .$$

(Here we use the Trace Theorem 1.1.)

ii) Proposition 1.4 leads to a *Poincaré type inequality:* under the hypotheses of that proposition,

(1.20) There exists a $c = c(\Omega)$ such that

$$\|u\|_{L^1(\Omega)^n} \leq c \sum_{i,j=1}^{n} \|\varepsilon_{ij}(u)\|_{L^1(\Omega)} \ ,$$

for all u in

(1.21) $$\{v \in LD(\Omega) \ , \ p(v) = 0\} \ .$$

In particular $\sum_{i,j=1}^{n} \|\varepsilon_{ij}(u)\|_{L^1(\Omega)}$ is a norm on (1.21) equivalent to that of $LD(\Omega)$. □

1.3 Proof of Theorem 1.1

Since Γ is bounded it is enough to define the trace $\gamma_0(u)$ in the neighbourhood of each point y of Γ .

i) Let $y \in \Gamma$. Since at least one of the components of $\nu(y)$ is non-zero, we can suppose that $\nu_n(y) > 0$. We now consider a neighbourhood Σ of y in Γ which can be represented in the form

(1.22) $$\Sigma = \{(x',x_n),x' \in \mathcal{O} \ , \ x_n = a(x')\} \ ,$$

where $x' = (x_1,\ldots,x_{n-1})$, $\mathcal{O}$ is a bounded open set of $\mathbb{R}^{n-1}$ and a is a function of class $\mathscr{C}^1$ which is strictly positive on $\bar{\mathcal{O}}$. We may also suppose that

(1.23) $$\operatorname*{Inf}_{x\in\Sigma} \nu_n(x) = \rho_1 > 0 \ .$$

For α small, say $0<\alpha<\alpha_1$ for an appropriate α_1, we introduce the sets

$$\Sigma_\alpha = \{(x',x_n),\ x' \in \mathcal{O}\ ,\ x_n = a(x')-\alpha\} \tag{1.24}$$

$$\mathcal{A}_\alpha = \{(x',x_n),\ x' \in \mathcal{O}\ ,\ a(x') - \alpha<x_n<a(x')\}\ . \tag{1.25}$$

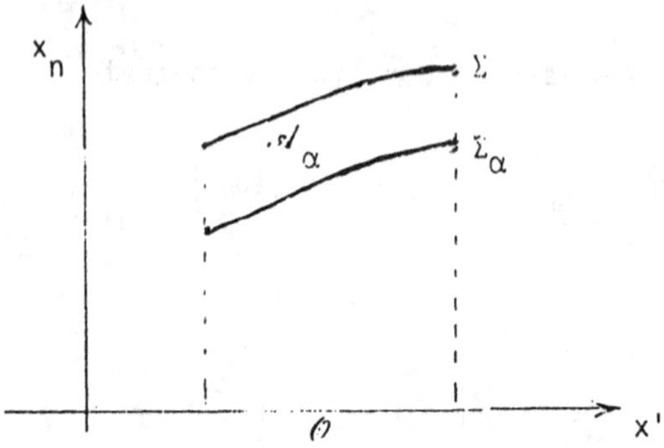

Fig. 1.1 The sets $\Sigma_\alpha, \mathcal{A}_\alpha$

Our first objective is to define the trace of u_n on Σ. After modifying u_n on a set of measure zero, the function $s \mapsto u_n(x',s)$ is, for almost all x' of $\mathcal{O}$, absolutely continuous for $s\in(a(x')-\alpha_1,a(x')$ and we have, for $0<\alpha<\alpha'<\alpha_0<\alpha_1$,

$$u_n(x',a(x')-\alpha') = u_n(x',a(x')-\alpha) + \int_{a(x')-\alpha}^{a(x')-\alpha'} \frac{\partial u_n}{\partial x_n}(x',s)ds\ . \tag{1.26}$$

We write this in the form

$$g_{\alpha'}(x) - g_\alpha(x) = \int_{a(x')-\alpha}^{a(x')-\alpha'} \frac{\partial u_n}{\partial x_n}(x',s)ds\ ,\ x \in \Sigma\ , \tag{1.27}$$

after having put, for almost all $x = (x',x_n)\in \Sigma$,

$$g_\alpha(x) = u_n(x',a(x')-\alpha)\ . \tag{1.28}$$

By integrating over Σ, we obtain

$$\int_\Sigma |g_{\alpha'}-g_\alpha|d\Gamma \leq \int_{\mathcal{O}} \left(\int_{a(x')-\alpha}^{a(x')-\alpha'} \left|\frac{\partial u_n}{\partial x_n}(x',s)\right|ds\right) \frac{dx'}{\nu_n(x',a(x'))}$$

$$(1.29)\qquad \int_\Sigma |g_{\alpha'}-g_\alpha|d\Gamma \leq \frac{1}{\rho_1}\int_{\mathcal{A}_{\alpha'}\setminus\mathcal{A}_\alpha} \left|\frac{\partial u_n}{\partial x_n}\right| dx .$$

When α and $\alpha' \to 0$, the integral on the right of (1.29) converges to zero, which proves that g_α is a Cauchy sequence in $L^1(\Sigma)$. We call g its limit in $L^1(\Sigma)$: g will be the trace of u_n on Σ. Clearly if $u \in LD(\Omega) \cap \mathscr{C}(\bar{\Omega})^n$ then $g = u_{|\Sigma}$.

The mapping $u \mapsto g$ is linear, and by letting α tend to zero in (1.29) and then integrating with respect to α' from 0 to α_0, we have

$$\int_\Sigma |g-g_{\alpha'}|d\Gamma \leq \frac{1}{\rho_1}\int_{\mathcal{A}_{\alpha'}} \left|\frac{\partial u_n}{\partial x_n}\right| dx \leq \frac{1}{\rho_1}\int_{\mathcal{A}_{\alpha_0}} \left|\frac{\partial u_n}{\partial x_n}\right| dx$$

$$\alpha_0 \int_\Sigma |g|d\Gamma \leq \int_0^{\alpha_0}\int_\Sigma |g_{\alpha'}|d\Gamma\, d\alpha' + \frac{\alpha_0}{\rho_1}\int_{\mathcal{A}_{\alpha_0}} \left|\frac{\partial u_n}{\partial x_n}\right| dx$$

$$(1.30)\qquad \int_\Sigma |g|d\Gamma \leq \frac{1}{\alpha_0}\int_{\mathcal{A}_{\alpha_0}} |u_n|dx + \frac{1}{\rho_1}\int_{\mathcal{A}_{\alpha_0}} \left|\frac{\partial u_n}{\partial x_n}\right| dx .$$

This last relation shows that the linear transformation $u \mapsto g$ is a continuous mapping from $LD(\mathcal{A}_{\alpha_0})$, (and hence from $LD(\Omega)$) into $L^1(\Sigma)$.

Similarly we can define the trace of u_i on a neighbourhood of y in Γ such that $\nu_i(y) \neq 0$, for each value of $i = 1,\dots,n-1$. If $\nu_i(y) = 0$ we note that $\nu_n(y) + \nu_i(y) \neq 0$, so that $\nu_n + \nu_i \neq 0$ in a neighbourhood of y in Σ. We choose a new orthonormal basis of $\mathbb{R}^n$ for which the nth direction is that of $e_n + e_i$. The last component of u in this base is, $v_n = (u_n + u_i)/\sqrt{2}$. In exactly the same way as before we define a trace of v_n in a neighbourhood Σ_i of y and this trace is in $L^1(\Sigma_i)$. This enables us, by taking the difference, to define the trace of u_i on $\Sigma_i \cap \Sigma$. This trace is in $L^1(\Sigma_i \cap \Sigma)$ and depends linearly and continuously on $u \in LD(\Omega)$.

ii) By collecting all the previous results, we see that we have defined, for all u of $LD(\Omega)$, a trace $\gamma_0(u) \in L^1(\Gamma)^n$, γ_0 being a linear continuous mapping of $LD(\Omega)$ into $L^1(\Gamma)^n$, and $\gamma_0(u) = u_{|\Gamma}$ for all $u \in LD(\Omega) \cap \mathscr{C}(\bar{\Omega})^n$.

Since by (1.8) $\mathscr{C}^1(\bar{\Omega})^n$ is dense in $LD(\Omega)$, γ_0 is the unique continuous extension of the mapping of the usual trace: $u \mapsto u_{|\Gamma}$ for regular functions.

In particular, as $W^{1,1}(\Omega)^n \subset LD(\Omega)$, the operator γ_0 coincides with the trace operator in $W^{1,1}(\Omega)^n$ defined by E. Gagliardo [1] (cf. I.(1.32)). As the operator γ_0 is already surjective from $W^{1,1}(\Omega)^n$ onto $L^1(\Gamma)^n$, it is *a fortiori* also surjective from $LD(\Omega)$) on $L^1(\Gamma)^n$.

To conclude the proof of Theorem 1.1, it only remains to prove Green's formula (1.11). This is elementary if $u \in \mathscr{C}^1(\bar{\Omega})^n$. By continuity, thanks to (1.8) it is still valid for all u of $LD(\Omega)$.

□

Remark 1.3. We deduce from Theorem 1.2 the following result:

(1.31) There exists a continuous linear operator Π from $LD(\Omega)$ into $LD(\mathbb{R}^n)$ such that $\Pi u(x) = u(x)$ for almost all x of Ω, $\forall u \in LD(\Omega)$.

Since $\gamma_0(u) \in L^1(\Gamma)^n$, there exists, as is shown in E. Gagliardo [1] a function $v \in W^{1,1}(\mathbb{R}^n \backslash \bar{\Omega})^n$ whose trace on Γ coincides with $\gamma_0(u)$, and this extension can be chosen in a linear and continuous fashion. It is then easy to see that the function Πu, equal to u on Ω and to v on $\mathbb{R}^n \backslash \bar{\Omega}$ is in $LD(\mathbb{R}^n)$ and that Π is a linear continuous operator from $LD(\Omega)$ into $LD(\mathbb{R}^n)$. By multiplying v by a function $\psi \in \mathscr{C}_0^\infty(\mathbb{R}^n)$ equal to 1 in a neighbourhood of $\bar{\Omega}$, we can suppose the support of Πu to be a subset of a fixed compact set (namely the support of ψ).

1.4. Proof of Theorem 1.2

Thanks to Remark 1.3 it will be sufficient to establish the result for the functions of $LD(\mathbb{R}^n)$ whose support is in a fixed compact set ([1]). The functions to be considered can be reduced to regular functions by the regularisation process, so that in the end we have only to establish the following result:

([1]) We shall give in Section 2 another proof of Theorem 1.2 which does not depend on Remark 1.3 and the results of E. Gagliardo [1].

(1.32) There is a constant $c = c(n)$, depending only on n, such that for all $u \in \mathscr{C}_0^\infty(\mathbb{R}^n)^n$

$$\|u\|_{L^{n^*}(\mathbb{R}^n)^n} \leq c(n) \sum_{i,j=1}^{n} \|\varepsilon_{ij}(u)\|_{L^1(\mathbb{R}^n)}$$

$(n^* = n/(n-1))$. We proceed in several steps.

i) Let α be a vector of $\mathbb{R}^n$ with components ± 1, and let $v = v_\alpha$ be the function $\alpha.u = \sum_{i=1}^{n} \alpha_i u_i$ (u given in $\mathscr{C}_0^\infty(\mathbb{R}^n)^n$). We then have

$$v_\alpha(x) = \int_{-\infty}^{0} \frac{d}{ds} v_\alpha(x+s\alpha)ds = \int_{-\infty}^{0} \sum_{i,j=1}^{n} \alpha_i \alpha_j \frac{\partial u_i}{\partial x_j}(x+\alpha s)ds$$

$$= \sum_{i,j=1}^{n} \alpha_i \alpha_j \int_{-\infty}^{0} \varepsilon_{ij}(u)(x+\alpha s)ds$$

and therefore, by Schwarz's inequality

(1.33) $$|v_\alpha(x)| \leq n \int_{-\infty}^{+\infty} |\varepsilon(u)(x+\alpha s)|ds$$

where of course

(1.34) $$|\varepsilon(u)(x)| = \{\sum_{i,j=1}^{n} |\varepsilon_{ij}(u)(x)|^2\}^{1/2}$$

We consider the vectors e_k of the canonical basis of $\mathbb{R}^n$ and the vectors $h_k = \alpha - \alpha_k e_k$, $k = 1,\dots,n-1$, and write

- for $i \neq k$, $$u_i(x) = \int_{-\infty}^{0} \frac{d}{ds} u_i(x+s\, h_k)ds = \int_{-\infty}^{0} \sum_{\substack{j=1 \\ j\neq k}}^{n} \alpha_j\, u_{i,j}(x+s\, h_k)ds ;$$

- for $i = k$, $$u_k(x) = \int_{-\infty}^{0} u_{k,k}(x+s\, e_k)ds .$$

We deduce from this

$$v_\alpha(x) = \sum_{i=1}^{n} \alpha_i\, u_i(x)$$

$$= \int_{-\infty}^{0} \sum_{\substack{i,j=1\\ i,j\neq k}}^{n} \alpha_i\, \alpha_j\, u_{i,j}(x+sh_k)ds + \int_{-\infty}^{0} \alpha_k\, u_{k,k}(x+se_k)ds$$

$$= \int_{-\infty}^{0} \sum_{\substack{i,j=1\\ i,j\neq k}} \alpha_i\, \alpha_j\, \varepsilon_{ij}(u)(x+sh_k)ds + \int_{-\infty}^{0} \alpha_k\, \varepsilon_{k,k}(u)(x+se_k)ds\ .$$

We thus have, for each k, an inequality similar to (1.33):

$$(1.35)\quad |v_\alpha(x)| \leqslant n\{\int_{-\infty}^{+\infty} |\varepsilon(u)(x+sh_k)|ds + \int_{-\infty}^{+\infty} |\varepsilon(u)(x+se_k)|\, ds\}\ .$$

Using (1.34) and the inequalities (1.35) for $k = 1,\dots,n-1$, we can write:

$$|v_\alpha(x)|^n$$
$$\leqslant n^n\{\int_{-\infty}^{+\infty} |\varepsilon(u)(x+s\alpha)|ds\}.\prod_{k=1}^{n-1} \int_{-\infty}^{+\infty} \{|\varepsilon(u)(x+sh_k)|+|\varepsilon(u)(x+se_k)|\}ds\ .$$

$$(1.36)\quad |v_\alpha(x)|^{n/n-1} \leqslant n^{n/(n-1)}(\sum_\sigma I_\sigma(x)^{(n-1)})^{1/(n-1)} \leqslant c_n \sum_\sigma I_\sigma(x)\ ,$$

where c_n depends only on n, σ runs through the subsets of $\{1,\dots,n\}$, and where, for each σ,

$$(1.37)\ \{I_\sigma(x)\}^{n-1}$$

$$= \{\int_{-\infty}^{+\infty} |\varepsilon(u)(x+s\alpha)|ds\}.\prod_{\substack{k=1\\ k\in\sigma}}^{n-1} \int_{-\infty}^{+\infty} |\varepsilon(u)(x+sh_k)|ds\}\ .$$

$$.\prod_{\substack{k=1\\ k\notin\sigma}}^{n-1} \{\int_{-\infty}^{+\infty} |\varepsilon(u)(x+se_k)|ds\}\ .$$

ii) To estimate the expression I_σ, we introduce, for each σ, a basis of $\mathbb{R}^n$ formed by the vectors $\delta_1,\dots,\delta_n$, where $\delta_i = h_i$ for $i \in \sigma$, $\delta_i = e_i$ for $i = 1,\dots,n-1$, $i \notin \sigma$ and $\delta_n = \alpha$. We write $\xi_1,\dots,\xi_n$ for the components of x with respect to this basis so that

$$x = \sum_{i=1}^{n} x_i\, e_i = \sum_{i=1}^{n} \xi_i\, \delta_i .$$

Then writing $\tilde{g}(\xi) = g(x)$, we see that:

$$I_\sigma(x) = \tilde{I}_\sigma(\xi) = \{ \prod_{i=1}^{n} \int_{-\infty}^{+\infty} |\varepsilon(u)(x+s\delta_i)|ds\}^{1/(n-1)}$$

$$\tilde{I}_\sigma(\xi) = \{ \prod_{i=1}^{n} \int_{\infty}^{\infty} |\widetilde{\varepsilon(u)}(\xi+s\delta_i)|ds\}^{1/(n-1)} . \tag{1.38}$$

The right-hand side of (1.38) is the product of the functions $\theta_1,\dots,\theta_n$, where θ_i is independent of ξ_i

$$\theta_i(\xi) = \int_{-\infty}^{+\infty} |\widetilde{\varepsilon(u)}(\xi+s\delta_i)|ds\} . \tag{1.39}$$

By Lemma 1.1 below:

$$\int_{\mathbb{R}^n} I_\sigma(x)dx = c(\sigma) \int_{\mathbb{R}^n} \tilde{I}_\sigma(\xi)d\xi = c(\sigma) \int_{\mathbb{R}^n} (\theta_1 \dots \theta_n)^{1/(n-1)} d\xi$$

$$\leq c(\sigma)\{ \prod_{i=1}^{n} \int_{\mathbb{R}^n} |\widetilde{\varepsilon(u)}(\xi)|d\xi\}^{1/(n-1)}$$

$$\leq c(\sigma)^{1/n}\{\int_{\mathbb{R}^n} |\varepsilon(u)(x)|dx\}^{n/(n-1)}$$

where $c(\sigma)$ is the Jacobian $D(x)/D(\xi)$, whose value is constant.

With these upper bounds (1.36) gives

$$\int_{\mathbb{R}^n} |v_\alpha(x)|^{n/(n-1)} dx \leq c_n \,(\sum_\sigma c(\sigma)^{1/n})\{\int_{\mathbb{R}^n} |\varepsilon(u)(x)|dx\}^{n/(n-1)} \tag{1.40}$$

iii) We conclude by making a number of successive choices of the vector α. For example by taking $\alpha = (1,1,\dots,1)$ and then $\alpha = (1,-1,\dots,-1)$, (1.40) gives

$$\|u_1+\dots+u_n\|_{L^{n^*}(\Omega)} \leq c\| \, |\varepsilon(u)| \, \|_{L^1(\Omega)}$$

$$\|u_1-u_2-\dots-u_n\|_{L^{n^*}(\Omega)} \leq c'\| \, |\varepsilon(u)| \, \|_{L^1(\Omega)} .$$

Hence by addition

$$\|u_1\|_{L^{n^*}(\Omega)} \leq c''\| \, |\varepsilon(u)| \, \|_{L^1(\Omega)} . \tag{1.41}$$

Proceeding in a similar way for the other components, we clearly obtain (1.32).

To conclude the proof of Theorem 1.2 it remains only to establish Lemma 1.1.

With this lemma in view we shall write $d\xi$ for $d\xi_1\dots d\xi_n$, $d\hat{\xi}_i$ for $d\xi_i = d\xi_1\dots d\xi_{i-1}\cdot d\xi_{i+1}\dots d\xi_n$, $d\xi'$ for $d\xi_1\dots d\xi_{n-1}$, and $d\hat{\xi}_i'$ for $d\xi_1\dots d\xi_{i-1}\cdot d\xi_{i+1}\dots d\xi_{n-1}$, and denote by π_i the projection in $\mathbb{R}^n$ on the hyperplane $\xi_i = 0$.

LEMMA 1.1.

Let $\theta_1,\dots,\theta_n$ be n measurable functions ≥ 0 on $\mathbb{R}^n$, such that for $i = 1,\dots,n$:

$$\theta_i \text{ does not depend on } \xi_i \tag{1.42}$$

$$\theta_i \in L^1(\Pi_i \, \mathbb{R}^n) \tag{1.43}$$

Then $\theta_1\dots\theta_n \in L^1(\mathbb{R}^n)$ and

$$\int_{\mathbb{R}^n} \left(\prod_{i=1}^n \theta_i\right)^{1/(n-1)} d\xi \leq \left\{\prod_{i=1}^n \int_{\Pi_i \mathbb{R}^n} \theta_i \, d\hat{\xi}_i\right\}^{1/(n-1)} . \tag{1.44}$$

<u>Proof</u>

Formula (1.44) is obviously true when $n = 2$. It can be proved for general n, by induction on n. We suppose that (1.44) has been proved up to $n-1$ and write Holder's inequality, for almost all ξ_n:

$$\int_{\Pi_n \mathbb{R}^n} \left(\prod_{i=1}^{n-1} \theta_i\right)^{\frac{1}{n-1}} (\theta_n)^{\frac{1}{n-1}} d\hat{\xi}_n$$

$$\leqslant \left(\int_{\Pi_n \mathbb{R}^n} \left(\prod_{i=1}^{n-1} \theta_i\right)^{\frac{1}{n-2}} d\hat{\xi}_n\right)^{\frac{n-2}{n-1}} \left(\int_{\Pi_n \mathbb{R}^n} \theta_n \, d\xi_n\right)^{\frac{1}{n-1}}$$

$$\leqslant \prod_{i=1}^{n-1} \left(\int_{\Pi_n \Pi_i \mathbb{R}^n} \theta_i \, d\hat{\xi}_i^{\prime}\right)^{\frac{1}{n-1}} \}.\left(\int_{\Pi_n \mathbb{R}^n} \theta_n \, d\xi_n\right)^{\frac{1}{n-1}}$$

by the inductive hypothesis.

To obtain (1.44) there remains only to integrate this last inequality with respect to ξ_n.

1.5. Proof of Theorem 1.3

After a few preliminaries in i), we shall prove in ii) that $u \in L^1_{loc}(\Omega)^n$, and establish the final result in iii).

i) The fundamental solution of the linear elasticity equations is a tempered distribution $E = E(\alpha) \in \mathcal{D}'(\mathbb{R}^n)^n$ such that, for all $\alpha \in \mathbb{R}^n$

$$\Delta E + \text{grad div } E = \alpha\delta \,, \quad (1.45)$$

[1]

where δ is the Dirac distribution at 0. It is well known (and easily verified) that

---

[1] In this proof E is not the space of symmetric tensors of order 2 on $\mathbb{R}^n$.

(1.46) $$E = \frac{3\alpha}{4} M - \frac{x}{4} (\alpha.\text{grad } M)$$

M denoting the fundamental solution of Laplace's equation ($\Delta M = \delta$). Thus

(1.47) $$E = \begin{cases} \frac{3\alpha}{8\pi} \log \frac{1}{r} + \frac{x}{8\pi} (\frac{\alpha.r}{r^2}) & \text{if } n = 2 , \\ c_n(\frac{3\alpha}{4r^{n-2}} + \frac{(n-2)}{4} \frac{x(\alpha.x)}{r^n} & \text{if } n \geq 3 ; \end{cases}$$

where c_n depends only on n and $r = |x|$.

We note for later on that

(1.48) The distribution derivatives of order 1 of E are functions which are locally Lebesgue-integrable ($\in L^1_{loc}(\mathbb{R}^n)$) .

In fact the derivatives are regular and can be taken as derivatives in the classical sense except at the origin. Explicit calculation shows that the first derivatives are majorised by a function $c|x|^{1-n}$ which is integrable over any bounded subset of $\mathbb{R}^n$. This implies (1.48).

ii) Our object is to show that $u \in L^1_{loc}(\mathbb{R}^n)$.

Let ω be a relatively compact subset of Ω , $\bar{\omega} \subset \Omega$, and let Ω_1 and Ω_2 be two other open subsets of Ω with $\bar{\omega} \subset \Omega_1 \subset \bar{\Omega}_1 \subset \Omega_2 \subset \bar{\Omega}_2 \subset \bar{\Omega}$. We suppose $\eta > 0$ small enough for the η-neighbourhood of ω (i.e. the union of the balls $B(x, \eta)$ with centre $x \in \omega$ and radius η) to be contained in Ω_1 (and similarly so that the η-neighbourhoods of Ω_1 and Ω_2 are contained in Ω_2 and Ω respectively).

Again let γ be a function of class $\mathscr{C}^\infty$ on $\mathbb{R}^n$, whose support is in the ball $B(0,\eta)$ and which has the value 1 in a neighbourhood of 0 . If we apply the elasticity operator to γE and if we write $E_1,\dots,E_n$ for the components of $E(\alpha)$ ($E_i = E_i(\alpha)$), we see that the jth component of $(\Delta + \text{grad div})(\gamma E)$ is

(1.49) $$\Delta(\gamma E_j) + D_j \,\text{div}(\gamma E) = \gamma(\Delta E_j + D_j \,\text{div } E) + 2 \,\text{grad } \gamma.\text{grad } E_j + \\ + \text{grad } \gamma.D_j E + (D_j\gamma) \,\text{div } E + (\Delta\gamma)E_j + (\text{grad } D_j\gamma).E .$$

Since γ is equal to 1 in a neighbourhood of 0, the sum of the first two terms on the right reduces to $\alpha_j\delta$. The other terms involving derivatives of γ vanish near 0 and outside the ball $B(0, \eta)$. Their sum is a function of class $\mathscr{C}^\infty$, say ζ_j, with compact support in $B(0, \eta)$.

Now let u be a distribution which satisfies the hypotheses of Theorem 1.3, and let $\phi \in \mathscr{C}_0^\infty(\Omega_2)$, $\phi = 1$ on Ω_1. We shall show that ϕu (= u on Ω_1) is a function of class L^1 on ω. The product ϕu is a well-defined distribution on $\mathbb{R}^n$; we take its convolution product with $(\alpha_j\delta + \zeta_j)$:

$$I_j = (\alpha_j\delta + \zeta_j) * (\phi u_j) = \phi(\alpha_j u_j) + \zeta_j*(\phi u_j) .$$

As the last function belongs to $\mathscr{C}^\infty$, we can conclude that $\alpha.u = \sum_j \alpha_j u_j$ is L^1 on ω, if we show that the same applies to $I = \sum I_j$. To show this we write

$$I = \sum_{j=1}^n \{\Delta(\gamma E_j) + D_j \operatorname{div}(\gamma E)\}*(\phi u_j)$$

$$\sum_{i,j=1}^n \{D_i(\gamma E_j) + D_j(\gamma E_i)\}*\{D_i(\phi u_j)\}$$

$$= \sum_{i,j=1}^n \{\varepsilon_{ij}(\gamma E)\}*\{D_i(\phi u_j) + D_j(\phi u_i)\}$$

$$= \sum_{i,j,=1}^n \{\varepsilon_{ij}(\gamma E)\}*\{\varepsilon_{ij}(\phi u)\}$$

$$= \sum_{i,j=1}^n \{\varepsilon_{ij}(\gamma E)\}*(\phi\varepsilon_{ij}(u)) + \frac{1}{2}\sum_{i,j=1}^n \{\varepsilon_{ij}(\gamma E)\}*\{u_jD_i\phi + u_iD_j\phi\} .$$

Now the derivatives of γE_j are zero outside $B(0,\eta)$ and those of ϕ are zero outside $\Omega_2\backslash\bar{\Omega}_1$, so that the convolution product

$$\varepsilon_{ij}(\gamma u)*\{u_j D_i\phi + u_i D_j\phi\}$$

whose support is contained in $\Omega_2\backslash\Omega_1 + B(0,\eta)$ vanishes on ω.

Thus, on ω , I reduces to

$$(1.50) \qquad \sum_{i,j=1}^{n} \{\varepsilon_{ij}(\gamma E)\} * (\phi\varepsilon_{ij}(u)) \ .$$

Since E and its first derivatives are locally integrable $\varepsilon_{ij}(\gamma E)$ belongs to $L^1(\mathbb{R}^n)$ for all i,j . By hypothesis $\phi\varepsilon_{ij}(u)$ is an integrable function so that each of the convolution products in (1.50) belong to $L^1(\mathbb{R}^n)$ and I is an L^1 function on ω .

It has been proved that $\alpha.u$ is L^1 on ω for all $\alpha \in \mathbb{R}^n$ and every ω which is relatively compact in Ω . It follows that $u \in L^1_{loc}(\Omega)^n$.

iii) In this last step of the proof we show that $u \in L^1(\Omega)^n$. For every $y \in \Gamma$ we have to find an open neighbourhood $\mathscr{A}$ of y throughout which u is L^1 .

Since at least one of the components of $\nu(y)$ is non-zero, we may suppose that $\nu_n(y) > 0$. As in the proof of Theorem 1.1 we then consider a neighbourhood Σ of y which satisfies (1.22)-(1.23). We also consider the sets (1.24)-(1.25) and the formula (1.26).

Our first objective is to prove that the restriction of u_n to η_0 is L^1 . For $0<\alpha<\alpha'<\alpha_0<\alpha_1$, we write

$$(1.51) \quad \int_{\mathscr{A}_{\alpha_0}\setminus\mathscr{A}_\alpha} |u_n(x)|dx = \int_\alpha^{\alpha_0}\left(\int_{\Sigma_{\alpha'}} |u_n(x)| \frac{d\Gamma}{\nu_n(x)}\right) d\alpha'$$

$$= \int_\alpha^{\alpha_0}\left(\int_\Sigma |u_n(x',a(x')-\alpha') \frac{d\Gamma}{\nu_n(x',a(x'))}\right) d\alpha'$$

$\leq$ (using (1.23) and (1.26))

$$\leq \frac{(\alpha_0-\alpha)}{\rho_1}\int_{\Sigma_{\alpha_0}} |u_n|d\Gamma + (\alpha_0-\alpha)\int_{\mathscr{A}_{\alpha_0}\setminus\mathscr{A}_\alpha} \left|\frac{\partial u_n}{\partial x_n}\right| d$$

When $\alpha \to 0$, there remains

$$(1.52) \qquad \int_{\mathscr{A}_{\alpha_0}} |u_n|\,dx \leq \frac{\alpha_0}{\rho_1}\int_{\Sigma_{\alpha_0}} |u_n|\,d\Gamma + \alpha_0 \int_{\mathscr{A}_{\alpha_0}} \left|\frac{\partial u_n}{\partial x_n}\right| dx \;.$$

Since $u_n \in L^1_{loc}(\Omega)$, the integral $\int_{\Sigma_\alpha} |u_n| d\Gamma$, by Fubini's Theorem, is finite for almost all α , $(0 < \alpha < \alpha_1)$; in particular we can choose α_0 so that $\int_{\Sigma_{\alpha_0}} |u_n| d\Gamma < \infty$ and (1.52) then implies that u_n is L^1 on $\mathscr{A}_{\alpha_0}$.

It could be shown in the same way that u_i is L^1 in a neighbourhood of y , for each $i = 1,\dots,n-1$, such that $\nu_i(y) \neq 0$. If $\nu_i(y) = 0$ (and ν_n satisfies (1.29)) then, by continuity, $\nu_n(x) + \nu_i(x) > 0$ in a neighbourhood of y in Γ . We choose a new orthonormal basis for which the nth direction coincides with $e_n + e_i$. The last component of u in this basis is $v_n = (u_n + u_i)/\sqrt{2}$. We prove, exactly as above, that v_n is L^1 in a neighbourhood of y , and as u_n is already L^1 around y , we obtain the desired result for u_i .

The proof of Theorem 1.3 is thus complete.

□

1.6. Proof of Theorem 1.4

We must show that a set B of functions which is bounded in $LD(\Omega)$ is relatively compact in $L^p(\Omega)^n$, $1 \leq p < n^*$. After Remark 1.3 it suffices to prove this when $\Omega = \mathbb{R}^n$, the functions under consideration having their support in a fixed compact subset $\mathcal{O}$ of $\mathbb{R}^n$ ([1]).

We know (cf. for example N. Dunford and J.T. Schwartz [1]) that we have to show that:

- For all $\eta > 0$, there is a compact $K \in \mathbb{R}^n$ such that

$$(1.53) \qquad \int_{\complement K} |u|^p \, dx \leq \eta \;, \quad \text{for all } u \in B \;.$$

([1]) We shall give, in Section 2, another proof of Theorem 1.4 which does not make use of Remark 1.3 or the results of E. Gagliardo [1].

- For all $\eta > 0$, there exists an $\alpha > 0$ such that for $|h| = (h_1^2 + \ldots + h_n^2)^{1/2} \leqslant \alpha$, we have

$$\|\tau_h u - u\|_{L^p(\mathbb{R}^n)^n} \leqslant \eta \text{ , for all } u \in B \text{ ,} \tag{1.54}$$

where τ_h is the translation operator

$$\tau_h f(x) = f(x_1 - h_1, \ldots, x_n - h_n) \text{ .}$$

The property (1.54) forms the subject of Lemma 1.2 below. The property (1.53) is a consequence of the Inclusion Theorem 1.2:

$$\int_{\complement K} |u|^p \, dx = \int_{\mathcal{O}\setminus K} |u|^p \, dx \leqslant \left\{\text{meas.}\, \mathcal{O}\setminus K\right\}^{\frac{n^*-p}{n^*}} \left(\int_{\mathbb{R}^n} |u|^{n^*} \, dx\right)^{\frac{n^*}{p}}$$

$$\leqslant c \left\{\text{meas}\, \mathcal{O}\setminus K\right\}^{\frac{n^*-p}{n^*}} \text{ , } \forall\, u \in B \text{ ,}$$

and, to satisfy (1.53), it is therefore sufficient to choose $K \subset \mathcal{O}$ such that $\mathcal{O}\setminus K$ has a small enough measure.

LEMMA 1.2

Let B be a set of functions which is bounded in $LD(\mathbb{R}^n)$, the functions in B having their support contained in a fixed compact subset of $\mathbb{R}^n$.

Then for all p with $1 \leqslant p < n^*$, all s with $0 \leqslant s < 1$, and all $u \in B$:

$$\|\tau_h u - u\|_{L^p(\mathbb{R}^n)^n} \leqslant c'|h|^s \, \|u\|_{LD(\mathbb{R}^n)} \tag{1.55}$$

where $c' < +\infty$ depends only on p, n, s, and the set $\mathcal{O} \subset \mathbb{R}^n$ containing the support of the functions $u \in B$.

The proof of this lemma due to L. Paris [1] is based on the lemmata 1.3 and 1.4 which follow:

LEMMA 1.3.

Let f *be a summable (L-integrable) function on* $\mathbb{R}^n$, *with compact support, and let* $0 \leq \lambda < n$. *Then the functions* $|x|^{-\lambda}$ *and* f *are convolvable and* $g = |x|^{-\lambda} * f$ *is a function which is* L^q *on every compact subset of* $\mathbb{R}^n$ *and for every* $q < n/\lambda$. *Moreover for every compact* $K \subset \mathbb{R}^n$,

$$(1.56) \qquad \|g\|_{L^q(K)} \leq c_1' \|f\|_{L^1(\mathbb{R}^n)}$$

where c_1' *depends only on* q, n, λ, K *and the support of* f.

This is a particular case of a lemma proved by S.L. Sobolev and we refer the reader for the proof to S.L. Sobolev [1]; cf. also L. Schwartz [1] Ch. VI, § 6.

LEMMA 1.4.

Let f *be a real function on* $\mathbb{R}^n$, *which is positively homogeneous of degree* $1-n$ *and of class* $\mathscr{C}^1$ *outside the origin. Then for all* s *such that* $0 \leq s \leq 1$, *there exists a constant* c' *depending only on* s, n *and* f *such that*

$$|f(x+h)-f(x)| \leq c_2' |h|^s \left\{\frac{1}{|x+h|^{n-1+s}} + \frac{1}{|x|^{n-1+s}}\right\}$$

for all $x, h \in \mathbb{R}^n$.

<u>Proof</u>

Because of homogeneity we can suppose that $|h| = 1$. Then for $|x| \leq 3/2$, we have $|x+h| \leq 5/2$ and we have

$$|f(x+h)-f(x)| \leq |f(x+h)| + |f(x)| \leq$$

$$\leq \{|x|^{1-n} + |x+h|^{1-n}\}.\{\sup_{|z|=1} |f(z)|\}$$

$$\leq c_3' \left(\frac{5}{2}\right)^s \left\{\frac{1}{|x|^{n-1+s}} + \frac{1}{|x+h|^{n-1+s}}\right\}.$$

If $|x| \geq 3/2$, then $|x+h| \geq 1/2$ and

$$|f(x+h)-f(x)| \leq \operatorname{Sup}_y |\nabla f(y)| ,$$

where y runs through the set of $x + th$, with $0 \leq t \leq 1$. As ∇f is positively homogeneous of degree $-n$, this last supremum is less than or equal to $c_4' \operatorname{Sup}_y |y|^{-n}$, where c_4' is the supremum of $|\nabla f(z)|$ for $|z| = 1$ and y runs through the same interval. Hence

$$|f(x+h)-f(x)| \leq c_4'\left\{\frac{1}{|x|^n} + \frac{1}{|x+h|^n}\right\} \leq c_4'(2^{s-1})\left\{\frac{1}{|x|^{n-1+s}} + \frac{1}{|x+h|^{n-1+s}}\right\} .$$

from which the result follows.

□

<u>Proof of Lemma 1.2</u>

As in the proof of Theorem 1.3 we use the fundamental solution E of the linear elasticity problem (cf. (1.45)(1.46)). For $\alpha \in \mathbb{R}^n$, we write, with (1.45)

$$\begin{aligned} \alpha.u &= \alpha(\delta * u) \\ &= (\Delta E + \text{grad div } E) * u \\ &= (E_{i,jj} + E_{i,ij}) * u_i \\ &= 2\, \varepsilon_{ij}(E) * \varepsilon_{ij}(u) \\ &= 2\, E_{i,j} * \varepsilon_{ij}(u) . \end{aligned}$$

Similarly if $\delta_{(h)}$ denotes the Dirac distribution at h ,

$$\begin{aligned} \tau_h(\alpha.u) &= \delta_{(h)} * (\alpha.u) \\ &= 2(\delta_{(h)} * E_{i,j}) * \varepsilon_{ij}(u) \\ &= 2(\tau_h E_{i,j}) * \varepsilon_{ij}(u) ; \end{aligned}$$

whence

$$(1.57) \qquad \begin{aligned} |\tau_h(\alpha.u)-(\alpha.u)| &= 2|(\tau_h(E_{i,j}) - E_{i,j}) * \varepsilon_{ij}(u)| \\ &\leq 2|\tau_h(E_{i,j}) - E_{i,j}| * |\varepsilon_{ij}(u)| . \end{aligned}$$

By (1.47) the functions $E_{i,j}$ are $\mathscr{C}^\infty$ except at 0 and positively homogeneous of degree $1-n$. It follows with the help of Lemma 1.4 that

$$|\tau_h(E_{i,j})-E_{i,j}| \leq c_4'|h|^s \{\tau_h \frac{1}{|x|^{n-1+s}} + \frac{1}{|x|^{n-1+s}}\} ;$$

(1.57) then implies

$$|\tau_h(\alpha.u)-(\alpha.u)| \leq 2c_4'|h|^s \sum_{i,j=1}^{n} \frac{1}{|x|^{n-1+s}} * \{\tau_{-h}|\varepsilon_{ij}(u)|+|\varepsilon_{ij}(u)|\} .$$

By applying Lemma 1.3 with $\lambda = n-1+s$, $q = p$ we obtain the desired result ($\alpha \in \mathbb{R}^n$ is arbitrary):

$$(1.58) \quad \|\tau_h(\alpha.u) - (\alpha.u)\|_{L^p(\mathbb{R}^n)} \leq$$

$$\leq 4c_1' c_4'|h|^s \sum_{i,j=1}^{n} \|\varepsilon_{ij}(u)\|_{L^1(\mathbb{R}^n)} ;$$

the required conditions on λ specified in Lemma 1.3 restrict us to $0 \leq s < 1$, and we obtain as stated $1 \leq p < n^*$. As we have indicated, the constant $c' = 4c_1' c_4'$ depends only on p, s, n and the support of the functions $u \in B$.

2. The space BD (Ω) (I)

We now begin the study of the space $BD(\Omega)$ which, as will appear, is essentially the space of functions whose energy of deformation in the Hencky plasticity model is finite. After some reminders and remarks on measure spaces, in Section 2.1, we give the definition of $BD(\Omega)$ and some of its simple properties in Section 2.2. In Sections 2.3 and 2.4 we state the main results for the space $BD(\Omega)$, analogous to the Theorems 1.1 to 1.4 established for $LD(\Omega)$, namely:

- a regularity result;
- the trace theorem: if $u \in BD(\Omega)$, one can define the trace of u on Γ, and this trace belongs to $L^1(\Gamma)^n$;
- a Sobolev type injection result;
- a theorem on compactness.

The proofs are largely based on the results already established for $LD(\Omega)$. The trace theorem and some of its applications are given in Section 2.3. The other theorems and their consequences are proved in Section 2.4.

2.1. The space of bounded measures

Let Ω, as before, be an open subset of $\mathbb{R}^n$. We shall denote by $M_1(\Omega)$ the space of bounded measures on Ω (cf. N. Bourbaki [3], N. Dinculeanu [1], P.R. Halmos [1]). This space coincides with the space of distributions μ on Ω such that [1]

(2.1) $$\sup_{\substack{\phi\in\mathscr{C}_0^\infty(\Omega)\\ |\phi(x)|\leq 1}} \langle\mu,\phi\rangle < \infty .$$

The number in (2.1) defines a norm on $M_1(\Omega)$, written as $\int_\Omega |\mu|$ or $\|\mu\|_{M_1(\Omega)}$. It is well known that $M_1(\Omega)$ is a Banach space for this norm; it is the dual of the space $\mathscr{C}_0(\Omega)$ of continuous functions on Ω, with compact support in Ω, where the norm is that of uniform convergence.

One can identify $L^1(\Omega)$ with a subspace of $M_1(\Omega)$ provided that one identifies (as we shall do) a measure $\mu = h\,dx$, $h \in L^1(\Omega)$, with the function h. The norm induced on $L^1(\Omega)$ by $M_1(\Omega)$ coincides with the usual norm of $L^1(\Omega)$, and $L^1(\Omega)$ is a closed subspace of $M_1(\Omega)$.

We shall say that a sequence of measures $\mu_j \in M_1(\Omega)$ converges weakly (or vaguely) to $\mu \in M_1(\Omega)$ if, for all ϕ of $\mathscr{C}_0(\Omega)$ (or for all ϕ of $\mathscr{C}_0^\infty(\Omega)$):

(2.2) $$\lim_{j\to\infty} \langle\mu_j,\phi\rangle = \langle\mu,\phi\rangle .$$

Since $M_1(\Omega)$ is the dual of a normed space, it is weakly sequentially compact, which means that for any bounded sequence μ_j of $M_1(\Omega)$, there exists a $\mu \in M_1(\Omega)$ and a subsequence $\mu_{j'}$ converging weakly to μ.

The norm $\|\mu\|_{M_1(\Omega)}$ is lower semicontinuous for the weak topology. More generally for any open $\mathscr{O} \subset \Omega$ and any sequence μ_j converging weakly to μ, we have

---

(1) We use the notation (without making any particular distinction) $\langle\mu,\phi\rangle$ or $\int_\Omega \phi\mu$ for the integral of the function ϕ with respect to the measure μ, an integral which is also written by some authors as $\int_\Omega \phi d\mu$.

(2.3) $$\int_{\mathcal{O}} |\mu| \leq \liminf_{j\to\infty} \int_{\mathcal{O}} |\mu_j| \, .$$

When $\Omega = \mathbb{R}^n$ a measure $\mu \in M_1(\mathbb{R}^n)$ can be convolved with a function ρ belonging to $\mathscr{C}_0^\infty(\mathbb{R}^n)$, the convolution product $\rho * \mu$ being a function in $\mathscr{C}^\infty(\mathbb{R}^n)$. We have

(2.4) $$\int_{\mathbb{R}^n} |\rho*\mu| \, dx \leq \int_{\mathbb{R}^n} |\rho| dx \, . \int_{\mathbb{R}^n} |\mu|$$

and if $\rho \geq 0$ and $\mu \geq 0$

(2.5) $$\int_{\mathbb{R}^n} \rho*\mu \, dx = \left(\int_{\mathbb{R}^n} \rho \, dx\right) . \int_{\mathbb{R}^n} \mu \, .$$

It is possible to regularise a measure μ by convolving it with functions ρ_η defined as in I.(1.8). By (2.5), since $\rho_\eta \geq 0$ and $\int \rho_\eta \, dx = 1$,

(2.6) $$\int_{\mathbb{R}^n} |\rho_\eta*\mu| dx \leq \int_{\mathbb{R}^n} |\mu|$$

and when $\eta \to 0$:

(2.7) $$\begin{cases} \rho_\eta * \mu \to \mu \quad \text{weakly in } M_1(\mathbb{R}^n) \, , \\ \int_{\mathbb{R}^n} |\rho_\eta*\mu| dx \to \int_{\mathbb{R}^n} |\mu| \, . \end{cases}$$

If $\Omega \neq \mathbb{R}^n$, then for all $\mu \in M^1(\Omega)$ and all $\phi \in \mathscr{C}_0^\infty(\Omega)$, $\phi\mu$ is defined as a bounded measure on $\mathbb{R}^n$ (with compact support on Ω) which one can convolve with functions of $\mathscr{C}_0^\infty(\mathbb{R}^n)$ or regularise. With the regularising functions ρ_η defined as above, if $\eta > 0$ is small enough (in fact less than the distance of the support of ϕ from Γ) then $\rho_\eta*(\phi\mu)$ has its support in Ω, and passing to the restrictions to Ω, we have:

(2.8) $$\rho_\eta * (\phi\mu) \to \phi\mu \, , \text{ weakly in } M_1(\Omega)$$

when $\eta \to 0$. If $0 \leq \phi \leq 1$ and $\phi = 1$ in the neighbourhood of ω, an open relatively compact subset of Ω , then for small enough positive η , $\rho_\eta * (\phi\mu)$ has its support in Ω , and by passage to the corresponding restrictions to ω ,

$$\rho_\eta * (\phi\mu)_{|\omega} \to \mu_{|\omega} \text{ weakly in } M_1(\omega) , \tag{2.9}$$

or in other words

$$\langle \rho_\eta * (\phi\mu), \theta \rangle \to \langle \mu, \theta \rangle , \tag{2.10}$$

when $\eta \to 0$, for all $\theta \in \mathscr{C}_0(\omega)$.

Another interesting topology on $M_1(\Omega)$ is the one associated with the family of seminorms

$$|(\mu,\phi)| , \quad \phi \in \mathscr{C}_0(\Omega)$$

and the pseudometric

$$\left| \int_\Omega |\mu| - \int_\Omega |\nu| \right| .$$

This defines a topology on $M_1(\Omega)$, intermediate between the weak topology and the norm topology. A sequence μ_j converges to μ for this topology if

$$\begin{cases} \mu_j \to \mu \text{ weakly in } M_1(\Omega) \\ \text{and } \int_\Omega |\mu_j| \to \int_\Omega |\mu| . \end{cases} \tag{2.11}$$

When the μ_j are ≥ 0 , (2.11) is simply convergence in measure in the strict sense (cf. N. Bourbaki [1]).

The following simple lemma will sometimes be useful:

LEMMA 2.1

If μ_j *converges to* μ *as* $j \to \infty$, *in the sense of (2.11), then for any Borel set* $\mathcal{O} \subset \Omega$, *such that*

$$(2.12) \qquad \int_{\Omega\cap\partial\mathcal{O}} |\mu| = 0$$

we have

$$(2.13) \qquad \int_{\mathcal{O}} |\mu_j| \to \int_{\mathcal{O}} |\mu| \,, \text{ when } j \to \infty \,.$$

Similarly, for all ϕ *belonging to* $\mathscr{C}(\bar{\Omega})$,

$$(2.14) \qquad \int_{\Omega} \phi\mu_j \to \int_{\Omega} \phi\mu$$

and

$$(2.15) \qquad \int_{\Omega} \phi|\mu_j| \to \int_{\Omega} \phi|\mu| \,,$$

when $j \to \infty$.

Proof

We write $\mathring{\mathcal{O}}$ for the interior of $\mathcal{O}$ and $\bar{\mathcal{O}}$ its closure in Ω . By (2.3) and (2.12),

$$(2.16) \qquad \int_{\mathcal{O}} |\mu| = \int_{\mathring{\mathcal{O}}} |\mu| \leq \liminf_{j\to\infty} \int_{\mathcal{O}} |\mu_j|$$

$$(2.17) \qquad \int_{\Omega\setminus\mathcal{O}} |\mu| = \int_{\Omega\setminus\bar{\mathcal{O}}} |\mu| \leq \liminf_{j\to\infty} \int_{\Omega\setminus\bar{\mathcal{O}}} |\mu_j|$$

$$\leq \int_{\Omega} |\mu| - \limsup_{j\to\infty} \int_{\mathcal{O}} |\mu_j| \;;$$

and (2.13) follows.

We shall prove (2.14), the proof of (2.15) being exactly similar. For any $\delta>0$, there is a compact $C \subset \Omega$ such that

$$(2.18) \qquad \int_{\Omega\setminus C} |\mu| \leq \delta \,, \text{ and } \int_{\partial C} |\mu| = 0 \,.$$

Let $\psi \in \mathscr{C}_0^\infty(\Omega)$, with $0\leq\psi\leq 1$ and $\psi = 1$ on C . We write

$$\left|\int_\Omega \phi\mu_j - \int_\Omega \phi\mu\right| \leq \left|\int_\Omega \phi\psi\, \mu_j - \int_\Omega \phi\psi\, \mu\right| + \int_\Omega |\phi|(1-\psi)|\mu_j| + \int_\Omega |\phi|(1-\psi)|\mu|$$

$$\leq \left|\int_\Omega \phi\psi\, \mu_j - \int_\Omega \phi\psi\, \mu\right| + \sup_{x\in\Omega}|\phi(x)|.\left\{\int_{\Omega\setminus C} |\mu_j| + \int_{\Omega\setminus C} |\mu|\right\}.$$

By (2.2) (2.3) (2.13) and (2.18) (and since $\partial C = \Omega \cap \partial(\Omega\setminus C)$) ,

$$\text{(2.19)} \qquad \limsup_{j\to\infty} \left|\int_\Omega \phi\mu_j - \int_\Omega \phi\mu\right| \leq 2\delta \sup_{x\in\Omega} |\phi(x)| \;.$$

As $\delta>0$ can be chosen to be arbitrarily small, this proves (2.14).

Remark 2.1 We will sometimes also consider spaces of bounded measure with values in an Euclidean space X (essentially $X = \mathbb{R}^n$, $X = E$ or $X = E^D$) . The space is then denoted by $M_1(\Omega;X)$; it is a Banach space for the norm

$$\int_\Omega |\mu| = \|\mu\|_{M_1(\Omega;X)} = \sup_{\substack{\phi\in\mathscr{C}_0^\infty(\Omega;X)\\ |\phi(x)|_X \leq 1}} \langle\mu,\phi\rangle \;.$$

Everything said about $M_1(\Omega)$ can easily be extended to this case.

□

The space of functions integrable on Ω whose first derivatives (in the sense of the Theory of Distributions) are bounded measures on Ω is the space $BV(\Omega)$, also called the space of functions of bounded variation in Ω . This space has been studied by numerous authors ([1]). It has served as a 'model' for the study of the space $BD(\Omega)$.

The space is defined as follows:

$$\text{(2.21)} \qquad BV(\Omega) = \{v \in L^1(\Omega),\ D_i v \in M_1(\Omega),\ i = 1,\dots,n\} \;.$$

([1]) Cf. in particular R. Caccioppoli [1], L. Cesari [1], E. de Giorgi [1] [2], E. Giusti [1], W.H. Fleming [1], M. Miranda [1], A.V. Volpert [1], L.C. Young [1-2-3].

It is a Banach space for the 'natural' norm

$$\|v\|_{BV(\Omega)} = \|v\|_{L^1(\Omega)} + \sum_{i=1}^{n} \|D_i v\|_{M_1(\Omega)} . \tag{2.22}$$

The space $BV(\Omega)$ contains $W^{1,1}(\Omega)$ but is not identical to it; for example if $n = 1$, the Heaviside function is in $BV(-1,+1)$ but not in $W^{1,1}(-1,+1)$. Nevertheless if Ω is of class $\mathscr{C}^1$, one can (cf. M. Miranda [1]) define the trace on Γ of a function $v \in BV(\Omega)$, and this trace is in $L^1(\Gamma)$ (as it is for $v \in W^{1,1}(\Omega)$, cf. E. Gagliardo [1]). The space $BV(\Omega)$ is contained (with continuous injection) in $L^{n*}(\Omega)$, $n* = n/(n-1)$ for $n \geq 2$, and the injection of $BV(\Omega)$ into $L^p(\Omega)$ is compact if Ω is bounded and $1 \leq p < n*$. These results are the same as those for $W^{1,1}(\Omega)$.

2.2. Definition of $BD(\Omega)$

The space $BD(\Omega)$ is the space of the $u \in L^1(\Omega)^n$ such that $\varepsilon_{ij}(u)$ is a bounded measure for all $i,j = 1,\dots,n$:

$$BD(\Omega) = \{u \in L^1(\Omega)^n,\ \varepsilon_{ij}(u) \in M_1(\Omega),\ i,j = 1,\dots,n\} . \tag{2.23}$$

We can define on this space a norm

$$\|u\|_{BD(\Omega)} = \|u\|_{L^1(\Omega)^n} + \sum_{i,j=1}^{n} \|\varepsilon_{ij}(u)\|_{M_1(\Omega)} . \tag{2.24}$$

It is easily seen that $BD(\Omega)$ is a Banach space for this norm.

If $u \in BD(\Omega)$ then, for all $\alpha \in \mathbb{R}^n$, $(\alpha.\nabla)(\alpha.u) \in M_1(\Omega)$; this expression can in fact be written

$$(\alpha.\nabla)(\alpha.u) = \sum_{i,j=1}^{n} \alpha_i\, \alpha_j\, \frac{\partial u}{\partial x_i} = \sum_{i,j=1}^{n} \alpha_i\, \alpha_j\, \varepsilon_{ij}(u) . \tag{2.25}$$

Conversely if $u \in L^1(\Omega)^n$ and if $(\alpha.\nabla)(\alpha.u) \in M_1(\Omega)$ for all $\alpha \in \mathbb{R}^n$ the function u is in $BD(\Omega)$: to see this we have only to note that

$(\alpha.\nabla)(\alpha.u) \in M_1(\Omega)$ putting successively $\alpha = e_i + e_j$, $i,j = 1,\ldots,n$ where e_i is the canonical basis of $\mathbb{R}^n$. This shows that the definition of $BD(\Omega)$ is intrinsic, it does not depend on the orthonormal frame of reference chosen in $\mathbb{R}^n$. We could of course also have established this by carrying out a change of orthonormal axes in $\mathbb{R}^n$.

The space $LD(\Omega)$ is a *proper* subset of $BD(\Omega)$ (cf. below); $BD(\Omega)$ therefore also contains the subspaces of $LD(\Omega)$: $W^{1,1}(\Omega)^n, \mathscr{C}^1(\bar{\Omega})^n,\ldots$ It also contains $BV(\Omega)^n$ but [1]

(2.26) $$BV(\Omega)^n \subset BD(\Omega),\ BV(\Omega)^n \neq BD(\Omega) .$$

The following lemma is useful and carries the implication that $BD(\Omega) \neq LD(\Omega)$.

LEMMA 2.2

Let Σ be an $(n-1)$-dimensional variety of class $\mathscr{C}^1$ which separates Ω into two open subsets Ω_1 and Ω_2 .

Let u be a function on Ω whose restriction $u^{(k)}$ to Ω_k is in $\mathscr{C}^1(\bar{\Omega}_k)^n$ for $k = 1,2$.

Then, for $i,j = 1,\ldots,n$,

(2.27) $$\varepsilon_{ij}(u) = \{\varepsilon_{ij}(u)\} + \mathscr{T}_{ij}(u^{(2)}-u^{(1)})\delta_\Sigma ,$$

where $\{\varepsilon_{ij}(u)\}$ is the function equal to $\varepsilon_{ij}(u^{(k)})$ on Ω_k , δ_Σ is the Dirac surface distribution on Σ , and ν denotes the unit normal on Σ directed towards Ω_2 , and

(2.28) $$\mathscr{T}_{ij}(p) = \frac{1}{2}(p_i\,\nu_j + p_j\,\nu_i) .$$

Proof

For $\varphi \in \mathscr{C}_0^\infty(\Omega)$ we write

---

[1] If $BV(\Omega)^n = BD(\Omega)$ were true, then by the closed-graph theorem the norms of these two spaces would be equivalent and there would be a c such that

$$\int_\Omega |D_i\,u_j| \leq c\{\int_\Omega |u|dx + \int_\Omega |\varepsilon(u)|\} ,\ \forall\, u \in BD(\Omega) ,\ \forall\, i,j .$$

It would follow that Korn's inequality would hold in L^1, which is not true.

$$\langle \varepsilon_{ij}(u),\phi\rangle = -\frac{1}{2}\langle u_i, D_j\phi\rangle - \frac{1}{2}\langle u_j, D_i\phi\rangle = -\frac{1}{2}\int_\Omega (u_i\, D_j\phi + u_j\, D_i\phi)\, dx .$$

We separate the integrals over Ω into integrals over Ω_1 and Ω_2 and apply the classical Green's formulae to Ω_1 and Ω_2. We arrive at

$$\langle \varepsilon_{ij}(u),\phi\rangle = \int_\Omega \{\varepsilon_{ij}(u)\}\phi\, dx + \int_\Sigma \mathscr{T}_{ij}(u^{(2)}-u^{(1)})\phi\, d\Gamma ,$$

which is precisely (2.27).

□

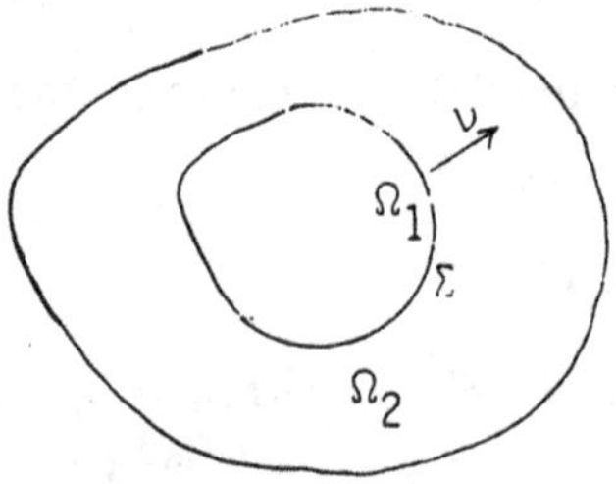

Fig. 2.1.

It follows from this lemma that

(2.29) $$\int_\Omega |\varepsilon_{ij}(u)| \leq \int_{\Omega_1} |\varepsilon_{ij}(u)|dx + \int_{\Omega_2} |\varepsilon_{ij}(u)|dx + c\int_\Sigma |u^{(2)}-u^{(1)}|\, d\Gamma ,$$

and that the function u indicated is in $BD(\Omega)$ (1). On the other hand it cannot belong to $LD(\Omega)$ unless

(2.30) $$\mathscr{T}_{ij}(u^{(2)}-u^{(1)}) = 0 \text{ on } \Sigma \text{ for all } i,j ,$$

which would imply $u^{(2)} = u^{(1)}$ on Σ (2).

---

(1) This function u is also in $BV(\Omega)^n$.

(2) To verify this one writes the components of $\mathscr{T}_{ij}(p)$ at an arbitrary point y of Σ, with reference to an orthonormal frame of reference in which one of the base vectors is $\nu = \nu(y)$.

Since $LD(\Omega)$ is a proper subset of $BD(\Omega)$ and the norm of $BD(\Omega)$ reduces to that of $LD(\Omega)$ on $LD(\Omega)$, we have

(2.31) $LD(\Omega)$ is a closed subspace of $BD(\Omega)$
$LD(\Omega) \neq BD(\Omega)$.

Approximation of functions of the class $BD(\Omega)$

Of course the regular functions are not dense in $BD(\Omega)$ for the norm topology. We shall establish in Section 3 some density results for a weaker topology. For the moment we make the following simple remark: let ω and ω' be two open subsets of Ω with $\bar{\omega} \subset \omega' \subset \bar{\omega}' \subset \Omega$ and let ϕ be a function of the class $\mathscr{C}_0^\infty(\Omega)$ with $0 \leq \phi \leq 1$, and $\phi = 1$ on ω'. If $u \in BD(\Omega)$ the function ϕu is in $BD(\mathbb{R}^n)$, and by regularising, and using (2.4)-(2.7), we see that, for $\eta \to 0$

$$\rho_\eta * (\phi u) \to \phi u \quad \text{in } L^1(\mathbb{R}^n)^n$$

$$\varepsilon_{ij}(\rho_\eta * (\phi u)) \to \varepsilon_{ij}(\phi u) \quad \text{weakly in } M_1(\Omega), \quad \forall\, i,j$$

$$\int_{\mathbb{R}^n} |\varepsilon_{ij}(\rho_\eta * (\phi u))| \to \int_{\mathbb{R}^n} |\varepsilon_{ij}(\phi u)|$$

By Lemma 2.1 and Remark 2.1 we also have

$$\int_{\mathcal{O}} |\varepsilon_{ij}(\rho_\eta * (\phi u))| \to \int_{\mathcal{O}} |\varepsilon_{ij}(\phi u)|,$$

for all i,j and every Borel set $\mathcal{O}$ in $\mathbb{R}^n$ such that

$$\int_{\partial\mathcal{O}} |\varepsilon_{ij}(\phi u)| = 0, \quad \forall\, i,j.$$

Let us call, for the moment, u_η the restriction of $\rho_\eta * (\phi u)$ to Ω. The restriction of u_η to ω is independent of ϕ as soon as η is small enough, say η less than the distance of ω to $\partial\omega'$: u_η then coincides on ω with the function $\rho_\eta * \tilde{u}$, where $\tilde{u}$ is the function equal to u on Ω and to zero outside Ω. Observing that the derivatives of ϕ vanish on ω' and hence on ω, we conclude that, when $\eta \to 0$

$$(2.32)\quad \begin{cases} u_\eta \to u \text{ in } L^1(\Omega)^n \\ \varepsilon_{ij}(u_\eta) \to \varepsilon_{ij}(u) \text{ weakly in } M_1(\Omega) \end{cases}$$

and

$$(2.33)\quad \begin{cases} \int_\omega |\varepsilon_{ij}(u_\eta)| \to \int_\omega |\varepsilon_{ij}(u)| , \\ \text{for all } \omega \subset \bar{\omega} \subset \Omega \text{ such that } \int_{\partial\omega} |\varepsilon_{ij}(u)| = 0, \end{cases}$$

$(i,j = 1,\dots,n)$.

2.3 Trace Theorem in $BD(\Omega)$

We now state a trace theorem for $BD(\Omega)$ analogous to Theorem 1.1 and deduce some consequences from it. We shall suppose that:

$$(2.34)\qquad \Omega \text{ is an open set of class } \mathscr{C}^1$$

THEOREM 2.1

There exists a continuous surjective linear operator γ_o *from* $BD(\Omega)$ *into* $L^1(\Gamma)^n$ *such that*

$$(2.35)\qquad \gamma_o(u) = u_{|\Gamma} \text{ for all } u \text{ of } BD(\Omega) \cap \mathscr{C}(\bar{\Omega})^n .$$

Furthermore for all i,j *and all* $\phi \in \mathscr{C}^1(\bar{\Omega})$ *we have the generalised Green's formula*

$$(2.36)\quad \frac{1}{2}\int_\Omega \left(u_i \frac{\partial\phi}{\partial x_j} + u_j \frac{\partial\phi}{\partial x_i}\right) dx + \int_\Omega \phi\, \varepsilon_{ij}(u) = \int_\Gamma \phi\, \mathscr{T}_{ij}(\gamma_o(u))d\Gamma .$$

Proof

i) The proof is basically a repetition of that of Theorem 1.1. At each point y of Γ, since one of the components of $\nu(y)$ must be non-zero, we may suppose for example that $\nu_n(y) > 0$, and we consider a neighbourhood Σ of y which admits of a representation of the type (1.22) (1.23). We also introduce the sets Σ_α and $\mathscr{A}_\alpha$ defined in (1.24) (1.25).

The formulae (1.26) (1.27) are not directly applicable but the inequality (1.29) holds for almost all α, α' with $0<\alpha<\alpha'<\alpha_1$: in fact by Fubini's theorem, since $u \in L^1(\Omega)^n$, the function g_α is defined and belongs to $L^1(\Sigma)$ for all $\alpha \in (0,\alpha_1) \setminus \mathscr{E}$ where $\mathscr{E}$ is a set of measure zero (we write $dx = dx_n\, d\Gamma/\nu_n$ in $\mathscr{A}_{\eta 1}$ and use (1.23)). We approximate the function u by regular functions u_η as in (2.32) (2.33). By letting η tend to zero and using once more Fubini's Theorem and (2.32), we see that for almost all $\alpha \in (0,\alpha_1)$ $g_{\alpha\eta} \to g_\alpha$ in $L^1(\Sigma)$, where $g_{\alpha\eta}$ is defined in terms of u_α by the relation analogous to (1.28). Now since $\partial u_n/\partial x_n \in M_1(\Omega)$, for almost all α , $\int_{\Sigma_\alpha} |\frac{\partial u_n}{\partial x_n}| = 0$, and it is therefore possible to apply (2.33) to $\mathscr{A}_{\alpha'} \setminus \mathscr{A}_\alpha$ (1). Under these conditions the formula (1.29) which holds for u_η , for all η , entails in the limit the validity of the formula for u , and for $\alpha, \alpha' \in (0,\alpha_1) \setminus \mathscr{E}'$ where $\mathscr{E}' \supset \mathscr{E}$ is a set of measure zero.

ii) As in the case of Theorem 1.1 we conclude that g_α is a Cauchy sequence in $L^1(\Sigma)$ when $\alpha \to 0$, $\alpha \in (0,\alpha_1) \setminus \mathscr{E}'$. Let g be its limit. The correspondence we have just established between u and g is not necessarily linear since the set $\mathscr{E}'$ depends on u . We note however that the formula (1.30) is none the less true and can be proved in the same way.

The trace of u_i on a neighbourhood of y in Γ , for any value of $i = 1,\dots,n-1$, for which $\nu_i(y) \neq 0$, is defined similarly. If $\nu_i(y) = 0$ then $\nu_n + \nu_i \neq 0$ at y and, by continuity, in a neighbourhood of y . We choose a new orthonormal basis of $\mathbb{R}^n$, for which the nth direction is $e_n + e_i$. The last component of u with respect to this basis is $v_n = (u_n+u_i)/\sqrt{2}$. Precisely as above we

---

(1) Σ and $\mathscr{O}$ must be chosen so that $\int_{\partial\mathscr{O} \times (0,\alpha_1)} |\varepsilon(u)| = 0$, which can always be done.

define a trace of v_n on a neighbourhood Σ_i of y and this trace is in $L^1(\Sigma_i)$. By taking the difference we define the trace of u_i on $\Sigma_i \cap \Sigma$ and this trace is in $L^1(\Sigma_i \cap \Sigma)$ and satisfies a relation analogous to (1.30).

iii) By joining these together we have thus defined a trace $\gamma_0(u)$ of u on Γ, which is in $L^1(\Omega)^n$ and satisfies

$$\|\gamma_0(u)\|_{L^1(\Gamma)^n} \leqslant c_2(\Omega)\ \|u\|_{BD(\Omega)} . \tag{2.37}$$

We have not yet established that γ_0 is linear nor proved that $\gamma_0(u)$ is uniquely defined (since there is no density of regular functions in $BD(\Omega)$). These two results will follow from Green's formula.

To prove Green's formula (2.36) we can again assume that the conditions (1.22) (1.23) hold and restrict ourselves to a function ϕ with compact support in $\Sigma \cup \mathscr{A}_{\alpha_0}$.

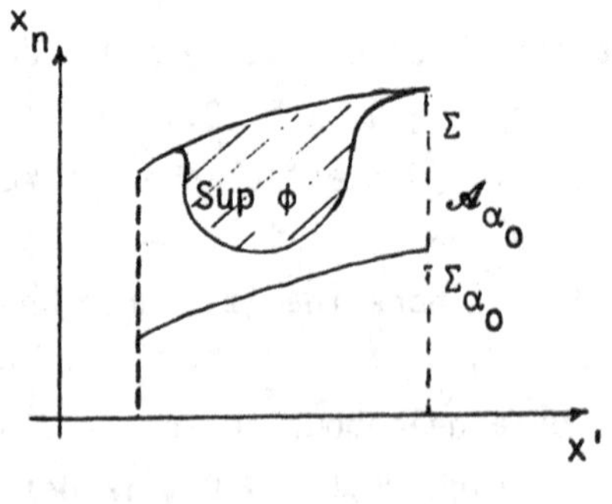

Fig. 2.2 : Support of ϕ

The Green's formula (2.36) is obvious if we replace Ω by $\mathscr{A}_{\alpha_0} \setminus \mathscr{A}_\alpha$, $0<\alpha<\alpha_0$, and u by a regularised u_η (as in (2.32) (2.33)). Letting η converge to 0 we obtain (2.36) with the function u, and with Ω replaced by $\mathscr{A}_{\alpha_0} \setminus \mathscr{A}_\alpha$. We now obtain the desired result by allowing α to tend to zero $(\alpha, \alpha_0 \in (0,\alpha_1) \setminus \mathscr{E}')$.

Having proved Green's formula (2.36) we thus conclude that $\mathscr{T}(\gamma_0(u))$

and hence $\gamma_0(u)$ are each uniquely defined. The formula (2.36) also shows that γ_0 is linear. Lastly (2.37) implies the continuity of γ_0 from $BD(\Omega)$ into $L^1(\Gamma)^n$.

This proves Theorem 2.1.

Remark 2.2. Let M be a variety of class $\mathscr{C}^1$ contained in Ω which separates Ω into two open sets Ω_1 and Ω_2. Applying Theorem 2.1 to Ω_1 and Ω_2 allows us to define for all $u \in BD(\Omega)$ two traces $\gamma_0^{(1)}(u)$ and $\gamma_0^{(2)}(u)$ which are in $L^1(M)^n$. *These two traces are not necessarily equal.* We shall call them the internal and external traces (for a given orientation of the normal to M); cf. also the Remark 2.3.

□

We deduce from Theorem 2.1,

PROPOSITION 2.1.

We suppose Ω to satisfy (2.34). If $u \in BD(\Omega)$ the function $\tilde{u}$, equal to u on Ω and to zero on $\mathbb{R}^n \setminus \Omega$ is in $BD(\mathbb{R}^n)$. Furthermore the mapping $u \mapsto \tilde{u}$ is a continuous linear mapping from $BD(\Omega)$ into $BD(\mathbb{R}^n)$.

Proof

A calculation similar to the one in Lemma 2.2 (but with the standard Green's formula replaced by (2.36)) shows that

$$\langle \varepsilon_{ij}(\tilde{u}), \phi \rangle = \int_\Omega \phi \, \varepsilon_{ij}(u) - \int_\Gamma \mathscr{T}_{ij}(\gamma_0(u))\phi \, d\Gamma \,, \tag{2.38}$$

for all ϕ in $\mathscr{C}_0^\infty(\mathbb{R}^n)$. It follows, with the help of (2.37) that

$$\int_{\mathbb{R}^n} |\varepsilon_{ij}(\tilde{u})| \leq \int_\Omega |\varepsilon_{ij}(u)| + c(\Omega)\|u\|_{BD(\Omega)} \,,$$

which shows that $\tilde{u} \in BD(\mathbb{R}^n)$ and that the linear mapping $u \mapsto \tilde{u}$ is continuous.

Remark 2.3. i) More generally if $w = u$ in Ω and $= v$ in $\mathbb{R}^n \setminus \bar{\Omega}$, $u \in BD(\Omega)$, $v \in BD(\mathbb{R}^n \setminus \bar{\Omega})$, then $w \in BD(\mathbb{R}^n)$ and by a calculation analogous to that used in Proposition 2.1 and Lemma 2.2 we have

$$(2.39) \quad \langle \varepsilon_{ij}(w), \phi \rangle = \int_\Omega \phi\, \varepsilon_{ij}(u) + \int_{\mathbb{R}^n \setminus \bar{\Omega}} \phi \varepsilon_{ij}(v) + \int_\Gamma \mathcal{T}_{ij}(\gamma_0(v-u))\phi \, d\Gamma ,$$

for all i,j and all $\phi \in \mathscr{C}_0^\infty(\mathbb{R}^n)$. This implies

$$(2.40) \quad \int_{\mathbb{R}^n} |\varepsilon_{ij}(w)| = \int_\Omega |\varepsilon_{ij}(u)| + \int_{\mathbb{R}^n \setminus \bar{\Omega}} |\varepsilon_{ij}(v)| + \int_\Gamma | \mathcal{T}_{ij}(\gamma_0(v-u))| d\Gamma$$

$$(2.41) \quad \|w\|_{BD(\mathbb{R}^n)} \leq c(\Omega) \{ \|u\|_{BD(\Omega)} + \|v\|_{BD(\mathbb{R}^n \setminus \bar{\Omega})} \} .$$

ii) If $u \in BD(\Omega)$ then, by reason of (2.39) and (2.40), the internal and external traces on the boundary of a ball $B(x,r) \subset \Omega$, are distinct if and only if $\int_{\partial B(x,r)} |\varepsilon(u)| \neq 0$.

Since $\int_\Omega |\varepsilon(u)| < \infty$, *the internal and external traces can be different only for an enumerable set of values of* r. This remark can easily be extended to apply to a more general family of hypersurfaces depending regularly on a parameter r.

□

2.4 Other properties of $BD(\Omega)$

We shall suppose for simplicity that Ω is of class $\mathscr{C}^1$. We then have

THEOREM 2.2. *(Sobolev embedding)*

The space $BD(\Omega)$ *is contained in* $L^{n^*}(\Omega)^n$, $n^* = n/(n-1)$ *and there is a continuous injection of the former into the latter.*

THEOREM 2.3. *(Regularity)*

If $u \in \mathcal{D}'(\Omega)^n$ *and if* $\varepsilon_{ij}(u) \in M_1(\Omega)$ *for all* $i,j = 1,\ldots,n$, *then* u *is in* $BD(\Omega)$.

THEOREM 2.4. *(Compactness Theorem)*

The injection of $BD(\Omega)$ *into* $L^p(\Omega)^n$ *is compact, for all* p *such that* $1 \leq p < n^* = n/(n-1)$.

Proof of Theorem 2.2

If $u \in BD(\Omega)$ then, by Proposition 2.1, the function $\tilde{u}$ equal to u on Ω and to 0 on $\mathbb{R}^n \backslash \bar{\Omega}$ is in $BD(\mathbb{R}^n)$. By regularising we obtain functions $\tilde{u}_\eta = \rho_\eta * \tilde{u}$ which are in $\mathscr{C}_0^\infty(\Omega)^n$. By utilising Theorem 1.2 and in particular the inequality (1.32), we see that

$$(2.42) \qquad \|\tilde{u}_\eta\|_{L^{n^*}(\mathbb{R}^n)^n} \leq c(n) \sum_{i,j=1}^{n} \|\varepsilon_{ij}(\tilde{u}_\eta)\|_{L^1(\mathbb{R}^n)}$$

$$\leq c(\Omega)\|u\|_{BD(\Omega)} \quad \text{(by (2.5) and (2.41))}$$

When $\eta \to 0$, $\tilde{u}_\eta$ converges to $\tilde{u}$ in $L^1(\mathbb{R}^n)^n$ (cf. (2.32)), and the inequality (2.42) shows that $\tilde{u} \in L^{n^*}(\mathbb{R}^n)^n$ and

$$(2.43) \qquad \|u\|_{L^{n^*}(\mathbb{R}^n)^n} \leq c(\Omega)\ \|u\|_{BD(\Omega)} .$$

Remark 2.4. As indicated earlier (cf. footnote ([1]) on page 130) a proof of Theorem 1.2 and 2.2 can be given which does not make use of Remark 1.3 and the Trace Theorem of E. Gagliardo [1]. In fact the proof of Theorem 2.2 is really based only on the inequality (1.32), which is independent of Remark 1.3. Incidentally Theorem 2.2 obviously implies Theorem 1.2, since $LD(\Omega) \subset BD(\Omega) \subset L^{n^*}(\Omega)^n$.

Proof of Theorem 2.3

The proof is essentially the same as that of Theorem 1.3 with the following slight modifications:

- in ii) when it has to be shown that $u \in L^1_{loc}(\Omega)^n$, we need to show that each of the terms in the sum (1.50) is L^1 on ω $(\subset \bar{\omega} \subset \Omega)$. In the proof of Theorem 1.3 we noted that $\varepsilon_{ij}(\gamma E) * (\phi\, \varepsilon_{ij}(u)) \in L^1(\mathbb{R}^n)$ since $\varepsilon_{ij}(\gamma E)$ belongs to $L^1(\mathbb{R}^n)$ as also does $\phi\, \varepsilon_{ij}(u)$. In the present case the conclusion is still true because $\varepsilon_{ij}(\gamma E) \in L^1(\mathbb{R}^n)$ and $\phi\, \varepsilon_{ij}(u) \in M_1(\mathbb{R}^n)$, and it is known (see, for example L. Schwartz [1]) that the convolution product of a function of class L^1 and a bounded measure is a function of class L^1.

- in iii) the inequality (1.52) is not obtained directly, but by approximating u by means of regular functions and passing to the limit using (2.32) (2.33) (an identical argument was used in proving Theorem 2.1).

Proof of Theorem 2.4.

Let u_m be a bounded sequence of functions in $BD(\Omega)$. By Proposition 2.1 the sequence $\tilde{u}_m$ is bounded in $BD(\mathbb{R}^n)$, and by (2.4) the family $\rho_\eta * \tilde{u}_m$ is in $LD(\mathbb{R}^n)$ and is bounded in that space, independently of m and η. By Theorem 1.4 (remembering that the functions $\rho_\eta * u_m$ have their support in a fixed compact set if, for example $0<\eta\leq 1$) the family $\rho_\eta * u_m$ is relatively compact in $L^p(\mathbb{R}^n)^n$, $\forall$ p, $1\leq p<n^*$, and the same applies to the sequence u_m, since for all m

$$\rho_\eta * \tilde{u}_m \to u_m \text{ in } L^{n^*}(\mathbb{R}^n)^n ,$$

when $\eta \to 0$ (cf. Theorem 2.2).

This proves Theorem 2.4 (1).

□

(1) In proving Theorem 2.4 we made use of Theorem 1.4 only for the particular case of a family of functions belonging to $LD(\mathbb{R}^n)$, having their support in a fixed compact set. The proof is independent of the extension operator of Remark 1.3 and thus Theorems 1.4 and 2.4 have been proved without using Remark 1.3 and the Trace Theorem of E. Gagliardo [1].

We draw from these theorems consequences which correspond exactly to Propositions 1.2 to 1.4, and whose proofs are so closely similar that they need not be reproduced.

PROPOSITION 2.2.

On the quotient space $BD(\Omega)/\mathscr{R}$ *, the expression*

$$\sum_{i,j=1}^{n} \int_\Omega |\varepsilon_{ij}(u)| \tag{2.44}$$

induces a norm equivalent to that induced by $BD(\Omega)$.

PROPOSITION 2.3

For all u *of* $BD(\Omega)$ *, there exists an* $r = r(u) \in \mathscr{R}$ *such that*

$$\|u-r\|_{BD(\Omega)} \leq c(\Omega) \sum_{i,j=1}^{n} \int_\Omega |\varepsilon_{ij}(u)| , \tag{2.45}$$

where the constant c *depends only on* Ω .

PROPOSITION 2.4.

If p *is a continuous seminorm on* $BD(\Omega)$ *which is a norm on* $\mathscr{R}$ *(so that* $p(u) = 0$ *,* $u \in \mathscr{R}$*, implies* $u = 0$*) , then*

$$p(u) + \sum_{i,j=1}^{n} \int_\Omega |\varepsilon_{ij}(u)| \tag{2.46}$$

is a norm on $BD(\Omega)$ *, equivalent to the initial norm.*

Remark 2.5. Statements analogous to Remarks 1.1 and 1.2 can be formulated:

i) Proposition 2.3 and Theorem 2.2 imply that for all $u \in BD(\Omega)$, there exists an $r = r(u)$ such that

$$\|u-r(u)\|_{L^{n^*}(\Omega)^n} \leq c(\Omega) \sum_{i,j=1}^{n} \int_\Omega |\varepsilon_{ij}(u)| , \tag{2.47}$$

where c depends only on Ω . An explicit choice of $r(u)$ can be one of those indicated in Remark 1.1. ii) which lead to continuous linear mappings $u \mapsto r(u)$ of $BD(\Omega)$ into $\mathscr{R}$.

ii) As in Remark 1.2, one can choose in Proposition 2.4

$$p(u) = \int_{\Gamma_0} |u(x)|dx \tag{2.48}$$

(using the Trace Theorem 2.1).

iii) Proposition 2.4 also implies a Poincaré inequality:

(2.49) There is a $c = c(\Omega)$ such that

$$\|u\|_{L^1(\Omega)^n} \leq c \sum_{i,j=1}^{n} \int_\Omega |\varepsilon_{ij}(u)| \ ,$$

for all u in

$$\{v \in BD(\Omega) \ , \ p(v) = 0\} \ . \tag{2.50}$$

In particular $\sum_{i,j=1}^{n} \int_\Omega |\varepsilon_{ij}(u)|$ is a norm on (2.50) equivalent to the norm on $BD(\Omega)$.

□

3. The space BD (Ω) (II)

We define two new topologies on $BD(\Omega)$ which are weaker than the norm topology (Section 3.1). We then go on to show that, for one of these topologies, the trace mapping is still continuous (Section 3.2) and that the set of regular functions is dense in $BD(\Omega)$ (Section 3.3). This last result will be used on several occasions in the sequel; however it is not absolutely necessary to read the proof, which is rather lengthy, in order to understand the subsequent sections.

3.1. Weak topologies on $BD(\Omega)$

It is possible to show that $BD(\Omega)$ *is the dual of a normed space*; cf. H. Matthies [1], G. Strang, R. Temam [2]. This enables us to define a weak-star topology on this space and the closed balls of $BD(\Omega)$ are then compact and sequentially compact for this topology.

Another way of defining this topology is the following: it is the topology defined by the family of norms and seminorms (cf. N. Bourbaki [2]):

(3.1) $$\|u\|_{L^1(\Omega)^n} ,$$

(3.2) $$\left|\int_\Omega \phi\, \varepsilon_{ij}(u)\right| , \quad \phi \in \mathscr{C}_0^\infty(\Omega) , \quad i,j = 1,\dots,n .$$

In particular a sequence u_m converges to u for this topology if, for $m \to \infty$

(3.3) $$\begin{cases} u_m \to u \text{ in } L^1(\Omega)^n \text{ and, for } i,j = 1,\dots,n\,, \\ \varepsilon_{ij}(u_m) \to \varepsilon_{ij}(u) \text{ weakly in } M_1(\Omega)\,. \end{cases}$$

We shall call this the *weak topology* of $BD(\Omega)$. It is easily seen that

(3.4) Every bounded sequence in $BD(\Omega)$ contains a weakly convergent subsequence.

For, if u_m is bounded in $BD(\Omega)$, it is relatively compact in $L^1(\Omega)^n$ by Theorem 2.4. There exists therefore a subsequence m' and a $u \in L^1(\Omega)^n$ such that $u_{m'}$ converges to u in $L^1(\Omega)^n$. As $\int_\Omega |\varepsilon_{ij}(u_m)| \leq \text{const.}$, $\varepsilon_{ij}(u) \in M_1(\Omega)$ for $i,j = 1,\dots,n$, and by a further extraction of a subsequence, we can obtain a subsequence $u_{m''}$ such that $\varepsilon_{ij}(u_{m''})$ converges weakly to $\varepsilon_{ij}(u)$, $\forall\, i,j$; whence (3.3) holds for $u_{m''}$.

Another natural topology on $BD(\Omega)$ is the one defined by the distance

(3.5) $$\|u-v\|_{L^1(\Omega)^n} + \left|\int_\Omega |\varepsilon(u)| - \int_\Omega |\varepsilon(v)|\right|\,.$$

It is easily established that this topology is intermediate between the weak topology and the topology of the norm on $BD(\Omega)$. We shall call it *the intermediate topology* ([1]). For example, a sequence u_m converges to u for the intermediate topology if, in addition to (3.3) we have :

(3.6) $$\int_\Omega |\varepsilon(u_m)| \to \int_\Omega |\varepsilon(u)|\,.$$

([1]) In fact we shall consider several intermediate topologies: see Remark 3.1.ii) and other intermediate topologies defined in Section 5.3.

There exist weakly convergent sequences u_m which do not converge for the intermediate topology; it follows that the balls of $BD(\Omega)$ are not relatively compact for the intermediate topology.

Remark 3.1. i) The weak topology defines a metrisable topology on the balls of $BD(\Omega)$ (cf. N. Bourbaki [2]).

ii) Instead of the distance (3.5) we could have considered the distance

$$\|u-v\|_{L^1(\Omega)^n} + \sum_{i,j=1}^{n} \left| \int_\Omega |\varepsilon_{ij}(u)| - \int_\Omega |\varepsilon_{ij}(v)| \right| .$$

We do not know whether the 'intermediate' topology defined in this way is the same as that defined above. One can define in the same way numerous other intermediate topologies (cf. in particular those introduced in Section 5.3).

Remark 3.2. It is now possible to answer the question raised in the Introduction to Chapter II : if u_m is a sequence of regular functions (say $u_m \in \mathscr{C}^1(\bar{\Omega})^n$) converging uniformly in $\bar{\Omega}$ to a function u with

$$\int_\Omega |\varepsilon(u_m)|dx \leq \text{const.} ,$$

What can be said about u?

From the preceding we can say that $u \in BD(\Omega)$ and that u_m converges weakly to u in $BD(\Omega)$. Incidentally, for this to be true, uniform convergence is unnecessary, it is sufficient that u_m should converge to u in $L^1(\Omega)^n$.

3.2. Continuity of the Trace

The trace operator γ_0 defined by Theorem 2.1 is continuous for the norm topology, but not for the weak topology [1]. We shall see, on the other hand, that γ_0 is continuous for the intermediate topology defined above.

THEOREM 3.1

The trace operator γ_0 *is a continuous operator from* $BD(\Omega)$ *into* $L^1(\Gamma)^n$ *, when* $BD(\Omega)$ *is endowed with the intermediate topology.*

In other words, if u_m converges to u in the sense defined by (3.3) and (3.6) then

$$\gamma_0(u_m) \to \gamma_0(u) \text{ in } L^1(\Gamma)^n . \tag{3.7}$$

To prove this theorem we introduce, for all small enough positive α $(0<\alpha<\alpha_0)$, the sets

$$\Omega_\alpha = \{x \in \Omega,\ d(x,\Gamma) > \alpha\} , \quad \Gamma_\alpha = \partial\Omega_\alpha . \tag{3.8}$$

If $u \in BD(\Omega)$, then by the Trace Theorem 2.2 and Remark 2.2, it is possible to define, for any α , two, possibly distinct, traces of u on Γ_α corresponding to those obtained by applying the trace theorem to Ω_α and $\Omega\setminus\bar{\Omega}_\alpha$ respectively. We shall denote these by $\gamma_0^-(u)$ and $\gamma_0^+(u)$ respectively, ($\gamma_0(u)$ being the trace on Γ) . In accordance with Remark 2.3, $\gamma_0^-(u) \neq \gamma_0^+(u)$ if and only if

$$\int_{\Gamma_\alpha} |\varepsilon(u)| \neq 0 , \tag{3.9}$$

and this can happen for, at most, an enumerable set $\mathcal{E}_1$ of values of α

---

(1) To see this it suffices to note that $BV(\Omega)^n \subset BD(\Omega)$ with continuous injection and that the trace is already discontinuous on $BV(\Omega)$: for example if $n = 1$ and $\Omega = (0,1)$, the sequence of functions $v_m(x) = x/m$ for $0<x<1/m$ and $=1$ for $1/m\leq x\leq 1$, converges weakly to 1 , whereas $v_m(0) = 0$ for all m .

We now state

LEMMA 3.1

There exists a constant $c = c(\Omega)$, *depending only on* Ω *such that for all sufficiently small* $\alpha>0$:

$$(3.10)\qquad \int_\Gamma |\gamma_0 u|\,d\Gamma \leq c\{\int_{\Omega\setminus\bar\Omega_\alpha} |\varepsilon(u)| + \int_{\Gamma_\alpha} |\gamma_0^+ u|\,d\Gamma + \int_{\Omega\setminus\bar\Omega_\alpha} |u|\,dx\} .$$

Proof

Since γ_0 is a continuous mapping from $BD(\Omega)$ into $L^1(\Gamma)^n$, there exists a $c' = c'(\Omega)$ such that

$$(3.11)\qquad \int_\Gamma |\gamma_0 v|\,d\Gamma \leq c'\{\int_\Omega |\varepsilon(v)| + \int_\Omega |v|\,dx\}$$

for all v of $BD(\Omega)$.

If $u \in BD(\Omega)$, let u^* be the function equal to 0 in Ω_α and to u in $\Omega\setminus\bar\Omega_\alpha$. By Remark 2.3 and (2.39), $u^* \in BD(\Omega)$ and, for all i,j ,

$$(3.12)\qquad \varepsilon_{ij}(u^*) = \theta_{\Omega\setminus\bar\Omega_\alpha} \cdot \varepsilon_{ij}(u) + \mathscr{T}_{ij}(\gamma_0^+(u))\delta_{\Gamma_\alpha} ,$$

where $\theta_{\mathscr{A}}$ is the characteristic function of $\mathscr{A}$ and δ_{Γ_α} is the Dirac (surface) distribution on Γ_α .

By applying (3.11) to u^* , we obtain (3.10).

□

We can now prove Theorem 3.1.

Proof of Theorem 3.1.

Let u_m be a sequence converging to u in the sense defined by (3.3) and (3.6). As we have already observed earlier, (3.9) holds for all $\alpha \in (0,\alpha_0)\setminus\mathscr{E}_1$ with $\alpha_0>0$ small enough, and $\mathscr{E}_1$ at most enumerable. By (3.6) and Lemma 2.1, we therefore have

$$\int_{\Omega_\alpha} |\varepsilon(u_m)| \to \int_{\Omega_\alpha} |\varepsilon(u)| ,$$

for all $\alpha \in (0,\alpha_0)\setminus\mathscr{E}_1$.

As $u_m - u$ converges to 0 in $L^1(\Omega)^n$, it follows from Fubini's Theorem that

$$\int_{\Gamma_\alpha} |u_m - u| d\Gamma \rightarrow 0 ,$$

for all $\alpha \in (0,\alpha_0) \setminus \mathscr{E}_2$, where $\mathscr{E}_2$ is of measure zero.

We now choose, for any given $\delta > 0$, as can always be done, a positive α with $\alpha \in (0,\alpha_0) \setminus (\mathscr{E}_1 \cup \mathscr{E}_2)$ and α small enough so that

$$\int_{\Omega \setminus \Omega_\alpha} |\varepsilon(u)| \leq \delta .$$

From what has been said, there will then exist an $m_1 = m_1(\delta)$ such that, for $m \geq m_1$,

$$\begin{cases} \displaystyle\int_{\Omega \setminus \Omega_\alpha} |\varepsilon(u_m)| \leq 2\delta \\ \displaystyle\int_{\Gamma_\alpha} |u_m - u| d\Gamma \leq \delta \\ \displaystyle\int_{\Omega \setminus \Omega_\alpha} |u_m - u| dx \leq \delta \end{cases}$$

The inequality (3.10) applied to $u_m - u$ implies that, for $m \geq m_1(\delta)$,

$$\int_\Gamma |\gamma_0(u_m - u)| d\Gamma \leq 4c(\Omega)\delta ,$$

which proves continuity, since δ can be chosen to be arbitrarily small.

□

3.3 Approximation of functions in $BD(\Omega)$

We give two results on the approximation of functions in $BD(\Omega)$ by regular functions.

THEOREM 3.2.

$\mathscr{C}^\infty(\bar{\Omega})$ is dense in the space $BD(\Omega)$ endowed with the intermediate topology.

Proof

i) We shall first prove that $LD(\Omega)$ is dense in $BD(\Omega)$ for the intermediate topology. The theorem will then follow from the fact that $\mathscr{C}^\infty(\bar{\Omega})^n$ is dense in $LD(\Omega)$ for the norm topology (cf. (1.8)).

Actually we shall show that $\mathscr{C}^\infty(\Omega)^n \cap LD(\Omega)$ is dense in $BD(\Omega)$. To do this we shall establish that for all $u \in BD(\Omega)$ and any $\delta>0$, there is a $u_\delta \in \mathscr{C}^\infty(\Omega)^n \cap LD(\Omega)$ such that

$$\|u_\delta - u\|_{L^1(\Omega)^n} \leqslant \delta \tag{3.13}$$

$$\left| \int_\Omega |\varepsilon(u_\delta)| - \int_\Omega |\varepsilon(u)| \right| \leqslant 4\delta \tag{3.14}$$

ii) We shall construct a u_δ satisfying (3.13) (3.14).

Since $|\varepsilon(u)|$ is a Radon measure, there exists an $r>0$ such that

$$\int_{\Omega \setminus \Omega_0} |\varepsilon(u)| \leqslant \delta , \tag{3.15}$$

where

$$\Omega_0 = \{x \in \Omega,\ d(x,\Gamma) > \frac{1}{r}\} . \tag{3.16}$$

We next define, for $j = 1,2,\dots$, the sets Ω_j,

$$\Omega_j = \{x \in \Omega,\ d(x,\Gamma) > \frac{1}{j+r}\} .$$

The open sets Ω_j satisfy the relations

$$\Omega_j \subset \bar{\Omega}_j \subset \Omega_{j+1} , \quad j = 0,1,\dots,$$

$$\bigcup_{j=0}^{\infty} \Omega_j = \Omega .$$

We now define the sets $\mathscr{A}_j$, for $j \geqslant 1$, by $\mathscr{A}_1 = \Omega_2$ and

$$\mathscr{A}_j = \Omega_{j+1} - \bar{\Omega}_{j-1} \quad \text{for } j>1$$

The $\mathscr{A}_j$ form an open covering of Ω and we consider a partition of unity, $\{\phi_j\}$, subordinated to this covering :

(3.17) $$\phi_j \in \mathscr{C}_0^\infty(\mathscr{A}_j) \ , \quad 0 \leqslant \phi_j \leqslant 1 \text{ and } \sum_{j=1}^{\infty} \phi_j = 1 \ .$$

We now consider a family of regularising functions ρ_{η_j} defined as in I.(1.8), where the sequence of the η_j is made satisfying certain conditions specified below, and we put :

(3.18) $$u_\delta = \sum_{j=1}^{\infty} \rho_{\eta_j} * (\phi_j u) \ .$$

The η_j must decrease monotonically to zero, that is to say must satisfy

$$\eta_{j+1} \leqslant \eta_j \ , \quad \eta_j \to 0 \ \text{ for } \ j \to \infty \ ,$$

and must also be small enough to ensure that the five following conditions are satisfied :

(3.19) $$\left| \int_\Omega |\rho_{\eta_1} * (\phi_1 \varepsilon(u))| dx - \int_\Omega |\phi_1 \varepsilon(u)| \right| \leqslant \delta$$

(3.20) $$\int_\Omega |\rho_{\eta_j} * (u\phi_j) - (u\phi_j)| dx \leqslant \delta 2^{-j}$$

(3.21) $$\int_\Omega |\rho_{\eta_j} * (u \,\bar{\varepsilon}(\phi_j)) - u \,\bar{\varepsilon}(\phi_j)| dx \leqslant \delta 2^{-j}$$

(3.22) $$(\Omega_2 \setminus \bar{\Omega}_1) + B(0,\eta_1) \subset \Omega_3 \setminus \bar{\Omega}_0$$

(3.23) $$\mathscr{A}_j + B(0,\eta_j) \subset \mathscr{A}_{j-1} \cup \mathscr{A}_j \cup \mathscr{A}_{j+1} \ , \quad j \geqslant 2 \ ,$$

where $B(0, \eta_j)$ is the ball of centre 0 and radius η_j (containing the support of ρ_{η_j}) , and where $u \,\bar{\varepsilon}\, (\phi)$ is defined by:

(3.24) $$u \,\bar{\varepsilon}(\phi) = \varepsilon(\phi u) - \phi\varepsilon(u) = \frac{1}{2} \{u \otimes \nabla\phi + \nabla\phi \otimes u\} \ .$$

The existence of an η_1 small enough to ensure that (3.19) holds is implied by (2.7). The existence of η_j small enough to ensure that (3.20) and (3.21) are both satisfied is a consequence of (1.9).

By (3.22) and as $\phi_1 = 1$ on Ω_1, it follows that $\rho_{n_1} * (\phi_1\, \varepsilon(u)) = \varepsilon(u)$ in a neighbourhood of $\bar{\Omega}_0$; (3.15) and (3.19) then imply

$$(3.25)\quad \left|\int_\Omega |\varepsilon(u)| - \int_\Omega |\rho_{n_1} * (\phi_1\, \varepsilon(u))|\right| \leq \delta + \int_\Omega (1-\phi_1)|\varepsilon(u)| \leq 2\delta\ .$$

From (3.17) we have

$$u = \sum_{j=1}^{\infty} \phi_j u$$

and, with (3.20) we obtain immediately (3.13) :

$$\int_\Omega |u_\delta - u|\,dx \leq \sum_{j=1}^{\infty} \int_\Omega |\rho_{n_j} * (\phi_j u) - \phi_j u|\,dx \leq \delta\ .$$

To show that u satisfies (3.14) we write :

$$\varepsilon(u_\delta) = \sum_{j=1}^{\infty} \rho_{n_j} * (\phi_j\, \varepsilon(u)) + \sum_{j=1}^{\infty} \rho_{n_j} * (u\, \bar{\varepsilon}(\phi_j))$$

$$\left(\text{since } \sum_{j=1}^{\infty} u\, \bar{\varepsilon}(\phi_j) = 0\right)$$

$$= \rho_{n_1} * (\phi_1\, \varepsilon(u)) + \sum_{j=2}^{\infty} \rho_{n_j} * (\phi_j\, \varepsilon(u)) + \sum_{j=1}^{\infty} \{\rho_{n_j} * (u\, \bar{\varepsilon}(\phi_j)) - u\, \bar{\varepsilon}(\phi_j)\}$$

$$\left|\int_\Omega |\varepsilon(u_\delta)| - \int_\Omega |\rho_{n_1} * (\phi_1\, \varepsilon(u))|\right|$$

$$\leq \sum_{j=2}^{\infty} \int_\Omega |\rho_{n_j} * (\phi_j\, \varepsilon(u))|\,dx$$

$$+ \sum_{j=1}^{\infty} \int_\Omega |\rho_{n_j} * (u\, \bar{\varepsilon}(\phi_j)) - u\, \bar{\varepsilon}(\phi_j)|\,dx$$

$$\leq \sum_{j=2}^{\infty} \int_\Omega \phi_j |\varepsilon(u)| + \delta \quad \text{(by (2.6) and (3.21))}$$

$$\leq 2\delta\ , \text{ by (3.15).}$$

This inequality, coupled with (3.25) implies (3.14), thus completing the proof of Theorem 3.2.

Remark 3.3. i) It may be noted, in connection with Remark 3.1, ii) that we could have chosen u_δ so that, in addition to satisfying (3.14),

$$\left| \int_\Omega |\varepsilon_{ij}(u_\delta)| - \int_\Omega |\varepsilon_{ij}(u)| \right| \leq \delta \ , \quad \forall\ i,j \ ,$$

which would prove that $\mathscr{C}^\infty(\bar{\Omega})^n$ is dense in $BD(\Omega)$ for the metric defined in Remark 3.1 ii). This point will be examined in more detail and extended in Section 5.

ii) It is also possible to impose further conditions on u_δ involving differential operators other than $\varepsilon(u)$, and $L^p(\Omega)$ spaces with $p \neq 1$. An example will be given in Section 3.4; general results are proved in R. Temam [10], F. Demengel - R. Temam [1]. For other density results of this type, cf. G. Anzelotti and M. Giaquinta [1] and E. Giusti [1] in the case of $BV(\Omega)$, and G. Anzelotti and M. Giaquinta [2] in the case of $BV(\Omega)$.

□

We can complete Theorem 3.2 by the following

THEOREM 3.3.

For any u *in* $BD(\Omega)$, *there exists a sequence* $u_m \in \mathscr{C}^\infty(\Omega)^n \cap LD(\Omega)$ *such that*

(3.26) $$\gamma_o(u_m) = \gamma_o(u) \ , \quad \text{for all } m \ ,$$

(3.27) $$u_m \to u \ , \text{ with the metric (3.5), when } m \to \infty$$

Proof

It is sufficient to prove that the u_δ which satisfy (3.13) (3.14) also satisfy

(3.28) $$\gamma_0(u_\delta) = \gamma_0(u) \ .$$

To do this we shall show (keeping the same notations and hypotheses as those of Theorem 3.2) that $u - u_\delta$ is the limit in $BD(\Omega)$ of the finite sum

$$u_N - u_{\delta N} = \sum_{j=1}^{N} \{\phi_j u - \rho_{n_j} * (\phi_j u)\}$$

when $N \to \infty$. Since the functions (3.29) vanish in the neighbourhood of Γ, their trace on Γ is zero; as the operator γ_0 is continuous on $BD(\Omega)$ for the norm topology, this will prove (3.28).

We have

$$(u-u_\delta) - (u_N-u_{\delta N}) = \sum_{j=N+1}^{\infty} \{\phi_j u - \rho_{n_j} * (\phi_j u)\}$$

$$\|(u-u_\delta)-(u_N-u_{\delta N})\|_{L^1(\Omega)^n} \leq \sum_{j=N+1}^{\infty} \int_\Omega |\phi_j u-\rho_{n_j} * (\phi_j u)|dx \tag{3.30}$$

$$\leq \delta 2^{-N-1} \text{ (by (3.20))}$$

and this tends to 0 as N tends to infinity. Similarly

$$\int_\Omega |\varepsilon(u-u_\delta+u_N-u_{\delta N})| \leq \sum_{j=N+1}^{\infty} \int_\Omega |u\bar{\varepsilon}(\phi_j)-\rho_{n_j} * (u\bar{\varepsilon}(\phi_j))|dx \tag{3.31}$$

$$+ \sum_{j=N+1}^{\infty} \int_\Omega \{\phi_j|\varepsilon(u)| + |\rho_{n_j} * (\phi_j \varepsilon(u))|\}$$

$$\leq \delta\, 2^{-N-1} + 2\int_{\Omega\setminus\Omega_N} |\varepsilon(u)| \text{ (by (3.21) and (2.4))}$$

and this tends to zero as N tends to infinity.

3.4 The space $U(\Omega)$

The search for weak solutions to the strain problem in plasticity (cf. I.(4.16)) will lead us to consider a space which is a little more restricted than $BD(\Omega)$, namely the space

$$U(\Omega) = \{v \in BD(\Omega) ,\ \operatorname{div} v \in L^2(\Omega) \quad . \tag{3.32}$$

This space is precisely the space of finite-energy displacements for I.(4.16). That this is so can be readily appreciated by asking one's self (as in Remark 3.2) the following question :

> If u_m is a sequence of functions in $\mathscr{C}^1(\bar{\Omega})^n$ which converge in $L^1(\Omega)^n$ (or resp. which converge uniformly) to u, and is such that
>
> $$\Psi(\varepsilon(u_m)) \leqslant \text{const.}$$
>
> What can be said about the function u?

The expression for Ψ given in I.(4.15) and I.(4.18) implies that

$$\frac{\kappa}{2}\int_\Omega (\operatorname{div} u_m)^2 dx + k_o \int_\Omega |\varepsilon^D(u_m)|\, dx \leqslant \text{const.} \tag{3.33}$$

and it is easily deduced that $u \in U(\Omega)$ (or resp. $u \in U(\Omega) \cap \mathscr{C}(\bar{\Omega})^n$). The space $U(\Omega)$ is therefore a natural subject of interest.

□

$U(\Omega)$ is a Banach space for the natural norm

$$\|u\|_{U(\Omega)} = |\operatorname{div} u|_{L^2(\Omega)} + \|u\|_{BD(\Omega)} . \tag{3.34}$$

It is possible to define on $U(\Omega)$ a weak topology and an intermediate topology analogous to those defined on $BD(\Omega)$. The weak topology is the one determined by

- the norm $\|u\|_{L^1(\Omega)^n}$

- and the family of seminorms

$$\left|\int_\Omega \theta \operatorname{div} u\, dx\right| , \quad \theta \in \mathscr{C}_o(\Omega) \tag{3.35}$$

$$\left|\int_\Omega \phi\, \varepsilon_{ij}(u)\right| , \quad \phi \in \mathscr{C}_o(\Omega) , \quad i,j=1,\dots,n . \tag{3.36}$$

As with $BD(\Omega)$ one finds that if u_m is a bounded sequence of functions in $U(\Omega)$ then there exists a subsequence (again denoted by u_m) which converges weakly to u in $U(\Omega)$, that is to say :

$$(3.37)\quad \begin{cases} u_m \to u \text{ in } L^1(\Omega)^n \text{ (for the norm)} \\ \operatorname{div} u_m \to \operatorname{div} u \text{ in } L^2(\Omega) \text{ weakly} \\ \varepsilon_{ij}(u_m) \to \varepsilon_{ij}(u) \text{ in } M_1(\Omega) \text{ weakly,} \end{cases}$$

$i,j = 1,\dots,n$.

The intermediate topology on $U(\Omega)$ is that associated with the metric

$$(3.38)\quad d(u,v) = \|u-v\|_{L^1(\Omega)^n} + |\operatorname{div}(u-v)|_{L^2(\Omega)} + \left|\int_\Omega |\varepsilon(u)| - \int_\Omega |\varepsilon(v)|\right|.$$

A sequence u_m converges to u for the intermediate topology if, in addition to (3.37)

$$(3.39)\quad \operatorname{div} u_m \to \operatorname{div} u \text{ in } L^2(\Omega) \text{ strongly}$$

$$(3.40)\quad \int_\Omega |\varepsilon(u_m)| \to \int_\Omega |\varepsilon(u)| .$$

It is worth noting that if $n = 2$, the space $U(\Omega)$ is *of local type*, that is to say that

$$(3.41)\quad \text{if } u \in U(\Omega) \text{ and } \phi \in \mathscr{C}^\infty(\bar{\Omega}) \text{ then } u\phi \in U(\Omega) .$$

On the other hand, if $n \geq 3$, we do not know whether this is so : by Theorem 2.2 $BD(\Omega) \subset L^2(\Omega)^2$ if $n = 2$, and $BD(\Omega) \subset L^p(\Omega)^n$, $p<2$ if $n \geq 3$. Since

$$(3.42)\quad \operatorname{div}(\phi u) = \phi \operatorname{div} u + u.\operatorname{grad} \phi ,$$

we cannot affirm, when $n \geq 3$, that $u.\operatorname{grad}\phi$ belongs to $L^2(\Omega)$, and the same thing applies therefore to $\operatorname{div}(\phi u)$.

We end by the following result analogous to Theorems 3.2 and 3.3.

THEOREM 3.4

For all u *in* $U(\Omega)$, *there exists a sequence* $u_m \in \mathscr{C}^\infty(\Omega)^n \cap LD(\Omega) \cap U(\Omega)$ *with the following properties* :

(3.43) $$\gamma_o(u_m) = \gamma_o(u) \ , \quad \forall\ m \ .$$

(3.44) *As* $m \to \infty$, u_m *converges to* u *for the intermediate topology (i.e. converges in the sense defined by (3.37) (3.39) and (3.40)).*

Proof

i) As a first step we shall prove the result (weaker when $n \geq 3$) obtained by replacing (3.39) by the weaker conclusion :

(3.45) $$\text{div } u_m \to \text{div } u \quad \text{in} \quad L^{n^*}(\Omega) \quad \text{strongly.}$$

By using (3.42) we note that $\text{div } (\phi u) \in L^{n^*}(\Omega)$ when $\phi \in \mathscr{C}^\infty(\bar{\Omega})$ and $u \in U(\Omega)$. The proof follows the lines of that of Theorems 3.2 and 3.3 whose hypotheses and notation we retain :

- we have to find a u_δ which satisfies, in addition to (3.13) (3.14):

(3.46) $$||\text{div } (u-u_\delta)||_{L^{n^*}(\Omega)} \leq \delta \ .$$

To do this we impose on ρ_{n_j} the following further conditions :

(3.47) $$||u.\nabla\phi_j - \rho_{n_j} * (u.\nabla\phi_j)||_{L^{n^*}(\Omega)} \leq \delta\ 2^{-j-1}$$

(3.48) $$||\phi_j.\nabla u - \rho_{n_j} * (\phi_j.\nabla u)||_{L^{n^*}(\Omega)} \leq \delta\ 2^{-j-1} \ .$$

We then have

$$\text{div}(u-u_\delta) = \sum_{j=1}^\infty \{\phi_j.\nabla u - \rho_{n_j} * (\phi_j.\nabla u)\} + \sum_{j=1}^\infty \{u.\nabla\phi_j - \rho_{n_j} * (u.\nabla\phi_j)\}$$

whence $\|\mathrm{div}(u-u_\delta)\|_{L^{n^*}(\Omega)} \leq 2\delta$ (by (3.47) and (3.48))

- in Theorem 3.3 we also need to prove that

$$(3.49) \qquad \|\mathrm{div}((u-u_\delta)-(u_N-u_{\delta N}))\|_{L^{n^*}(\Omega)} \to 0 ,$$

when $N \to \infty$. But

$$\|\mathrm{div}(u-u_\delta-u_N+u_{\delta N})\|_{L^{n^*}(\Omega)} \leq$$

$$\leq \sum_{j=N+1}^{\infty} \|\phi_j \nabla.u - \rho_{n_j} * (\phi_j \nabla.u)\|_{L^{n^*}(\Omega)} + \sum_{j=N+1}^{\infty} \|u.\nabla\phi_j - \rho_{n_j} * (u.\nabla\phi_j)\|_{L^{n^*}(\Omega)}$$

$$\leq \delta\, 2^{-N-1} , \text{ by virtue of (3.47) and (3.48),}$$

and this tends to zero when $N \to \infty$.

ii) It follows from the first step that there does exist a sequence u_m having the required properties (3.37) (3.40) (3.43) and (3.45). The second step of the proof follows an idea of A. Anzelotti and M. Giaquinta [2].

Since $\mathscr{C}_0^\infty(\bar{\Omega})$ is dense in $L^2(\Omega)$ a sequence of functions can be found which converges to div u in $L^2(\Omega)$ with, for all m

$$(3.50) \qquad \int_\Omega \theta_m \, dx = \int_\Omega \mathrm{div}\, u \, dx \quad (^1)$$

---

(1) There exists a sequence of functions ϕ_m in $\mathscr{C}_0^\infty(\Omega)$ which converges to div u in $L^2(\Omega)$. If $\int_\Omega \mathrm{div}\, u \, dx \neq 0$, we take $\theta_m = \phi_m(\int_\Omega \mathrm{div}\, u \, dx)/\int_\Omega \phi_m \, dx$ which is well-defined for large enough m. If $\int_\Omega \mathrm{div}\, u\, dx = 0$, we take

$$\theta_m = \phi_m - \phi(\int_\Omega \phi_m \, dx)/\int_\Omega \phi \, dx ,$$

where $\phi \in \mathscr{C}_0^\infty(\Omega)$ is arbitrarily chosen.

As is shown in M.E. Bogovski [1] or M. Giaquinta and G. Modica [1], there exists for all m, a function $f_m \in W^{1,n^*}(\Omega)^n \cap \mathscr{C}^\infty(\Omega)^n$, ($n^* = n/(n-1)$), such that

(3.51) $$\operatorname{div} f_m = \theta_m - \operatorname{div} u_m \quad \text{in } \Omega$$

(3.52) $$f_m = 0 \quad \text{on } \partial\Omega \quad (^1)$$

and

(3.53) $$|f_m|_{W^{1,n^*}(\Omega)^n} \leq c\|\theta_m - \operatorname{div} u_m\|_{L^{n^*}(\Omega)}$$

where c depends only on Ω.

We therefore put

(3.54) $$v_m = u_m + f_m ,$$

and we shall see that v_m has all the required properties : $v_m \in LD(\Omega) \cap \mathscr{C}^\infty(\Omega)^n \cap U(\Omega)$ and

$$\int_\Omega |v_m - u|dx \leq \int_\Omega |u_m - u|dx + \int_\Omega |f_m|dx ,$$

which tends to 0 by (3.37) and (3.53). Similarly

$$\operatorname{div}(v_m - u) = \theta_m - \operatorname{div} u \to 0 \quad \text{in } L^2(\Omega)$$

when $m \to \infty$. Lastly

(^{1}) For $n = 2$ or 3 this also follows from well-known results for the Stokes problem, cf. L. Cattabriga [1], O.A. Ladyzhenskaya [1] or R. Temam [5] : there exists a function $f_m \in W^{1,n^*}(\Omega)^n \cap \mathscr{C}^\infty(\Omega)^n$ satisfying (3.51) (3.52) (3.53) and a function $p_m \in L^{n^*}(\Omega) \cap \mathscr{C}^\infty(\Omega)$ such that

$$- \Delta f_m + \operatorname{grad} p_m = 0$$

$$\varepsilon(v_m) = \varepsilon(u_m) + \varepsilon(f_m)$$

$$\left| \int_\Omega |\varepsilon(v_m)| - \int_\Omega |\varepsilon(u_m)| \right| \leq \int_\Omega |\varepsilon(f_m)|$$

and this tends to 0 as $m \to \infty$, since the first derivatives of f_m converge to 0 in $L^{n^*}(\Omega)$, by (3.53).

The Theorem is thus proved. □

Remark 3.4. i) We are unable to characterise the space $\gamma_0(U(\Omega))$, i.e. the space of the $\gamma_0(u)$ with $u \in U(\Omega)$, though Theorem 3.4 implies that this space is the same as the space $\gamma_0(U(\Omega) \cap LD(\Omega))$ (which is likewise unknown (1)).

ii) It can be shown, exactly as in Theorem 3.4 ii) that $\mathscr{C}^\infty(\bar{\Omega})^n$ is dense in

$$U(\Omega) \cap LD(\Omega) = \{v \in LD(\Omega) \ , \ \operatorname{div} v \in L^2(\Omega)\} \ ,$$

which does not automatically follow from Proposition I.1.3, since the space is not of local type when the dimension n exceeds 2.

Remark 3.5. By Theorem 2.2 and Theorem 2.4, $U(\Omega) \subset L^p(\Omega)^n$, $p \leq n^* = n/(n-1)$ the injection being continuous and even compact if $p<n^*$. The sequence u_m given by Theorems 3.3 and 3.4 thus satisfies, in addition to (3.37) the condition

(3.55) $$u_m \to u \text{ in } L^p(\Omega)^n \text{ strongly}$$

for all $p<n^*$. Moreover we can choose the sequence u_m in such a way that, when $m \to \infty$.

(3.56) $$u_m \to u \text{ in } L^{n^*}(\Omega)^n \ .$$

---

(1) It has however, been established in G. Anzelotti and M. Giaquinta [2] that when u describes $U(\Omega)$, $\gamma_\tau(u)$ describes the whole of the space consisting of the vectors of $L^1(\Gamma)^n$ tangent to Γ.

To ensure that the sequence u_m in Theorem 3.3 should satisfy (3.56) it is sufficient to impose, instead of (3.21), the condition :

$$||\rho_{n_j} * (u.\bar{\varepsilon}(\phi_j)) - u.\bar{\varepsilon}(\phi_j)||_{L^{n*}(\Omega)^n} \leq \delta\, 2^{-j} .$$

To verify that the sequence u_m in Theorem 3.4 satisfies (3.56), we have to write, in place of (3.54) :

$$||v_m - u_m||_{L^{n*}(\Omega)^n} \leq ||u_m - u||_{L^{n*}(\Omega)^n} + ||f_m||_{L^{n*}(\Omega)^n}$$

and, since $u_m \to u$ in $L^{n*}(\Omega)^n$, it suffices to verify that $f_m \to 0$ in that space, and this follows easily from (3.53) $(n^* \leq 2)$.

4. Convex functions of a measure

Our aim in this Section and those which follow it, is to extend the definition of the functional

$$\Psi(\varepsilon^D(u)) = \int_\Omega \psi(\varepsilon^D(u))dx \ ,$$

(see I.(4.15)) to the case where $u \in BD(\Omega)$. This will be done at the beginning of Section 6. In the present section we define $f(\mu)$ when μ is a bounded measure on Ω and f a convex function with at most linear growth at infinity : $f(\mu)$ is then a bounded measure on Ω , and is moreover equal to the function $f \circ h \, dx$ when μ is a function defined by $\mu = h\,dx$, $h \in L^1(\Omega)$.

After specifying the hypotheses on f (Section 4.1) we shall define $f(\mu)$ in Section 4.2 and give its main properties. Next we shall give the Lebesgue decomposition of $f(\mu)$ (knowing that of μ) - in Section 4.3; finally in Section 4.4 we give various additional results. The presentation follows that given in F. Demengel - R. Temam [1] and R. Temam [11].

4.1. Hypotheses on f

Let X be a finite-dimensional Euclidean space. As before we write $|\xi|$ for the norm, and $\xi.\eta$ for the scalar product (cf. Ch.I, Section 1), and we assume that we are given a convex function f mapping X into $\mathbb{R}$, which satisfies

$$f(\xi) \leq k_1(1+|\xi|) \ , \quad \forall \ \xi \in X \ ,$$

where $k_1 \geq 0$. Its domain is thus the whole space X , and it is known that f is, in these circumstances, continuous and indeed locally Lipschitzian throughout X , and that it is sub-differentiable at every point of X (cf. Ch.I, Section 2.1). Its conjugate f^* is a l.s.c. proper convex function on X , and with values in $\mathbb{R} \cup \{+\infty\}$:

$$f^*(\eta) = \operatorname*{Sup}_{\xi \in X} \{\eta.\xi - f^*(\xi)\} \ , \ \forall \ \eta \in X \ . \tag{4.2}$$

The domain (of finiteness) of f^* , given by

$$K = \text{dom } f^* = \{\eta \in X \ , \ f^*(\eta) < +\infty\} \ , \tag{4.3}$$

is not empty since f^* is not identically $+\infty$; (4.1) implies that K is bounded since

$$f^*(\eta) \geq \operatorname*{Sup}_{\xi \in X} [\eta.\xi - k_1(1+|\xi|)]$$

$$\geq \operatorname*{Sup}_{r=|\xi|\geq 0} [r(|\eta|-k_1)-k_1] = \begin{cases} -k_1 & \text{if } |\eta|\leq k_1 \\ +\infty & \text{otherwise.} \end{cases}$$

Thus, if $B(0,k_1)$ denotes the ball in X of centre 0 and radius k_1

$$K \subset B(0,k_1) \tag{4.4}$$

$$f^*(\eta) \geq - k_1 \ , \quad \forall \ \eta \in K \ . \tag{4.5}$$

We shall suppose, furthermore, that f^* is bounded above in K :

$$f^*(\eta) \leq k_2 \ , \ \forall \ \eta \in K \ . \tag{4.6}$$

By lower semicontinuity this implies that f is finite on ∂K , $\partial K \subset K$ hence

$$\text{(4.7)} \qquad K \text{ is closed.}$$

We shall sometimes make some further hypotheses on f, and assume that:

$$\text{(4.8)} \qquad f \geq 0 \;, \quad f(0) = 0 \;.$$

It is clear that this implies the same properties for f^*

$$\text{(4.9)} \qquad f^* \geq 0 \;, \quad f^*(0) = 0 \quad (\text{and so } 0 \in K).$$

When f satisfies the conditions (4.1) and (4.6) but not (4.8), it is possible to find an affine function ℓ on X such that

$$\text{(4.10)} \qquad f = g + \ell \;.$$

with the function g satisfying the conditions (4.1) (4.6) and (4.8). Since f is everywhere subdifferentiable there exists, for every ξ_0 of X, an $a_0 \in \partial f(\xi_0)$, such that

$$f(\xi) \geq f(\xi_0) + a_0.(\xi-\xi_0) \;, \quad \forall\, \xi \in X$$

and thus

$$\text{(4.11)} \quad g(\xi) = f(\xi+\xi_0) - f(\xi_0) - a_0.\xi = f(\xi+\xi_0) - a_0.(\xi+\xi_0) + f^*(a_0)$$

obviously satisfies the conditions (4.1) and (4.8); as for (4.6) we note that

$$\text{(4.12)} \qquad g^*(\eta) = f^*(\eta+a_0) - \eta\xi_0 - f^*(a_0) \;, \quad \forall\, \eta \in X \;,$$

so that

$$\text{(4.13)} \qquad \operatorname{dom} g^* = \operatorname{dom} f^* - a_0 = K - a_0$$

$$\text{(4.14)} \quad g^*(\eta) \leqq k_2 + k_1(1 + 2|\xi_0|) \;, \quad \forall\, \eta \in \operatorname{dom} g^* = K - a_0 \;.$$

To arrive at (4.10), it is enough to restrict ourselves to $\xi_0 = 0$ for which $f^*(a_0) = -f(0)$.

In conclusion, we shall introduce the 'principal part' of f, denoted by f_∞ : this is the support function of K,

$$f_\infty(\xi) = \operatorname*{Sup}_{\eta\in K} \xi.\eta \ , \quad \forall\, \xi \in X \ . \tag{4.15}$$

It is a positively homogeneous convex function mapping X into $\mathbb{R}$ which satisfies

$$f_\infty(\xi) \leq k_1|\xi| \ , \quad \forall\, \xi \in X \ . \tag{4.16}$$

It is positive if f satisfies (4.8), because in that case $0 \in K$ by (4.9).

4.2 Definition of $f(\mu)$

Let Ω be a bounded open subset of $\mathbb{R}^n$. As there is no risk of misunderstanding we (throughout this Section) shall simply write $\mathscr{C}_0, \mathscr{C}_0^\infty, M_1, L^1,\ldots,$ for the spaces $\mathscr{C}_0(\Omega;X), \mathscr{C}_0^\infty(\Omega;X), M_1(\Omega;X), L^1(\Omega;X),\ldots$ respectively.

We take as given a convex function f from X into $\mathbb{R}$ which satisfies (4.1) and (4.6) and also, for the time being, (4.8). For a given $\mu \in M_1$, we wish to define $f(\mu)$ as a bounded (scalar) measure on Ω : we therefore have to define $\langle f(\mu),\phi\rangle$ for all $\phi \in \mathscr{C}_0(\Omega)$, and we shall begin by defining this expression when $\phi \geq 0$.

For all $\mu \in M_1 = M_1(\Omega;X)$ and all $\phi \in \mathscr{C}_0(\Omega)$, $\phi \geq 0$ we define

$$\langle f(\mu),\phi\rangle = \operatorname*{Sup}_{v} \left\{\int_\Omega (v\phi)\mu - \int_\Omega f^*(v)\phi \, dx\right\} , \tag{4.17}$$

where the supremum is taken with respect to the v belonging to $\mathscr{D}_f(\mathscr{C}_0)$ defined by

$$\mathscr{D}_f(\mathscr{C}_0) = \{v \in \mathscr{C}_0(\Omega;X) \ , \ f^*\circ v \in L^1(\Omega)\} \ . \tag{4.18}$$

The definition (4.17) is motivated by the fact, which we shall prove later (see Theorem 4.2), that

$$\langle f(\mu),\ \phi\rangle = \int_\Omega f(h(x))\ \phi(x)\ dx \ , \tag{4.19}$$

when μ is a function, $\mu = h\,dx$ with $h \in L^1(\Omega;X)$.

It is easily seen that the right-hand side of (4.17) is a finite number ≥ 0 ; indeed $0 \in \mathscr{D}_f(\mathscr{C}_0)$ since $0 \in K$ (cf.(4.9)) and as $f^* \geq 0$,

$$0 \leq \langle f(\mu),\phi\rangle \leq \sup_{v\in\mathscr{D}_f(\mathscr{C}_0)} \int_\Omega \mu\, v\phi \leq k_1 \int_\Omega \phi|\mu| \ . \tag{4.20}$$

We define

$$\mathscr{D}_f(\mathscr{C}_0^\infty) = \{v \in \mathscr{C}_0^\infty(\Omega;X) \ , \ f^* \circ v \in L^1(\Omega)\} \tag{4.21}$$

$$\mathscr{D}_f(L^1_\mu) = \{v \in L^1(\Omega,\mu;X) \ ; \ f^* \circ v \in L^1(\Omega)\} \ , \tag{4.22}$$

where $L^1(\Omega,\mu;X) = L^1_\mu$ is the set of (classes of) functions from Ω into X which are L^1 for the measure μ . We now have the following Lemma which will prove useful :

LEMMA 4.1.

$\langle f(\mu),\phi\rangle$ remains unchanged if, in (4.17) the supremum over $\mathscr{D}_f(\mathscr{C}_0)$ is replaced by the supremum over $\mathscr{D}_f(\mathscr{C}_0^\infty)$ or over $\mathscr{D}_f(L^1_\mu)$.

<u>Proof</u>

As $\mathscr{D}_f(\mathscr{C}_0^\infty) \subset \mathscr{D}_f(\mathscr{C}_0) \subset \mathscr{D}_f(L^1_\mu)$, it will be sufficient to prove that the two last-mentioned suprema are equal. To do this let v be given in $\mathscr{D}_f(L^1_\mu)$; we must approximate

$$\Lambda(v) = \int_\Omega (v\phi)\mu - \int_\Omega f^*(v)\phi\, dx \ , \tag{4.23}$$

by expressions $\Lambda(w)$, $w \in \mathscr{D}_f(\mathscr{C}_0^\infty)$. Let θ_r be an increasing sequence of functions of class $\mathscr{C}_0^\infty(\Omega)$, satisfying $0\leq\theta_r\leq 1$, and converging (simply) to 1 as $r \to \infty$. By convexity and from (4.9) ,

$$0 \leq f^*(\theta_r v) \leq \theta_r\, f^*(v) + (1-\theta_r)\, f^*(0) \leq f^*(v) \ .$$

Lebesgue's theorem then shows that $\Lambda(\theta_r v) \to \Lambda(v)$, when $r \to \infty$ and we can therefore suppose that $v \in \mathscr{D}_f(L^1_\mu)$ has compact support in Ω. The function v being extended to vanish outside Ω we now consider the regularised functions $\rho_\eta * v$, where

$$\rho_\eta(x) = \frac{1}{\eta^n}\rho\left(\frac{x}{\eta}\right)$$

$\rho \in \mathscr{C}_0^\infty(\mathbb{R}^n)$, $0 \leq \rho \leq 1$, $\int_{\mathbb{R}^n} \rho(x)dx = 1$, $\rho(x) = 0$ for $|x| \geq 1$ and, to simplify, $\rho(-x) = \rho(x)$. The functions $\rho_\eta * v$ vanish outside Ω as soon as η is smaller than the distance between $\partial\Omega$ and the support of v. Thus by convexity and thanks to (1.9),

$$0 \leq f^*(\rho_\eta * v) \leq \rho_\eta * f^*(v) . \tag{4.24}$$

It follows that $\rho_\eta * v \in \mathscr{D}_f(\mathscr{C}_0^\infty)$ and we have

$$\Lambda(\rho_\eta * v) = \int_\Omega (\rho_\eta * v)\phi\, \mu - \int_\Omega f^*(\rho_\eta * v)\phi\, dx$$

$$\geq \int_\Omega v(\rho_\eta * (\mu\phi)) - \int_\Omega \rho_\eta * f^*(v)\phi\, dx .$$

When $\eta \to 0$, this minorant of $\Lambda(\rho_\eta * v)$ converges to $\Lambda(v)$, whence

$$\liminf_{\eta \to 0} \Lambda(\rho_\eta * v) \geq \Lambda(v) . \tag{4.25}$$

This shows that the infima over $\mathscr{D}_f(\mathscr{C}_0^\infty)$ and $\mathscr{D}_f(L^1_\mu)$ are equal, thus proving the lemma.

□

We shall now show that $\langle f(\mu), \phi \rangle$ is an additive function of ϕ

LEMMA 4.2.

If f satisfies (4.1) (4.6) (4.8), $\mu \in M_1$ and $\phi_1, \phi_2 \in \mathscr{C}_o(\Omega)$, $\phi_i \geq 0$, $i = 1,2$ then

$$\langle f(\mu),\ \phi_1+\phi_2\rangle = \langle f(\mu),\ \phi_1\rangle + \langle f(\mu),\ \phi_2\rangle\ .$$

Proof

The inequality

$$\langle f(\mu),\phi_1+\phi_2\rangle \leq \langle f(\mu),\phi_1\rangle + \langle f(\mu),\phi_2\rangle\ .$$

is obvious. We shall show that the opposite inequality holds with a correction term $\eta>0$, which can be made arbitrarily small. By the definition of $\langle f(\mu),\ \phi_i\rangle$, $i = 1,2$ there exist v_1 and v_2 in $\mathscr{C}_0$, such that, for $i = 1,2$:

$$(4.26)\qquad \langle f(\mu),\phi_i\rangle - \eta \leq \int_\Omega (v_i\ \phi_i)\mu - \int_\Omega f^*(v_i)\phi_i\ dx\ .$$

Accordingly let v be the function from Ω into X which equals 0 when $\phi_1 + \phi_2 = 0$ and which equals

$$\frac{v_1(x)\ \phi_1(x) + v_2(x)\ \phi_2(x)}{\phi_1(x) + \phi_2(x)} \quad \text{if} \quad \phi_1(x) + \phi_2(x) \neq 0\ .$$

Clearly v is μ-measurable, and since $|v(x)| \leq |v_1(x)| + |v_2(x)|$ for all $x \in \Omega$, it follows that $v \in L^1(\Omega,\mu;X)$ and that

$$\int_\Omega (\phi_1+\phi_2)v\ \mu = \sum_{i=1}^{2} \int_\Omega v_i\ \phi_i\ \mu\ .$$

Let $\Omega_0 = \{x \in \Omega,\ \phi_1(x) + \phi_2(x) > 0\}$. For all x of Ω_0, we have, since f^* is convex and $f^* \geq 0$.

$$0 \leq f^*(v(x)) \leq \frac{\phi_1(x)}{\phi_1(x)+\phi_2(x)}\ f^*(v_1(x)) + \frac{\phi_2(x)}{\phi_1(x)+\phi_2(x)}\ f^*(v_2(x))$$

$$\leq f^*(v_1(x)) + f^*(v_2(x))\ .$$

On $\Omega\setminus\Omega_0$, $f^*(v(x)) = 0$ since $f^*(0) = 0$. It follows that

$f^*\circ v \in L^1(\Omega)$ and thus $v \in \mathscr{D}_f(L^1_\mu)$. The preceding inequalities show that

$$\int_\Omega f^*(v)(\phi_1+\phi_2)dx = \int_{\Omega_0} f^*(v)(\phi_1+\phi_2)dx \leq \sum_{i=1}^{2} \int_{\Omega_0} f^*(v_i)\ \phi_i\ dx$$

$$= \sum_{i=1}^{2} \int_\Omega f^*(v_i)\phi_i\ dx\ ,$$

and, bearing in mind Lemma 4.1, we obtain the stated inequality

$$\langle f(\mu),\phi_1+\phi_2\rangle \geq \int_\Omega v(\phi_1+\phi_2)\mu - \int_\Omega f^*(v)(\phi_1+\phi_2)dx$$

$$\geq \sum_{i=1}^{2} \{\int_\Omega v\ \phi_i\mu - \int_\Omega f^*(v_i)\phi_i\ dx\}$$

$$\geq \sum_{i=1}^{2} \langle f(\mu),\phi_i\rangle - 2\eta\ .$$

□

If now $\phi \in \mathscr{C}_0(\Omega)$, we can express ϕ in the form $\phi = \phi^+ - \phi^-$ with

$$\phi^+ = \max(\phi,0)\ ,\quad \phi^- = \max(-\phi,0)\ ,$$

and define $\langle f(\mu), \phi\rangle$ by

$$\langle f(\mu),\phi\rangle = \langle f(\mu),\phi^+\rangle - \langle f(\mu),\phi^-\rangle\ . \tag{4.27}$$

Note that, in view of Lemma 4.1, we can replace ϕ^+ and ϕ^- in (4.27) by any pair of non-negative functions ϕ_1 and ϕ_2 in $\mathscr{C}_0(\Omega)$, such that $\phi = \phi_1 - \phi_2$: in fact we then have $\phi_1 + \phi^- = \phi_2 + \phi^+$ and by Lemma 4.1

$$\langle f(\mu),\phi_1+\phi^-\rangle = \langle f(\mu),\phi_1\rangle + \langle f(\mu),\phi^-\rangle = \langle f(\mu),\phi_1+\phi^+\rangle$$

$$= \langle f(\mu),\phi_2\rangle + \langle f(\mu),\phi^+\rangle\ ,$$

(4.28) $$\langle f(\mu),\phi\rangle = \langle f(\mu),\phi_1\rangle - \langle f(\mu),\phi_2\rangle \ .$$

THEOREM 4.1.

Let f be a convex function from X into $\mathbb{R}$ which satisfies (4.1) (4.6) (4.8) and let $\mu \in M_1(\Omega:X)$. Then $f(\mu)$, defined by (4.17) and (4.27) is a positive bounded Radon measure on Ω, which is absolutely continuous with respect to μ.

Proof.

We have to show that $\phi \mapsto \langle f(\mu),\phi\rangle$ is a positive linear form on $\mathscr{C}_0(\Omega)$. If $\phi_1,\phi_2 \in \mathscr{C}_0(\Omega)$, $\phi = \phi_1 + \phi_2$, then

$$\phi_i = \phi_i^+ - \phi_i^- \ , \ i = 1,2 \ ;$$

$$\phi = (\phi_1^+ + \phi_2^+) - (\phi_1^- + \phi_2^-)$$

and by (4.28) and Lemma 4.2,

$$\begin{aligned}\langle f(\mu),\phi\rangle &= \langle f(\mu),\phi_1^+ + \phi_2^+\rangle - \langle f(\mu),\phi_1^- + \phi_2^-\rangle \\ &= \sum_{i=1}^{2} \{\langle f(\mu),\phi_i^+\rangle - \langle f(\mu),\phi_i^-\rangle\} \\ &= \langle f(\mu),\phi_1\rangle + \langle f(\mu),\phi_2\rangle \ .\end{aligned}$$

On the other hand if $\lambda \in \mathbb{R}$ and $\phi \in \mathscr{C}_0(\Omega)$, we need to show that

(4.29) $$\langle f(\mu),\lambda\phi\rangle = \lambda\langle f(\mu),\phi\rangle \ .$$

By writing $\phi = \phi^+ - \phi^-$ and treating separately the cases $\lambda \geq 0$ and $\lambda < 0$, we have only to prove (4.29) for the case $\lambda \geq 0$, $\phi \geq 0$, and here the proof is obvious.

We know from (4.20) that $f(\mu) \geq 0$. This implies that the linear form $\phi \mapsto \langle f(\mu),\phi\rangle$ is continuous on $\mathscr{C}_0(\Omega)$, and so $f(\mu)$ is a Radon measure. Furthermore, by (4.20) and (4.28) we have, for all $\phi \in \mathscr{C}_0(\Omega)$,

(4.30) $$|\langle f(\mu),\phi\rangle| \leq \langle f(\mu),|\phi|\rangle \leq k_1 \int_\Omega |\phi||\mu| \ ;$$

which shows that $f(\mu)$ is absolutely continuous with respect to μ, thus completing the proof of the theorem.

□

Remark 4.1. The function $f(\xi) = |\xi|$ satisfies the condition (4.1) (4.6) (4.8). As f^* is the characteristic function of the ball $|\xi| \leq 1$, it follows from (4.17) that $f(\mu)$ is, in this case, the measure $|\mu|$ in the ordinary sense of measure theory.

Similarly if $X = \mathbb{R}$ and $f(s) = s^+$ or s^-, the conditions (4.1) (4.6) and (4.8) are all satisfied and the measures $f(\mu)$ coincide with the usual measures μ^+ and μ^-. Thus for example in the case of s^+ we note that $f^*(s)$ is the indicator function of $[0,1]$ and, by (4.20) we have, for all ϕ in $\mathscr{C}_0(\Omega)$, with $\phi \geq 0$,

$$\langle f(\mu),\phi\rangle = \sup_{\substack{v\in\mathscr{C}_0(\Omega) \\ 0\leq v(x)\leq 1}} \int_\Omega v\phi\ \mu = \int_\Omega \phi\ \mu^+ .$$

□

We now turn to the definition of $\langle f(\mu),\phi\rangle$ when f satisfies (4.1) and (4.6) but not (4.8) : we use a decomposition of f of the type (4.10), $f = g + \ell$, where ℓ is a continuous affine function on X and g a convex function satisfying (4.1) (4.6) (4.8) (for example the function given by (4.11) with $\xi_0 = 0$) . We then put

(4.31) $$f(\mu) = g(\mu) + \ell(\mu) .$$

Clearly $f(\mu)$ is a bounded Radon measure on Ω which is absolutely continuous with respect to μ ; it only remains to be verified that (4.31) intrinsically defines $f(\mu)$, or in other words that $f(\mu)$ does not depend on the choice of g and ℓ in (4.10). To this end we shall show that, for all ϕ in $\mathscr{C}_0(\Omega)$, with $\phi \geqq 0$:

(4.32) $$f(\mu) = \sup_{v\in\mathscr{D}_f(L^1_\mu)} \{\int_\Omega (v\phi)\mu - \int_\Omega f^*(v)\phi\, dx\} .$$

Writing $\ell(\xi) = a.\xi + b$, $a \in X$, $b \in \mathbb{R}$ we have by (4.31), (4.17) and Lemma 4.1 :

$$\langle f(\mu),\phi\rangle = \sup_{v \in \mathscr{D}_g(L^1_\mu)} \{\int_\Omega (v\phi)\mu - \int_\Omega g^*(v)\phi \, dx\}$$

$$+ a \int_\Omega \phi\mu + b \int_\Omega \phi \, dx .$$

But $g^*(\eta) = f^*(a+\eta) + b$, $\forall\ \eta \in X$ and so $\mathscr{D}_g(L^1_\mu) = \mathscr{D}_f(L^1_\mu) - a$ and $\langle f(\mu),\phi\rangle$ is also equal to

$$\sup_{v+a \in \mathscr{D}_f(L^1_\mu)} \{\int_\Omega (v+a)\phi \ \mu - \int_\Omega f^*(v+a)\phi \, dx\} ,$$

which is precisely the same as the right-hand side of (4.32).

□

It follows from (4.32) that if f_1 and f_2 are two convex functions satisfying (4.1) and (4.6) and

$$f_1 \leqq f_2 , \tag{4.33}$$

then

$$f_1(\mu) \leqq f_2(\mu) . \tag{4.34}$$

This follows immediately from (4.32) and the fact that $f_1^* \geqq f_2^*$. If $f_1 = f$ and $f_2 = k_1(1+|\xi|)$, then by (4.1)

$$f(\mu) \leqq k_1(1+|\mu|) . \tag{4.35}$$

Now by (4.31) and as $g(\mu) \geqq 0$, we have

$$f(\mu) \geqq \ell(\mu) = a_0\mu - f^*(a_0) . \tag{4.36}$$

But since $a_0 \in \text{dom}\ f^*$, we have because of (4.4) and (4.6)

(4.37) $$f(\mu) \geq -k_1|\mu| - k_2 .$$

4.3 Lebesgue decomposition of $f(\mu)$

Let f be a convex function satisfying (4.1) and (4.6) and μ belong to M_1 ; we wish to determine the Lebesgue decomposition of $f(\mu)$, knowing that of μ :

(4.38) $$\mu = h\,dx + \mu^S ,$$

where $h \in L^1$ and μ^S is singular. We begin with the case where $\mu^S = 0$.

LEMMA 4.3

If μ is a function, $\mu = h\,dx$, then

(4.39) $$f(\mu) = f\circ h\,dx .$$

Proof

We may suppose that f satisfies (4.8) and we have to show that

(4.40) $$\langle f(\mu),\phi\rangle = \{\int_\Omega f(h(x))\phi(x)\,dx\} ,$$

when $\phi \in \mathscr{C}_0(\Omega)$, $\phi \geq 0$.

Since $f^* \geq 0$, $\int_\Omega f^*(v)\phi\,dx = +\infty$, if $v \in L^1 = L^1(\Omega;X)$ and $v \in \mathscr{D}_f(L^1_\mu)$. We can then replace in (4.32), the supremum over $\mathscr{D}_f(L^1_\mu)$ by the supremum over L^1 and write

$$\langle f(\mu),\phi\rangle = \sup_{v\in L^1} \{\int_\Omega (vh - f^*(v))\phi\,dx\} .$$

If one considers the function θ defined in $\Omega \times X$, by

$$\theta(x,\xi) = \begin{cases} (f^*(\xi)-\xi h(x))\ \phi(x) & \text{if} \quad \xi \in K \\ +\infty & \text{if} \quad \xi \notin K \end{cases}$$

then

(4.41) $$\langle f(\mu),\phi\rangle = \operatorname*{Sup}_{v\in L^1} \{-\int_\Omega \theta(x,v(x))dx\} \ .$$

The function θ is a Borel function from $\Omega \times X$ into $\mathbb{R}$ ([1]) and, for almost all x in Ω, $\xi \mapsto \theta(x,\xi)$ is l.s.c. It therefore satisfies the conditions (2.58) and (2.59) of Chapter I, and so is a normal integrand on $\Omega \times X$. The condition I.(2.57) is satisfied thanks to (4.4) and (4.5) :

$$\theta(x,\xi) \geq -k_1(1+|h(x)|)\ \phi(x) \ .$$

Thus it follows from (2.60) that the supremum in (4.41) is equal to

$$\int_\Omega \theta^*(x,0)\ dx \ ,$$

where θ^* is the function conjugate to θ with respect to ξ :

$$\theta^*(x,\eta) = \operatorname*{Sup}_{\xi} \{\eta.\xi - \theta(x,\xi)\} \ ,$$

and in particular

$$\theta^*(x,0) = \operatorname*{Sup}_{\xi\in K} \{(\xi\ h(x) - f^*(\xi))\ \phi(x)\} \ .$$

This supremum has the value $\phi(x)\ f(h(x))$ and hence (4.40) follows. □

We now turn to the general case. We shall make use of the function f_∞ associated with f by (4.15) ; this function satisfies the conditions

---

([1]) Cf. for example I. Ekeland - R. Temam [1], Theorem 1.1, Ch.VIII.

(4.1) and (4.6) and one can therefore define $f_\infty(\mu)$.

THEOREM 4.2

Let f be a convex function mapping X into $\mathbb{R}$ which satisfies (4.1) (4.6) and let $\mu \in M_1(\Omega;X)$ whose Lebesgue decomposition can be written in the form (4.38). Then the Lebesgue decomposition of $f(\mu)$ can be written

$$f(\mu) = f \circ h\, dx + f_\infty(\mu^S)\ . \tag{4.42}$$

Proof

i) We first prove the result assuming that f satisfies (4.8). It will be sufficient to show that

$$\langle f(\mu),\phi\rangle = \langle f\circ h\ dx,\phi\rangle + \langle f_\infty(\mu^S),\phi\rangle\ , \tag{4.43}$$

for all $\phi \geq 0$ in $\mathscr{C}_0(\Omega)$.

The inequality :

$$\langle f(\mu),\phi\rangle \leq \langle f\circ h\ dx,\phi\rangle + \langle f_\infty(\mu^S),\phi\rangle\ .$$

is easily proved, because by (4.17) and (4.38)

$$\begin{aligned}\langle f(\mu),\phi\rangle &\leq \operatorname*{Sup}_{v\,\in\,\mathscr{D}_f(\mathscr{C}_0)}\{\int_\Omega (v\phi)\mu^S\} + \operatorname*{Sup}_{v\,\in\,\mathscr{D}_f(\mathscr{C}_0)}\{\int_\Omega (vh-f^*(v))\phi\ dx\}\\ &\leq \operatorname*{Sup}_{v\,\in\,\mathscr{D}_{f_\infty}(\mathscr{C}_0)}\{\int_\Omega (v\phi)\mu^S - \int_\Omega f_\infty^*(v)\phi\ dx\} + \langle f(h\ dx),\phi\rangle\\ &\leq \langle f_\infty(\mu^S),\phi\rangle + \langle f(h\ dx),\phi\rangle\ \ ;\end{aligned}$$

we have here used the fact that $\operatorname{dom} f_\infty^* = \operatorname{dom} f^* = K$ and that $f_\infty^*(\xi) = 0$ for $\xi \in K$.

We next prove that the inequality in the opposite direction holds to within η . For any arbitrarily small $\eta > 0$ there exist two functions v_1 in $\mathscr{D}_{f_\infty}(\mathscr{C}_0)$ and v_2 in $\mathscr{D}_f(\mathscr{C}_0)$ such that

$$\langle f_\infty(\mu^S),\phi\rangle - \eta \leq \int_\Omega (v_1\phi)\mu^S$$

$$\langle f(h\,dx),\phi\rangle - \eta \leq \int_\Omega (v_2 h - f^*(v_2))\phi\,dx\ .$$

The sets $\mathscr{D}_f(\mathscr{C}_0)$ and $\mathscr{D}_{f_\infty}(\mathscr{C}_0)$ are in fact the same,

$$\mathscr{D}_f(\mathscr{C}_0) = \mathscr{D}_{f_\infty}(\mathscr{C}_0) = \{v \in \mathscr{C}_0(\Omega;X), v(x) \in K,\ \ \forall\ x\}\ .$$

Since μ^S is singular, there exists an open $\mathscr{O} \subset \Omega$ of arbitrarily small measure on which μ^S is concentrated. Let

$$\theta = (v_1-v_2)h - (f^*(v_1)-f^*(v_2)) \in L^1(\Omega,X)\ .$$

We choose the open set $\mathscr{O}$ with a measure small enough so that

$$|\int_{\mathscr{O}} \theta\phi\,dx| \leq \eta\ , \tag{4.44}$$

and define the function $\bar{v}$ by :

$$\bar{v} = \begin{cases} v_1 & \text{on } \mathscr{O} \\ v_2 & \text{on } \Omega\backslash\mathscr{O} \end{cases}$$

Clearly $\bar{v}$ is μ-measurable and belongs to $\mathscr{D}_f(L^1_\mu)$ and so by Lemma 4.1

$$\int_\Omega (\bar{v}\phi)\mu^S + \int_\Omega \bar{v}\phi h\,dx - \int_\Omega f^*(v)\phi\,dx \leq \langle f(\mu),\phi\rangle\ .$$

Now the left-hand side of this inequality is equal to

$$\int_\Omega v_1\,\phi\,\mu^S + \int_{\mathscr{O}} v_1\,\phi h\,dx + \int_{\Omega\backslash\mathscr{O}} v_2\,\phi h\,dx - \int_{\mathscr{O}} f^*(v_1)\phi\,dx - \int_{\Omega\backslash\mathscr{O}} f^*(v_2)\phi\,dx$$

$$= \int_\Omega v_1\,\phi\,\mu^S + \int_\Omega (v_2 h - f^*(v_2))\phi\,dx + \int_{\mathscr{O}} \theta\phi\,dx$$

$$\geq \langle f_\infty(\mu^S),\phi\rangle + \langle f(h\,dx),\phi\rangle - 3\eta\ ,$$ by the definition of v_1 and v_2 and (4.44).

The equation (4.43) has thus been proved.

ii) When f does not satisfy (4.8) we use the decomposition (4.10) and (4.11) of f. By (4.31) and what precedes,

$$f(\mu) = g(\mu) + a_0\mu - f^*(a_0)$$
$$= g(h\,dx) + a_0 h\,dx + g_\infty(\mu^S) + a_0\mu^S - f^*(a_0)$$

But, by (4.13)

$$g_\infty(\xi) = \operatorname*{Sup}_{\eta\in \operatorname{dom} g^*} \xi.\eta = -\xi a_0 + \operatorname*{Sup}_{\eta\in\operatorname{dom} f^*} \xi.\eta$$
$$= -\xi a_0 + f_\infty(\xi) .$$

and hence (4.42) is proved in this case.

To conclude we note that $f(\mu^S)$ is singular (in every case) and is concentrated on the same set as μ^S since, by (4.32) $\langle f_\infty(\mu^S), \phi\rangle = 0$ whenever $\phi \in \mathscr{C}_0(\Omega)$ and $\phi = 0$ on the support of μ^S.

The theorem has therefore been proved.

□

4.4 Complements

We end this Section by a few further results.

Since $f(\mu)$ is a bounded Radon measure, the integral $\int_\Omega \phi\, f(\mu)$ has a meaning for any ϕ in $\mathscr{C}(\bar\Omega)$ (or even in $L^\infty(\Omega,\mu)$) and when $\phi \in \mathscr{C}_0(\Omega)$, $\phi \geq 0$, this integral, equal to $\langle f(\mu),\phi\rangle$ is given by (4.17). We shall show that the formula (4.17) and those resulting from Lemma 4.1 extend to any nonnegative function in $\mathscr{C}(\bar\Omega)$.

LEMMA 4.4

For any nonnegative function ϕ *in* $\mathscr{C}(\bar\Omega)$,

$$\int_\Omega \phi\, f(\mu) = \operatorname*{Sup}_{v} \left\{\int_\Omega (v\phi)\mu - \int_\Omega f^*(v)\phi\, dx\right\}, \tag{4.45}$$

where it is indifferent whether the supremum be taken over all v *in* $\mathcal{D}_f(\mathcal{C}_0^\infty)$, *or all* v *in* $\mathcal{D}_f(\mathcal{C}_0)$, *or all* v *in* $\mathcal{D}_f(L^1_\mu)$.

Proof

i) We first establish the result for $\mathcal{D}_f(L^1_\mu)$. Let $\psi \in \mathcal{C}_0^\infty(\Omega)$, with $0 \leqslant \psi \leqslant 1$ and $\psi = 1$ on a relatively compact subset Ω' of Ω . We have

$$\int_\Omega \phi\, f(\mu) = \int_\Omega \psi\phi\, f(\mu) + \int_\Omega (1-\psi)\phi\, f(\mu)$$

$$= \sup_{v\in\mathcal{D}_f(L^1_\mu)} \{\int_\Omega (v\psi\phi)\mu - \int_\Omega f^*(v)\psi\phi\, dx\} + \int_\Omega (1-\psi)\phi\, f(\mu)$$

(by Lemma 4.1)

Hence, writing

$$I = \int_\Omega \phi\, f(\mu) - \sup_{v\in\mathcal{D}_f(L^1_\mu)} \{\int_\Omega v\phi\mu - \int_\Omega f^*(v)\phi\, dx\} ,$$

$$I \leqslant \int_\Omega (1-\psi)\phi\, f(\mu) + \sup_{v\in\mathcal{D}_f(L^1_\mu)} \{\int_\Omega (\psi-1)\, v\phi\mu - \int_\Omega (\psi-1)\, f^*(v)\phi\, dx\} .$$

We have $v(x) \in K$ and $|v(x)| \leqslant k_1$ p.p. when $v \in \mathcal{D}_f(L^1_\mu)$ (cf. (4.4)); by using (4.5) and (4.35) we can write

$$I \leqslant 2k_1 \int_\Omega (1-\psi)\phi\, |\mu| + 2k_1 \int_\Omega (1-\psi)\phi\, dx .$$

The right-hand side of this inequality can be made arbitrarily small by choosing a large enough Ω' and we conclude that $I \leqslant 0$.

We could show in exactly the same way that $I \geqslant 0$ and thus (4.45) holds for the $\mathcal{D}_f(L^1_\mu)$ case.

ii) To establish that the supremum on $\mathcal{D}_f(\mathcal{C}_0^\infty)$ (and hence on $\mathcal{D}_f(\mathcal{C}_0)$) is the same as on $\mathcal{D}_f(L^1_\mu)$, we have only to repeat the argument of Lemma 4.2 which is just as valid for ϕ in $\mathcal{C}_0(\Omega)$ as it is for ϕ in $\mathcal{C}(\bar{\Omega})$.

The Lemma 4.4 has been proved. □

We write $\theta_{\mathcal{O}}$ for the characteristic function of a set $\mathcal{O} \subset \Omega$. If $\mathcal{O}$ is a Borel subset of Ω and $\mu \in M_1 = M_1(\Omega;X)$, we can define two Radon measures $f(\theta_{\mathcal{O}}\mu)$ and $\theta_{\mathcal{O}}f(\mu)$. If $f(0) = 0$ and μ is a function these two measures coincide. We extend this to any measure μ.

PROPOSITION 4.1

Let f *be a convex function satisfying (4.1) (4.6) and*

$$f(0) = 0 . \tag{4.46}$$

Then for any measure $\mu \in M_1$ *and any Borel set* $\mathcal{O} \subset \Omega$,

$$f(\theta_{\mathcal{O}}\mu) = \theta_{\mathcal{O}} f(\mu) . \tag{4.47}$$

By (4.31) it will be sufficient to prove this when f also satisfies (4.8) which we shall now assume.

LEMMA 4.5.

For a given μ *in* M_1 *and* f *satisfying (4.1) (4.6) (4.8), the function* $\mathcal{O} \mapsto f(\theta_{\mathcal{O}}\mu)$ *is an additive increasing function of* $\mathcal{O}$ *(* $\mathcal{O}$ *being a Borel subset of* Ω*)*.

Proof

Since $f \geq 0$ it will be sufficient to prove that f is additive. Let $\mathcal{O}_1$ and $\mathcal{O}_2$ be two disjoint Borel subsets of Ω and let $\mathcal{O} = \mathcal{O}_1 \cup \mathcal{O}_2$. We need to verify that for any $\phi \in \mathscr{C}_0(\Omega)$, with $\phi \geq 0$.

$$\int_\Omega \phi\, f(\theta_{\mathcal{O}}\, \mu) = \int_\Omega \phi\, f(\theta_{\mathcal{O}_1}\mu) + \int_\Omega \phi\, f(\theta_{\mathcal{O}_2}\, \mu) . \tag{4.48}$$

For a fixed $\eta > 0$, there exists a $v \in \mathscr{C}_0(\Omega)$ such that

$$\int_\Omega \phi f(\theta_{\mathcal{O}}\, \mu) - \eta \leq \int_\Omega v\phi\, \theta_{\mathcal{O}}\, \mu - \int_\Omega f^*(v)\phi\, dx .$$

As $f^*(\theta_{\mathcal{O}}\, v) \leq f^*(v)$, the right-hand side of the above inequality is less than or equal to

$$\int_\Omega v_1\phi\, \theta_{\mathcal{O}_1}\mu + \int_\Omega v_2\phi\, \theta_{\mathcal{O}_2}\mu - \int_\Omega f^*(v_1)\phi\, dx - \int_\Omega f^*(v_2)\phi\, dx ,$$

(where $v_i = v\theta_{\mathcal{O}_i}$) and this is majorised by the right-hand side of (4.48).

For the inequality in the opposite direction, we note that there exist $v_i \in \mathcal{D}_f(\mathcal{C}_0)$, for $i = 1,2$, such that

$$\int_\Omega \phi\, f(\theta_{\mathcal{O}_i}\mu) - \eta \leq \int_\Omega v_i\phi\, \theta_{\mathcal{O}_i}\mu - \int_\Omega f^*(v_i)\phi\, dx ,$$

for $i = 1,2$. We write v for the function $\theta_{\mathcal{O}_1} v_1 + \theta_{\mathcal{O}_2} v_2$ which belongs to the class $\mathcal{D}_f(L^1_\mu)$ so that :

$$\int_\Omega \phi\, f(\theta_{\mathcal{O}_1}\mu) + \int_\Omega \phi\, f(\theta_{\mathcal{O}_2}\mu) - 2\eta \leq \int_\Omega v\phi\mu - \int_\Omega f^*(v)\phi\, dx \leq \int_\Omega \phi f(\theta_{\mathcal{O}}\,\mu) ,$$

and the result follows since η may be chosen arbitrarily small.

LEMMA 4.6.

With the hypotheses of Lemma 4.5, for any Borel subset $\mathcal{O} \subset \Omega$:

$$f_\infty(\theta_{\mathcal{O}}\,\mu) = \theta_{\mathcal{O}}\, f_\infty(\mu) . \tag{4.49}$$

Proof

We have to prove that for any ϕ in $\mathcal{C}_0(\Omega)$, with $\phi \geqq 0$,

$$\int_{\mathcal{O}} \phi\, f_\infty(\mu) = \operatorname*{Sup}_{v\in\mathcal{D}_{f_\infty}(\mathcal{C}_0)} \int_\Omega v\,\phi\,\theta_{\mathcal{O}}\,\mu .$$

We begin by examining the case where $\mathcal{O}$ is open. We saw, in proving Theorem 4.2 that

$$\mathscr{D}_{f_\infty}(\mathscr{C}_0) = \mathscr{D}_f(\mathscr{C}_0) = \{v \in \mathscr{C}_0(\Omega;X) \ , \ v(x) \in K \ , \ \forall\, x \in \Omega\} \ .$$

Since $\mathscr{O}$ is open, there exists an increasing sequence of functions θ_j in $\mathscr{C}_0^\infty(\mathscr{O})$ with $0 \leqq \theta_j \leqq 1$, which converge to $\theta_{\mathscr{O}}$ when $j \to \infty$. Thus $\int_\Omega \phi\, \theta_j f_\infty(\mu)$ converges to $\int_\Omega \phi\, \theta_{\mathscr{O}} f_\infty(\mu)$ when $j \to \infty$ and

$$\left|\int_\Omega \phi\, \theta_j\, f_\infty(\mu) - \sup_{v \in \mathscr{D}_f(\mathscr{C}_0)} \int_\Omega v\phi\, \theta_{\mathscr{O}}\, \mu\right| \leqslant$$

$$\leqslant \sup_{v \in \mathscr{D}_f(\mathscr{C}_0)} \left\{\int_\Omega \phi |v|\, |\theta_j - \theta_{\mathscr{O}}|\, |\mu|\right\}$$

$$\leqslant k_1 \int_\Omega \phi |\theta_j - \theta_{\mathscr{O}}|\, |\mu| \ ,$$

and this last term tends to 0 as $j \to \infty$. .

With (4.49) having been established for an open $\mathscr{O}$, Lemma 4.5 applied to f_∞ implies that (4.49) also holds good when $\mathscr{O}$ is a closed subset of Ω :

$$f_\infty(\theta_{\mathscr{O}}\, \mu) = f_\infty(\mu) - f_\infty(\theta_{\Omega\setminus\mathscr{O}}\mu) = \theta_{\mathscr{O}}\ f_\infty(\mu) \ .$$

Now let $\mathscr{O}$ be a Borel subset of Ω, A a compact set, B an open set, such that $A \subset \mathscr{O} \subset B$. From the foregoing and from Lemma 4.5 applied to f_∞, we have

$$\theta_A\ f_\infty(\mu) = f_\infty(\theta_A\mu) \leqq f_\infty(\theta_{\mathscr{O}}\, \mu) \leqq f_\infty(\theta_B\mu) = \theta_B\ f_\infty(\mu) \ .$$

For any $\phi \geqq 0$ in $\mathscr{C}_0(\Omega)$ we have therefore

$$\int_A \phi\ f_\infty(\mu) \leqq \int_\Omega \phi\ f_\infty(\theta_{\mathscr{O}}\, \mu) \leqq \int_B \phi\ f_\infty(\mu) \ .$$

By taking the supremum in A and the infimum in B, we obtain

$$\int_{\Omega} \phi \, f_\infty(\theta_{\mathcal{O}} \mu) = \int_{\mathcal{O}} \phi \, f_\infty(\mu) \;, .$$

which proves (4.49).

Proof of Proposition 4.1

We now complete the proof of Proposition 4.1.

If $\mathcal{O}$ is a Borel subset of Ω then by Theorem 4.2 and Lemma 4.6,

$$
\begin{aligned}
f(\theta_{\mathcal{O}} \mu) &= f(\theta_{\mathcal{O}} h \, dx) + f_\infty(\theta_{\mathcal{O}} \mu^S) \\
&= \theta_{\mathcal{O}} f \circ h \, dx + \theta_{\mathcal{O}} \; f_\infty(\mu^S) \\
&= \theta_{\mathcal{O}} f(\mu) \;,
\end{aligned}
\tag{4.50}
$$

whence (4.47).

5. Convex functionals of a measure

In this section we study convex functions defined on a measure space with values in $\mathbf{R}$ or $\mathbf{R} \cup \{+\infty\}$. Such convex functions have been studied in the past by various authors: see in particular H. Brezis [3], C. Goffman - J. Serrin [1], R.T. Rockafellar [3][4], R. Temam [5] [11]. In the above-mentioned references these convex functions are defined, either directly, or by duality. The presentation which we give here is different: it involves, as in F. Demengel - R. Temam [1], defining a functional in the form of the integral over Ω (or a Borel subset of Ω) of the measure $f(\mu)$ defined in Section 4. These functionals will be defined and studied in Section 5.1.

In Sections 5.2 and 5.3 we shall be concerned with approximation problems: for example the problem of approximating to a measure μ by regular functions u_m in such a way that $f(u_m)$ converges to $f(\mu)$ and $\int_\Omega f(u_m)\,dx$ to $\int_\Omega f(\mu)$; this is not a standard problem because the convergence of the u_m to μ is necessarily weak when μ is not a function. Results of this type form the object of Section 5.3 and we give similar results for functions defined on the space $BD(\Omega)$ studied in Sections 2 and 3.

5.1. Definition of convex functions

As in Section 4, X denotes a finite-dimensional Euclidean space.

Let f be a convex function mapping X into $\mathbb{R}$ which satisfies (4.1) and (4.6) and let $\mu \in M_1(\Omega;X)$. We write

$$(5.1) \qquad F(\mu) = \int_\Omega f(\mu) \ ,$$

where $f(\mu)$ is the Radon measure defined in Section 4. It follows immediately from Lemma 4.4 that

$$(5.2) \qquad F(\mu) = \operatorname*{Sup}_v \left\{ \int_\Omega v\mu - \int_\Omega f^*(v)\, dx \right\}$$

where the supremum can be taken indifferently over any one of the three sets $\mathcal{D}_f(\mathcal{C}_0^\infty)$, $\mathcal{D}_f(\mathcal{C}_0)$, $\mathcal{D}_f(L^1_\mu)$. By Lemma 4.3 if μ is a function, $\mu = h\,dx$, $h \in L^1$, and we then have simply

$$(5.3) \qquad F(\mu) = \int_\Omega f(h(x))dx \ .$$

In the general case, μ admits of a Lebesgue decomposition of the form

$$(5.4) \qquad \mu = h\,dx + \mu^S \ ,$$

where $h \in L^1$ and μ^S is singular. Then by Theorem 4.2, $F(\mu)$ can be written in the form

$$(5.5) \qquad F(\mu) = F(h\,dx) + F_\infty(\mu^S) = \int_\Omega f(h(x))dx + \int_\Omega f_\infty(\mu^S) \ ,$$

where f_∞ is defined in (4.15) and F_∞ is the functional associated with it by (5.1).

Clearly $F(\mu) \in \mathbb{R}$ for all μ of M_1 and by (4.35) and (4.37)

$$(5.6) \qquad -k_1 \int_\Omega (|\mu|-1) \leq F(\mu) \leq k_1 \int_\Omega (|\mu|+1) \ .$$

The function $\mu \mapsto F(\mu)$ is , by (5.2), convex and l.s.c. on $M_1(\Omega;X)$; as its domain is the whole space M_1 , it is also continuous on that space (cf. Ch.I, Section 2.1). By (5.2) it is also l.s.c. for the vague topology of the measure space. □

If now $\mathcal{O}$ is a Borel subset of Ω, we can associate with f the functional $F_{\mathcal{O}}$ defined by

$$F_{\mathcal{O}}(\mu) = \int_{\mathcal{O}} f(\mu) . \tag{5.7}$$

If we further suppose that

$$f(0) = 0 , \tag{5.8}$$

then by Proposition 4.1,

$$F_{\mathcal{O}}(\mu) = F(\theta_{\mathcal{O}}\mu) , \tag{5.9}$$

where $\theta_{\mathcal{O}}$ is the characteristic function of $\mathcal{O}$, and we have a formula analogous to (5.2)

$$F_{\mathcal{O}}(\mu) = \operatorname*{Sup}_{v} \{\int_{\mathcal{O}} v\mu - \int_{\Omega} f^{*}(v)\, dx\} , \tag{5.10}$$

where it is immaterial whether the supremum is taken with respect to all v in $\mathcal{D}_f(\mathcal{C}_0^{\infty})$, in $\mathcal{D}_f(\mathcal{C}_0)$, or in $\mathcal{D}_f(L^1_{\mu})$.

Clearly $F_{\mathcal{O}}(\mu) \in \mathbb{R}$ for all $\mu \in M_1$ and by (4.35) and (4.37)

$$- k_1 \int_{\mathcal{O}}(|\mu|-1) \leq F_{\mathcal{O}}(\mu) \leq k_1 \int_{\mathcal{O}}(|\mu|+1) . \tag{5.11}$$

The function $\mu \mapsto F_{\mathcal{O}}(\mu)$ is convex and continuous on $M_1(\Omega;X)$; on the other hand, since the mapping $\mu \mapsto \theta_{\mathcal{O}}\mu$ is not in general continuous for the vague topology, $F_{\mathcal{O}}(\mu)$ is, by contrast, not in general l.s.c. for this topology.

When $\mathcal{O}$ is open and f satisfies (4.8), formula (5.10) shows that

(5.12) $$F_{\mathcal{O}}(\mu) = \operatorname*{Sup}_{v} \{\int_{\mathcal{O}} \mu v - \int_{\mathcal{O}} f^*(v)\, dx\} ,$$

where the supremum is taken over all $v \in \mathscr{C}_0(\mathcal{O};X)$, with $f^* \circ v \in L^1(\mathcal{O})$; this shows that Ω plays no part and that $F_{\mathcal{O}}$ is then l.s.c. for the vague topology of $M_1(\Omega;X)$; of course this remains true, thanks to (4.31), if f does not satisfy (4.8).

Lemma 4.5 enables us to add a few remarks: suppose f satisfies (4.8), then

(5.13) $$F_{\mathcal{O}}(\mu) \geq 0 , \forall \mathcal{O}, \forall \mu ,$$

(5.14) $$F_{\mathcal{O}}(\mu) \leq F_{\mathcal{O}'}(\mu) , \forall \mu, \forall \mathcal{O}, \mathcal{O}' , \mathcal{O} \subset \mathcal{O}' ,$$

(5.15) $F_{\mathcal{O}}(\mu)$ is an additive function of $\mathcal{O}$.

Finally Theorem 4.2 shows that if μ is of the form (5.4)

(5.16) $$F_{\mathcal{O}}(\mu) = \int_{\mathcal{O}} f(h(x))dx + \int_{\mathcal{O}} f_\infty(\mu^S) = \int_{\mathcal{O}} f(h(x))dx + \int_{\Omega} f_\infty(\theta_{\mathcal{O}}\mu^S) .$$

Thus the contribution from μ^S vanishes if $\mathcal{O}$ does not meet a (negligible) set on which μ^S is concentrated.

5.2 Approximation of convex functions of measure

We recalled in Section 2 that $\mathscr{C}_0^\infty(\Omega;X)$ is dense in the space $M_1(\Omega;X)$ endowed with the topology of weak convergence of measures, each measure μ being capable of being approximated, by truncation and regularisation, by a sequence of functions in $\mathscr{C}_0^\infty(\Omega;X)$ (see (2.8)-(2.10)). The space $\mathscr{C}_0^\infty(\Omega;X)$ *is not dense* in the space $M_1(\Omega;X)$ endowed with the topology of the norm (if it were, M_1 would be identical to L^1). Our object here is to show that $\mathscr{C}_0^\infty$ is dense in M_1 , when M_1 is endowed with a topology intermediate between these two (the weak and the norm).

Given a convex function f mapping X into $\mathbb{R}$, and which satisfies (4.1) (4.6) as before, we write M_f for the space $M_1(\Omega;X)$ endowed with the topology associated with the pseudo-metrics defined below (cf. N. Bourbaki [1]):

- those corresponding to the seminorms $|\langle\mu,v\rangle|$, $v\in\mathscr{C}_0^\infty(\Omega;X)$, i.e.

$$e_v(\mu,\nu) = |\langle\mu-\nu,v\rangle| , \tag{5.17}$$

- the pseudo-metric

$$e_f(\mu,\nu) = \left|\int_\Omega f(\mu) - \int_\Omega f(\nu)\right| . \tag{5.18}$$

It is clear that this topology is intermediate between that of weak convergence, and that of the norm; for example a sequence μ_j converges to μ in M_f , if, as $j\to\infty$

$$\mu_j \to \mu \text{ weakly and} \tag{5.19}$$

$$\int_\Omega f(\mu_j) \to \int_\Omega f(\mu) . \tag{5.20}$$

We shall prove

THEOREM 5.1.

If the convex function f *satisfies (4.1) and (4.6) then* $\mathscr{C}_0^\infty(\Omega;X)$ *is dense in the pseudo-metric space* $M_1(\Omega;X)$ *carrying the topology associated with the pseudo-metrics (5.17) (5.18).*

The proof depends on several lemmata.

LEMMA 5.1.

i) Let μ_j *be a sequence of elements of* $M_1(\Omega;X)$ *converging weakly to* μ *, while* $f(\mu_j)$ *converges weakly to a measure* ν *as* $j\to\infty$ *. Then we have*

$$f(\mu) \leq \nu .$$

ii) If in addition $f\geq 0$ *and* $\limsup_{j\to\infty}\int_\Omega f(\mu_j) \leq \int_\Omega f(\mu)$ *, then*

$$f(\mu) = \nu \quad \text{and} \quad \int_\Omega f(\mu) = \lim_{j\to\infty} \int_\Omega f(\mu_j) \ . \tag{5.22}$$

Proof

To establish (5.21) we have to show that

$$\langle f(\mu),\phi\rangle \leq \langle \nu,\phi\rangle \ ,$$

for all $\phi\geq 0$ in $\mathscr{C}_0(\Omega)$. Now for a fixed $\eta>0$ there exists a $v \in \mathscr{D}_f(\mathscr{C}_0)$, such that

$$\langle f(\mu),\phi\rangle - \eta \leq \int_\Omega (v\phi)\mu - \int_\Omega f^*(v)\phi \ dx \ .$$

The right-hand side of this inequality is equal to:

$$\lim_{j\to\infty} \{\int_\Omega (v\phi)\mu_j - \int_\Omega f^*(v)\phi \ dx\} \leq$$

$$\leq \liminf_{j\to\infty} \ \sup_{w\in\mathscr{D}_f(\mathscr{C}_0)} \{\int_\Omega (w\phi)\mu_j - \int_\Omega f^*(w)\phi \ dx\}$$

$$\leq \lim_{j\to\infty} \langle f(\mu_j),\phi\rangle = \langle \nu,\phi\rangle \ .$$

Hence

$$\langle f(\mu),\phi\rangle - \eta \leq \langle \nu,\phi\rangle \ ,$$

and since $\eta>0$ can be chosen to be arbitrarily small we obtain the required inequality.

For (5.22) we note that by reason of (5.21), and since $f(\mu_j) \geq 0$:

$$\int_\Omega f(\mu) \leq \int_\Omega \nu \leq \liminf_{j\to\infty} \int_\Omega f(\mu_j) \leq \limsup_{j\to\infty} \int_\Omega f(\mu_j)$$

$$\leq \int_\Omega f(\mu) \ , \text{ by hypothesis.}$$

These inequalities are therefore equations. Since $\nu - f(\mu) \geq 0$ and

$$\int_\Omega (\nu - f(\mu)) = 0 \ ,$$

we therefore have $\nu = f(\mu)$ and

$$\lim_{j\to\infty} \int_\Omega f(\mu_j) = \int_\Omega f(\mu) . \qquad \square$$

Let $\rho \in \mathscr{C}_0^\infty(\mathbb{R}_n)$, with $\rho \geq 0$, $\rho(x) = \rho(-x)$, $\int_{\mathbb{R}^n} \rho(x)dx = 1$, and for a fixed $\eta > 0$ let $\rho_\eta(x) = \frac{1}{\eta^n}\rho(\frac{x}{\eta})$.

LEMMA 5.2.

Suppose $\Omega = \mathbb{R}^n$ (1) *and* $\mu \in M_1(\mathbb{R}^n;X)$; *suppose also that* f *satisfies (4.1) (4.6) (4.8) and that* ρ_η *is defined as above. Then when* $\eta \to 0$, $f(\rho_\eta * \mu)$ *converges weakly to* $f(\mu)$, *and*

$$\int_{\mathbb{R}^n} f(\rho_\eta * \mu) \to \int_{\mathbb{R}^n} f(\mu) . \tag{5.23}$$

Furthermore for all $\eta > 0$:

$$f(\rho_\eta * \mu) \leq \rho_\eta * f(\mu) . \tag{5.24}$$

Proof

i) We first show that

$$\int_{\mathbb{R}^n} f(\rho_\eta * \mu) \leq \int_{\mathbb{R}^n} f(\mu) . \tag{5.25}$$

For $v \in \mathscr{D}_f(\mathscr{C}_0^\infty)$, we have, since ρ is even

$$\int_{\mathbb{R}^n} v(\rho_\eta * \mu) - \int_{\mathbb{R}^n} f^*(v)\,dx = \int_{\mathbb{R}^n} (v * \rho_\eta)\mu - \int_{\mathbb{R}^n} f^*(v)\,dx$$

(1) When f satisfies (4.1) (4.6) (4.8) the theory developed in Section 4 applies also to the case where $\Omega = \mathbb{R}^n$. If f does not satisfy (4.8) they no longer apply inasmuch as the decomposition (4.10) (4.11) is no longer valid, since the constants do not define bounded measures on $\mathbb{R}^n$.

$$= \int_{\mathbb{R}^n} (v*\rho_\eta)\mu - \int_{\mathbb{R}^n} f^*(v*\rho_\eta)dx + \int_{\mathbb{R}^n} (f^*(v*\rho_\eta) - f^*(v))dx$$

$$\leq \int_{\mathbb{R}^n} f(\mu) + \int_{\mathbb{R}^n} (f^*(v*\rho_\eta)-f^*(v))dx \ .$$

By convexity, we have

$$f^*(\rho_\eta * v) \leq \rho_\eta * f^*(v) \tag{5.26}$$

whence by integration

$$\int_{\mathbb{R}^n} f^*(\rho_\eta * v)dx \leq \int_{\mathbb{R}^n} \rho_\eta * f^*(v)dx = \int_{\mathbb{R}^n} f^*(v)dx \ .$$

It follows that for all $v \in \mathcal{D}_f(\mathscr{C}_0^\infty)$,

$$\int_{\mathbb{R}^n} v(\rho_\eta * \mu) - \int_{\mathbb{R}^n} f^*(v)dx \leq \int_{\mathbb{R}} f(\mu)$$

and (5.25) is established, in view of (5.2).

ii) As $f \geq 0$, the same applies to the measures $f(\rho_\eta * \mu)$ and (5.25) shows that these measures remain bounded in M_1 when $\eta \to 0$. There is therefore a subsequence η_j tending to zero, and a $\nu \in M_1$ such that

$$f(\rho_{\eta_j} * \mu) \to \nu \ , \ \text{weakly} \ , \tag{5.27}$$

when η_j tends to zero. Lemma 5.1 i) shows that

$$f(\mu) \leq \nu \ . \tag{5.28}$$

Furthermore, in view of (5.25)

$$\limsup_{j\to\infty} \int_\Omega f(\rho_{\eta_j} * \mu) \leq \int_{\mathbb{R}^n} f(\mu)$$

and Lemma 5.1 ii) shows that $f(\mu) = \nu$ and (5.23) holds for the sequence η_j .

A standard argument by contradiction then shows that it is the entire family $f(\rho_\eta * \mu)$ which converges weakly to $f(\mu)$ when η tends to 0, and that (5.23) holds for the whole family η .

iii) To prove (5.24) we approximate μ by a sequence of functions $u_j \in L^1(\Omega;X)$, for example $u_j = \rho_{1/j} * \mu$. The convexity condition (5.26) with v replaced by u_j can be written:

$$f(\rho_\eta * u_j) \leq \rho_\eta * f(u_j) .$$

By (5.23) $f(u_j)$ converges weakly to $f(\mu)$ as j tends to ∞ . Similarly

$$\rho_\eta * u_j = \rho_\eta * \rho_{1/j} * \mu = \rho_{1/j} * (\rho_\eta * \mu)$$

converges weakly to $\rho_\eta * \mu$ and $f(\rho_\eta * u_j)$ converges weakly to $f(\rho_\eta * \mu)$ so that in the limit we have (5.24). □

Remark 5.1. We have in fact rather more than (5.23), namely:

$$\int_{\mathcal{O}} f(\rho_\eta * \mu) \to \int_{\mathcal{O}} f(\mu) ,$$

for any open $\mathcal{O}$. The reasoning is the same as that given above for $\mathbb{R}^n$, remembering that the integral $\int_{\mathcal{O}} f(\mu)$ is l.s.c. for the weak topology when $\mathcal{O}$ is an open set.

□

We now return to the case where Ω is a bounded open subset of $\mathbb{R}^n$ and establish the following lemma which implies Theorem 5.1.

LEMMA 5.3.

For all $\mu \in M_1(\Omega;X)$, *there exists a sequence* $u_j \in \mathscr{C}_0^\infty(\Omega;X)$ *converging to* μ *under the weak topology and such that*

(5.29) $$\int_\Omega |u_j|dx \leq \int_\Omega |\mu| \ , \quad \forall\, j \ ,$$

and

(5.30) $$\int_\Omega |\mu_j|dx \to \int_\Omega |\mu| \quad \text{as} \quad j \to \infty \ .$$

Furthermore for any function f *satisfying (4.1) (4.6)*

(5.31) $f(u_j)$ *converges weakly to* $f(\mu)$,

(5.32) $$\int_\Omega f(u_j) \to \int_\Omega f(\mu) \ , \quad \text{when} \quad j \to \infty \ .$$

Proof

i) We begin by approximating to μ by a sequence of measures with compact support in Ω : let Ω_j be a monotone increasing sequence of open sets, such that

$$\Omega_j \subset \overline{\Omega}_j \subset \Omega_{j+1} \ , \quad \forall\, j \ ,$$

$$\bigcup_{j=1}^{\infty} \Omega_j = \Omega \ ,$$

and suppose that for all j , $\phi_j \in \mathscr{C}_0^\infty(\Omega)$, $0 \leq \phi_j \leq 1$, $\phi_j = 1$ on Ω_j . Lebesgue's theorem implies that $\phi_j\mu$ converges to μ in $M_1(\Omega;X)$ as $j \to \infty$.

We next consider for all j , $u_j = \rho_{\eta_j} * (\phi_j\mu)$ where ρ_η is defined as above and $\eta_j > 0$ is less than the distance of the support of ϕ_j from $\partial\Omega$. Clearly u_j converges weakly to μ when $j \to \infty$. In fact, for all $\theta \in \mathscr{C}_0^\infty(\Omega)$,

$$\langle u_j, \theta \rangle = \int_\Omega (\rho_{\eta_j} * \theta) . \phi_j \mu \ ;$$

$\rho_{\eta_j} * \theta$ converges uniformly to θ while $\phi_j\mu$ converges to μ in $M_1(\Omega;X)$; hence $\lim_{j\to\infty} \langle u_j, \theta \rangle = \int_\Omega \theta\mu$.

On the other hand, by (2.6), and since $\phi_j \leq 1$:

$$(5.33) \qquad \int_\Omega |u_j|dx = \int_\Omega |\rho_{n_j} * (\phi_j\mu)| = \int_\Omega \phi_j|\mu| \leqslant \int_\Omega |\mu| \ ;$$

so that (5.29) follows. Moreover by semicontinuity, when j tends to ∞

$$\int_\Omega |\mu| \leqslant \liminf_{j\to\infty} \int_\Omega |u_j|dx \ ,$$

which, taken in conjunction with (5.33) implies (5.30).

ii) We now prove (5.31) (5.32) in the case of a function which also satisfies (4.8). By Lemma 5.2,

$$(5.34) \qquad 0 \leqslant f(u_j) = f(\rho_{n_j} * (\phi_j\mu)) \leqslant \rho_{n_j} * f(\phi_j\mu)$$

and by Lemma 5.4 to be proved below, $f(\phi_j\mu) \leqslant f(\mu)$ so that

$$(5.35) \qquad \int_\Omega f(u_j)dx \leqslant \int_{\mathbf{R}^n} \rho_{n_j} * f(\phi_j\mu)dx = \int_{\mathbf{R}^n} f(\phi_j\mu) \leqslant \int_\Omega f(\mu) \ .$$

The sequence of functions $f(u_j)$ is bounded in $L^1(\Omega)$; hence there exists a sequence j' tending to ∞ and a $\nu \in M_1(\Omega)$ such that $f(u_{j'})$ converges weakly to ν in $M_1(\Omega)$. It is easily deduced from (5.34) that $\nu \leqslant f(\mu)$, while Lemma 5.1 gives the reverse inequality $\nu \geqslant f(\mu)$. Thus $\nu = f(\mu)$ and the whole sequence $f(u_j)$ converges weakly to $f(\mu)$. Furthermore, by semicontinuity

$$(5.36) \qquad \int_\Omega f(\mu) \leqslant \liminf_{j\to\infty} \int_\Omega f(u_j)dx$$

which, in conjunction with (5.35), implies (5.34).

iii) It now remains only to prove (5.31) (5.32) in the case where f does not satisfy (4.8).

By applying (5.32) to the functions $f(\xi) = \xi_i^+$ and $f(\xi) = \xi_i^-$, where the ξ_i are the components of ξ in a basis of X , we see that

$$(5.37) \qquad \int_\Omega u_j \, dx \to \int_\Omega \mu \ , \ \text{as } j \to \infty \ .$$

We then obtain the required results by using the decomposition (4.10) (4.11) of f and the properties (5.31) and (5.32) for the function g appearing in (4.10).

□

To conclude we prove

LEMMA 5.4.

Suppose f satisfies (4.1) (4.6) and (4.8), $\mu \in M_1(\Omega;X)$ and let ϕ_1, ϕ_2 be functions in $\mathscr{C}(\bar{\Omega})$ such that

(5.38) $$0 \leq \phi_1 \leq \phi_2 .$$

Then

(5.39) $$0 \leq f(\phi_1\mu) \leq f(\phi_2\mu) .$$

<u>Proof</u>

For all $\theta \in \mathscr{C}_0^\infty(\Omega)$, with $\theta \geq 0$ and for all $v \in \mathscr{D}_f(\mathscr{C}_0)$, we write

$$\int_\Omega \phi_1(v\theta)\mu - \int_\Omega f^*(v)\theta \, dx = \int_\Omega \phi_2(hv\theta)\mu - \int_\Omega f^*(hv)\theta \, dx + r ,$$

where

$$r = \int_\Omega (f^*(hv) - f^*(v))\theta \, dx ,$$

and h is the function equal to $\phi_1(x)/\phi_2(x)$ when $\phi_2(x) \neq 0$ and to 0 when $\phi_1(x) = \phi_2(x) = 0$. The function h is of course measurable and since $|hv| \leq |v|$, we have $hv \in L^1(\Omega;\mu)$. Noting that $f^* \geq 0$ and $f^*(0) = 0$ (see (4.9)), and using convexity of f^*, we have (p.p.)

(5.40) $$f^*(h(x)v(x)) \leq h(x)f^*(v(x)) \leq f^*(v(x))$$

which implies that $hv \in \mathscr{D}_f(L^1_\mu)$. Noting also that $\mathscr{D}_f(L^1_\mu) \subset \mathscr{D}_f(L^1_{\phi_2\mu})$

and using Lemma 4.1, we see that

$$\int_\Omega (hv\theta)(\phi_2\mu) - \int_\Omega f^*(hv)\theta \, dx \leq \int_\Omega \theta f(\phi_2\mu) .$$

On the other hand, $r\leq 0$ by (5.40) and we have

$$\int_\Omega (v\theta)(\phi_1\mu) - \int_\Omega f^*(v)\theta \, dx \leq \int_\Omega \theta \, f(\phi_2\mu) ,$$

for all $v \in \mathscr{D}_f(\mathscr{C}_0)$ and all $\theta \in \mathscr{C}_0^\infty(\Omega)$, with $\theta\geq 0$; it follows that the inequality (5.39) holds.

This completes the proof of Theorem 5.1.

5.3. Approximation of convex functions on BD(Ω)

We shall now establish a density result for $BD(\Omega)$. We consider a convex function, defined on $X = E^D$, which satisfies the conditions (4.1) (4.6) and (4.8) and which could, in particular, be the function ψ^D defined in (3.42) of Chapter I. We associate with f a distance d_f defined on $BD(\Omega)$ by (compare (3.5)):

$$(5.41) \qquad d_f(u,v) = \|u-v\|_{L^1(\Omega)^n} + \left|\int_\Omega |\varepsilon(u)| - \int_\Omega |\varepsilon(v)|\right| + \\ + \left|\int_\Omega f(\varepsilon^D(u)) - \int_\Omega f(\varepsilon^D(v))\right| .$$

This metric defines on $BD(\Omega)$ a topology intermediate between the weak topology and that of the norm; for example (cf. (3.3) and (3.6)), a sequence u_m converges to u under this metric, if, when $m \to \infty$,

$$(5.42) \qquad \begin{cases} u_m \to u \text{ in } L^1(\Omega)^n , \\ \varepsilon_{ij}(u_m) \to \varepsilon_{ij}(u) \text{ weakly in } M_1(\Omega) , \ (i,j=1,\dots,n) , \\ \int_\Omega |\varepsilon(u_m)| \to \int_\Omega |\varepsilon(u)| , \\ \int_\Omega f(\varepsilon^D(u_m)) \to \int_\Omega f(\varepsilon^D(u)) . \end{cases}$$

We shall prove

THEOREM 5.2.

$\mathscr{C}^\infty(\bar{\Omega})^n$ is dense in the space $BD(\Omega)$ endowed with the topology defined by the metric (5.41).

More precisely, for every u of $BD(\Omega)$, there exists a sequence $u_m \in \mathscr{C}_o^\infty(\Omega)^n \cap LD(\Omega)$ such that

$$\gamma_o(u_m) = \gamma_o(u) \;,\quad \text{for all } m \;, \tag{5.43}$$

$$d_f(u_m,u) \to 0 \;,\quad \text{when } m \to \infty \;. \tag{5.44}$$

<u>Proof</u>

i) We saw in Section 4 that the function $\mu \mapsto \int_\Omega f(\mu)$ is continuous on $M_1(\Omega;X)$ under the norm topology. It follows that the function $u \mapsto \int_\Omega f(\varepsilon^D(u))$ is continuous on $BD(\Omega)$ and hence on $LD(\Omega)$. Then since $\mathscr{C}^\infty(\bar{\Omega})^n$ is dense in $LD(\Omega)$ (cf. (1.8)), $\mathscr{C}^\infty(\bar{\Omega})^n$ is also dense in $LD(\Omega)$ endowed with the metric d_f and the first part of the theorem is therefore an immediate consequence of (5.43) and (5.44).

The proof of (5.43) and (5.44) follows closely the lines of that of Theorems 3.2 and 3.3, and we resume the hypotheses and notations used in their proof.

ii) To establish (5.44) we shall show, as in Theorem 3.2, that for a given u in $BD(\Omega)$, there exists for any $\delta>0$, a function $u_\delta \in \mathscr{C}^\infty(\Omega)^n \cap LD(\Omega)$ such that

$$\|u_\delta - u\|_{L^1(\Omega)} \leq \delta \tag{5.45}$$

$$\left|\int_\Omega |\varepsilon(u_\delta)| - \int_\Omega |\varepsilon(u)|\right| \leq 4\delta \tag{5.46}$$

$$\left|\int_\Omega f(\varepsilon^D(u_\delta)) - \int_\Omega f(\varepsilon^D(u)) \leq c\delta \tag{5.47}$$

where the constant c depends only on f; (5.44) will then follow from this by taking $u_m = u_{\delta_m}$ with $\delta_m = 1/m$.

The construction of u_δ carried out in Theorem 3.2 gives (5.45) (5.46) (identical to (3.13) (3.14)); a slight change in the construction enables us to obtain (5.47) as well. For this purpose we choose Ω_0 in such a way that, in addition to (3.15) and (3.16) it satisfies the condition

$$\text{measure } (\Omega\setminus\Omega_0) \leq \delta \tag{5.48}$$

Now by (5.23) and Remark 5.1 we can choose η_1 small enough to ensure that, apart from (3.19) - (3.21).

$$\left|\int_{\Omega_0} f(\rho_{\eta_1}*(\phi_1\varepsilon^D(u))) - \int_{\Omega_0} f(\phi_1\varepsilon^D(u))\right| \leq \delta . \tag{5.49}$$

We now have

$$\left|\int_\Omega f(\varepsilon^D(u_\delta))-\int_\Omega f(\varepsilon^D(u))\right| \leq \tag{5.50}$$

$$\leq \left|\int_{\Omega_0} f(\varepsilon^D(u_\delta))-\int_{\Omega_0} f(\varepsilon^D(u))\right| + \int_{\Omega_0} |f(\varepsilon^D(u_\delta))|$$

$$+ \int_{\Omega_0} |f(\varepsilon^D(u))| .$$

Since all the functions ϕ_j , $j\geq 2$, and their derivatives vanish in a neighbourhood of $\bar{\Omega}_0$, $\varepsilon^D(u_\delta) = \rho_{\eta_1} * \varepsilon^D(\phi_1 u_\delta)$ on Ω_0 and $\varepsilon^D(u) = \phi_1 \varepsilon^D(u)$ on Ω_0 . Thus by Theorem 4.2,

$$\left|\int_{\Omega_0} f(\varepsilon^D(u_\delta))-\int_{\Omega_0} f(\varepsilon^D(u))\right| = \left|\int_{\Omega_0} f(\rho_{\eta_1}*(\phi_1\varepsilon^D(u)))-\int_{\Omega_0} f(\phi_1\varepsilon^D(u))\right| ,$$

and this is less than or equal to δ by reason of the hypothesis (5.49). The other terms on the right-hand side of (5.50) can be majorised with the help of (4.35) and the inequality $|\varepsilon^D(v)| \leq |\varepsilon(v)|$:

$$\int_{\Omega\setminus\Omega_0} |f(\varepsilon^D(u_\delta))|+ \int_{\Omega\setminus\Omega_0} |f(\varepsilon^D(u))| \leq k_1 \int_{\Omega\setminus\Omega_0} (2+|\varepsilon(u_\delta)|+|\varepsilon(u)|).$$

A calculation analogous to that of (3.25) shows us that

(5.51) $$\int_{\Omega\setminus\Omega_0} |\varepsilon(u_\delta)| \leq 2\delta$$

and then thanks to (3.15) (5.48) and what precedes, the right-hand side of (5.50) is majorised by $(1+5k_1)\delta$, which implies (5.47) with $c = 1+5k_1$.

We have proved (5.44); as for (5.43), the result has already been proved in Theorem 3.3, the sequence u_m there being the same as the one considered here.

Remark 5.2. We can choose η_1 so that (5.49) holds for any finite family of functions f satisfying (4.1) (4.6) (4.8) : the sequence u_m then approximates u for all the corresponding distances d_f . This clears up the point raised in Remark 3.3.

On the other hand, we do not know whether it is possible to find, as it was for Theorem 5.1, a sequence u_m which approximates u for *all* the distances d_f . □

We conclude this Section by giving an analogous result for the space $U(\Omega)$ (cf. (3.32)) :

(5.52) $$U(\Omega) = \{v \in BD(\Omega) \ , \ \operatorname{div} v \in L^2(\Omega)\} \quad .$$

Given a convex function f , mapping $X = E^D$ into $\mathbb{R}$, which satisfies (4.1) (4.6) (4.8), we can define on $U(\Omega)$ a topology intermediate between the weak topology (cf. (3.37)) and the topology of the norm; namely that defined by the following metric, which should be compared with that given in (3.38):

(5.53) $$d_f(u,v) = \|u-v\|_{L^1(\Omega)^n} + |\operatorname{div}(u-v)|_{L^2(\Omega)}$$
$$\left|\int_\Omega |\varepsilon(u)| - \int_\Omega |\varepsilon(v)|\right| + \left|\int_\Omega f(\varepsilon^D(u)) - \int_\Omega f(\varepsilon^D(v))\right| .$$

A sequence u_m converges to u in $U(\Omega)$ for this metric if (cf. (3.37) (3.39) (3.40)):

(5.54)
$$\begin{cases} u_m \to u \quad \text{in} \quad L^1(\Omega)^n \quad \text{(strong convergence)} \\ \operatorname{div} u_m \to \operatorname{div} u \quad \text{in} \quad L^2(\Omega) \quad \text{(strong convergence)} \end{cases}$$

(5.55)
$$\varepsilon_{ij}(u_m) \to \varepsilon_{ij}(u) \quad \text{in} \quad M_1(\Omega) \quad \text{weakly for all} \quad i,j$$

(5.56)
$$\begin{cases} \int_\Omega |\varepsilon(u_m)| \to \int_\Omega |\varepsilon(u)| \\ \int_\Omega f(\varepsilon^D(u_m)) \to \int_\Omega f(\varepsilon^D(u)) \,. \end{cases}$$

By combining the results of Theorems 3.4 and 5.2 we obtain (1)

THEOREM 5.3.

For every u *of* $U(\Omega)$, *there exists a sequence* $u_m \in \mathscr{C}^\infty(\Omega)^n \cap LD(\Omega) \cap U(\Omega)$ *with the following properties* :

(5.57)
$$\gamma_o(u_m) = \gamma_o(u) \,, \quad \forall\, m \,.$$

(5.58) *When* $m \to \infty$, u_m *converges to* u *under the topology defined by the metric (5.53) (that is to say converges in the sense of (5.54)-(5.56)).*

Remark 5.3. i) By Lemma 5.1 and (5.56), $f(\varepsilon^D(u_m))$ converges vaguely to $f(\varepsilon^D(u))$ when $m \to \infty$ and

$$\int_\Omega f(\varepsilon^D(u_m)) \to \int_\Omega f(\varepsilon^D(u)) \,.$$

ii) Remark 5.2 can be generalised: it is possible to choose u_m so that (5.57) (5.58) and the property just stated in i) above all hold good for any finite family of functions f .

Remark 5.4. It follows from Remark 3.5 that the sequence u_m given by

(1) As the sequence u_m considered in these two theorems are the same there is nothing more to be proved.

Theorem 5.3 satisfies

$$u_m \to u \text{ in } L^p(\Omega)^n , \tag{5.59}$$

as $m \to \infty$, for all p with $1 \le p < n^* = n/(n-1)$, and it is even possible to choose the sequence u_m so as to have (5.59) with $p = n^*$.

6. Example of a convex function of a measure: relaxation of the strain problem

Until the end of this Chapter, we shall be dealing again with the strain problem for the perfectly-plastic elastic model, i.e. the problem (4.16) (4.17) of Chapter I. We have seen that there is little hope in general of proving the existence of a regular solution to this problem because of the discontinuities which can arise in the field of displacements inside Ω or on its boundary. The introduction of the relaxed problem I. (6.4) took into account the discontinuities which could appear at the boundary (the boundary condition $u = u_0$ not satisfied on Γ_0). By utilising the space $BD(\Omega)$ we shall now introduce a new version of the relaxed problem which takes account of possible discontinuities of the displacement field within Ω . The definition of the relaxed problem uses the concept of a convex function of a measure studied in Sections 4 and 5.

In Section 6.1 we formulate the relaxed problem in $BD(\Omega)$ and in Section 6.2 we show that *its infimum is the same as that of the problems I. (4.16) and I. (6.4).* Next in Section 6.3 we make a few remarks on the minimising sequences of the relaxed problem and show that a solution exists in a simple particular case. The existence of a solution to the relaxed problem in the general case will be established in Section 8.

Throughout the final part of this Chapter Ω is an open subset of $\mathbf{R}^n$, where n is any integer ≥ 2. This entails no additional complications and enables us also to cover the useful case $n = 2$. The extension of the results of Chapter I to the case $n \neq 3$ is very easy.

6.1. Definition of the relaxed problem

We shall again use the notation of Sections 3, 4, 5, 6 of Chapter I, without repeating the definitions here. The function ψ which appears in I.(6.10) was defined in I.(4.9), I.(4.10); it is a convex function mapping E into $\mathbb{R}$. Its restriction ψ^D to E^D satisfies I.(4.18); the conjugate of ψ^D is, by I.(4.10), the function mapping E^D into $\mathbb{R} \cup \{+\infty\}$ defined by [1]:

$$(6.1) \qquad (\psi^D)^*(\eta^D) = \begin{cases} \frac{1}{4\mu}\, \eta^D.\eta^D & \text{if } \eta^D \in K^D \\ +\infty & \text{if } \eta^D \notin K^D \end{cases} .$$

Its domain (of finiteness) is thus the convex set K^D, and $(\psi^D)^*$ is of course bounded above on K^D: by putting $X = E^D$, $f = \psi^D$, we see that conditions (4.1) and (4.6) of Section 4 are satisfied. By (6.1) $(\psi^D)^* \geq 0$ and $(\psi^D)^*(0) = 0$; this implies by conjugation that ψ^D has these same properties and that condition (4.8) is also satisfied. To summarise (cf. (4.4) (4.5) (4.6))

$$(6.2) \qquad k_o(|\xi^D|-1) \leq \psi^D(\xi^D) \leq k_1|\xi^D| , \; \forall\, \xi^D \in E^D$$

$$(6.3) \qquad \sup_{\eta^D \in K^D} (\psi^D)^*(\eta^D) \leq k_2 = \frac{k_1^2}{4\mu}$$

$$(6.4) \qquad \psi^D \geq 0 , \quad \psi^D(0) = 0$$

$$(6.5) \qquad (\psi^D)^* \geq 0 , \quad (\psi^D)^*(0) = 0 .$$

If now $u \in BD(\Omega)$, then $\varepsilon^D(u)$ is a bounded measure on Ω with values in E^D and $\psi^D(\varepsilon^D(u))$ is a bounded measure on Ω with real values ≥ 0, which is completely defined by Theorem 4.1. In particular (cf. (4.34) and (6.2))

---

[1] The coefficient $\mu > 0$ in (6.1) is one of the Lamé constants and does not represent a measure.

(6.6) $\psi^D(\varepsilon^D(v))$ is a bounded Radon measure, ≥ 0, on Ω ,

$$\psi^D(\varepsilon^D(v)) \leq k_1|\varepsilon^D(v)| \ . \tag{6.7}$$

If $\varepsilon^D(v) = h\,dx + \varepsilon^D(v)^S$ is the Lebesgue decomposition of $\varepsilon^D(v)$ then, by Theorem 4.2

$$\psi^D(\varepsilon^D(v)) = \psi^D(h)dx + \psi^D_\infty(\varepsilon^D(v)^S) \ , \tag{6.8}$$

where ψ^D_∞ is the support function of K^D in E^D ; it is the restriction to E^D of the function ψ_∞ which appears in I.(5.1) and I.(5.2). As in the general case, the support of the singular part of $\psi^D(\varepsilon^D(v))$ is included in the support of the singular part of $\varepsilon^D(v)^S$.

All this allows us, as in Section 5, to define the functional

$$\Psi^D(\varepsilon^D(v)) = \int_\Omega \psi^D(\varepsilon^D(v)) \ . \tag{6.9}$$

This represents a well-defined finite nonnegative number for any $v \in BD(\Omega)$.

We next consider the space $U(\Omega)$ (cf. (3.32))

$$U(\Omega) = \{v \in BD(\Omega) \ , \ \text{div } v \in L^2(\Omega)\} \ , \tag{6.10}$$

and, for any $v \in U(\Omega)$, we consider the bounded measure (cf. I.(4.9))

$$\psi(\varepsilon(v)) = \frac{\kappa}{2}(\text{div } v)^2 + \psi^D(\varepsilon^D(v)) \ , \tag{6.11}$$

and the functional (compare with I.(4.15)):

$$\Psi(\varepsilon(v)) = \int_\Omega \psi(\varepsilon(v)) = \frac{\kappa}{2}\int_\Omega (\text{div } v)^2 dx + \int_\Omega \psi^D(\varepsilon^D(v)) \ . \tag{6.12}$$

Thanks to the trace theorem in $BD(\Omega)$ (Theorem 2.1), the expression $L(v)$ which appears in I.(4.14) has a meaning for every $v \in BD(\Omega)$,

(6.13)
$$L(v) = \int_\Omega f\ v\ dx + \int_{\Gamma_1} g\ v\ d\Gamma ,$$

provided that

(6.14)
$$f \in L^\infty(\Omega)^n ,\ g \in L^1(\Gamma_1)^n .$$

The same also applies to the term I.(6.6),

(6.15)
$$\int_{\Gamma_0} \psi_\infty(\mathscr{T}^D(\gamma_\tau(u_0-v)))d\Gamma ,$$

where $\gamma_\tau(u_0-v) = u_{0\tau}-v_\tau$ is the tangential component of the trace on Γ of u_0-v, and (cf. I.(6.2), I.(6.3))

(6.16)
$$\begin{cases} \mathscr{T}_{ij}(p) = \frac{1}{2}\,(p_i\ \nu_j + p_j\ \nu_i) \\ \mathscr{T}^D_{ij}(p) = \mathscr{T}_{ij}(p) - \frac{1}{n}\,p.\nu\ \delta_{ij} , \end{cases}$$

where $i,j = 1,\ldots,n$ for all $p = (p_1,\ldots,p_n) \in L^1(\Gamma_0)^n$, and where $\nu = (\nu_1,\ldots,\nu_n)$ denotes the outward normal on Γ.

Finally we define the set of admissible v which generalises I.(6.5),

(6.17)
$$\tilde{\mathscr{C}}_a = \{v \in U(\Omega) ,\ v.\nu = u_0.\nu \text{ on } \Gamma_0\} .$$

With all the preceding definitions, the relaxed problem $\mathscr{Q}$ can now be written exactly as I.(6.4)

(6.18)
$$\operatorname*{Inf}_{v\in\tilde{\mathscr{C}}_a} \{\Psi(\varepsilon(v)) + \int_{\Gamma_0} \psi_\infty(\mathscr{T}^D(\gamma_\tau(u_0-v))d\Gamma - L(v)\} .$$

It is clear that (6.18) is a generalisation of the problem I.(6.4), since the same functional is being minimised but over a larger set, namely the set (6.17) which contains the set I.(6.5). Let us remember incidentally that the problem I.(6.4) was itself an extension of the problem I.(4.16):

$$\text{I.}\underset{\mathscr{P}}{(4.16)} \subset \text{I.}\underset{\mathscr{PR}}{(6.4)} \subset \underset{\mathscr{Q}}{(6.18)} \ . \tag{6.19}$$

It follows that

$$\operatorname{Inf}\mathscr{P} = \operatorname{Inf}\mathscr{PR} \geq \operatorname{Inf}\mathscr{Q} \ , \tag{6.20}$$

the equality of the infima of $\mathscr{P}$ and $\mathscr{PR}$ having been proved in Theorem 6.1 Ch. I. We shall prove that the infimum of $\mathscr{Q}$ is the same as that of the other two problems.

6.2. Equality of the Infima

THEOREM 6.1.

The infimum of problem (6.18) is the same as that of the problems I.(6.4) and I.(4.16).

Proof

By Theorem 6.1 Ch.I it is sufficient to prove equality with the infimum of problem I.(6.4). We have to show that if $v \in \tilde{\mathscr{C}}_a$, there is a sequence v_m of elements of the admissible set I.(6.5) of the problem $\mathscr{PR}$ such that

$$\Psi(\varepsilon(v_m)) + \int_{\Gamma_0} \psi_\infty(\mathscr{T}^D(\gamma_\tau(u_0 - v_m)))d\Gamma - L(v_m) \tag{6.21}$$

$$\to \Psi(\varepsilon(v)) + \int_{\Gamma_0} \psi_\infty(\mathscr{T}^D(\gamma_\tau(u_0 - v_m)))d\Gamma - L(v) \ .$$

To do this we consider the sequence v_m provided by Theorem 5.3. Since the trace of v_m on Γ is equal to that of v, it follows that $v_m.\nu = u_0.\nu$ on Γ_0 for all m and

$$\int_{\Gamma_0} \psi_\infty(\mathscr{T}^D(\gamma_\tau(u_0 - v_m)))d\Gamma = \int_{\Gamma_0} \psi_\infty(\mathscr{T}^D(\gamma_\tau(u_0 - v)))d\Gamma \ ,$$

$$\int_{\Gamma_1} g\ v_m\ d\Gamma = \int_{\Gamma_1} g\ v\ d\Gamma\ .$$

The convergence of v_m to v in $L^1(\Omega)^n$ thus implies the convergence of $L(v_m)$ to $L(v)$. Since $\operatorname{div} v_m \to \operatorname{div} v$ strongly in $L^2(\Omega)$ (cf. (5.54)),

$$\frac{\kappa}{2}\int_\Omega (\operatorname{div} v_m)^2\,dx \to \frac{\kappa}{2}\int_\Omega (\operatorname{div} v)^2\,dx\ .$$

Finally the convergence of $\Psi^D(\varepsilon^D(v_m))$ to $\Psi^D(\varepsilon^D(v))$ is simply the second property in (5.56) with f replaced by ψ^D.

We have thus obtained (6.21)in this case, but the sequence v_m is composed not of elements of $H^1(\Omega)^n$ but of elements of $\mathscr{C}^\infty(\Omega)^n \cap LD(\Omega) \cap U(\Omega)$. To complete the proof we need only note that, as in Lemma I.6.2, for each m, one can approximate to v_m-u_0, as closely as one wishes in terms of the norm of $LD(\Omega) \cap U(\Omega)$, by a function w_m of $\mathscr{C}^2(\bar\Omega)^n$ which vanishes in the neighbourhood of Γ_0. As all the functionals in (6.18) are continuous for the (strong) topology of $LD(\Omega) \cap U(\Omega)$, we deduce that (6.21) holds with v_m replaced by u_0+w_m, and, for all m, u_0+w_m belongs to the admissible set I.(6.5).

6.3. Existence of a solution to the relaxed problem in a simple case

We shall make a few simple remarks about the minimising sequences of the problem $\mathscr{Q}$ and then prove the existence of a solution to this problem in a particular case.

Theorem 5.1, Ch.I gives us a necessary and sufficient condition for the problems $\mathscr{P}$, $\mathscr{PR}$, $\mathscr{Q}$ to have a finite infimum $(>-\infty)$; this condition can be written (cf. I.(5.12), I.(5.13)):

$$\text{(6.22)}\qquad \operatorname*{Inf}_{v} \Psi_\infty(\varepsilon^D(v)) \geq 1\ ,$$

where the infimum is taken over all v satisfying :

$$(6.23)\quad \begin{cases} v \in H^1(\Omega)^3 \ , \ \operatorname{div} v = 0 \ , \ v = 0 \ \text{ on } \Gamma_0 \\ L(v) = 1 \ . \end{cases}$$

On the other hand, by Proposition 4.1 and Remark 5.2 of Ch.I if the inequality in (6.22) is strict, then every minimising sequence for the problem $\mathscr{P}$ is bounded in the sense of I.(4.21), I.(4.22). We can now put this remark into a more precise form and extend it to the problem $\mathscr{Q}$.

PROPOSITION 6.1.

If the inequality in (6.22) is strict, then every minimising sequence for $\mathscr{P}$, $\mathscr{PR}$ *or* $\mathscr{Q}$ *is bounded in* $U(\Omega)/\mathscr{R}$ *if* Γ_0 *is empty and in* $U(\Omega)$ *if not.*

Proof

Just as in Proposition 4.1 Ch.I we introduce the problems $\mathscr{Q}_\lambda$, $\lambda \geq 0$,

$$(6.24)\qquad \operatorname*{Inf}_{v \in \tilde{\mathscr{C}}_a} \{\Psi(\varepsilon(v)) + \int_{\Gamma_0} \psi_\infty(\mathscr{T}^D(\gamma_\tau(u_0 - v)))d\Gamma - \lambda \, L(v)\} \ .$$

If $\bar{\lambda}$ is the infimum of problem (6.22) (6.23) then by Proposition 4.1 and Remark 5.2 of Ch.I. and also Theorem 6.1 above, the infimum of problem (6.24) is finite if and only if $0 \leq \lambda \leq \bar{\lambda}$. Since, by hypothesis, $\bar{\lambda} > 1$, there exists a λ such that $1 < \lambda < \bar{\lambda}$. Now consider a minimising sequence for $\mathscr{Q}$ $(= \mathscr{Q}_1)$:

$$J(v_m) - L(v_m) \to \inf \mathscr{Q} \ , \ \text{ for } m \to \infty \quad ,$$

with

$$(6.25)\qquad J(v) = \Psi(\varepsilon(v)) + \int_{\Gamma_0} \psi_\infty(\mathscr{T}^D(\gamma_\tau(u_0 - v)))d\Gamma \ .$$

We have

$$J(v_m) - L(v_m) \geq (1 - \tfrac{1}{\lambda}) \ J(v_m) + \tfrac{1}{\lambda}(J(v_m) - \lambda L(v_m)) \geq (1 - \tfrac{1}{\lambda}) \ J(v_m) + \tfrac{1}{\lambda} \inf \mathscr{Q}_\lambda .$$

It follows that $J(v_m)$ remains bounded as $m \to \infty$. Hence by (6.2):

(6.26) $\varepsilon^D(v_m)$ remains bounded in $M_1(\Omega;E^D)$,

(6.27) $\operatorname{div} v_m$ remains bounded in $L^1(\Omega)$.

We also deduce from this and the inequality (5.4), Ch.I, that $\mathscr{T}^D(\gamma_\tau(v_m-u_o))$ remains bounded in $L^1(\Gamma_o;E^D)$, as $m \to \infty$; this implies by I.(6.26) that $\gamma_\tau(v_m-u_o)$ remains bounded in $L^1(\Gamma_o)^n$. Since $v_m.\nu = u_o.\nu$,

(6.28) $v_{m|\Gamma_o}$ remains bounded in $L^1(\Gamma_o)^n$ when $m \to \infty$..

It was proved in Section 2 (Proposition 2.4 and Remark 2.6 ii)) that, if Γ_o is any nonvacuous subset of Γ ,

$$\int_\Omega |\varepsilon(v)| + \int_{\Gamma_o} |v| d\Gamma$$

is a norm on $BD(\Omega)$ equivalent to the initial norm. It follows from this with (6.26)-(6.28) that

(6.29) v_m remains bounded in $L^1(\Omega)^n$ and hence in $U(\Omega)$, as $m \to \infty$.

In the case where Γ_o is empty, Proposition 2.2 shows directly with (6.26) and (6.27) that

(6.30) v_m remains bounded in $L^1(\Omega)^n/\mathscr{R}$ and hence in $U(\Omega)/\mathscr{R}$, as $m \to \infty$.

This proves the proposition.

□

Remark 6.1. Condition (6.22) is equivalent, let us recall, to the existence of a σ admissible for $\mathscr{P}^*_\lambda$ with a $\lambda \geq 1$ (and hence satisfying (5.28) Ch.I). It is a necessary and sufficient condition for the finiteness of the infimum of $\mathscr{P}, \mathscr{PR}$ and $\mathscr{Q}$.

The strict inequality condition in (6.22) is equivalent (cf. Section 5, Ch.I) to the existence of a σ , admissible for $\mathscr{P}^*_\lambda$ with a $\lambda>1$. When stated in this form, the condition is known in the literature under the name of the *safe load condition.*

Remark 6.2. By (3.4) and (3.37) the minimising sequences v_m contains a subsequence, also denoted by v_m , which converges weakly in $U(\Omega)$:

$$\text{(6.31)} \quad \begin{cases} v_m \to u \text{ in } L^1(\Omega)^n \text{ (for the norm)} \\ \operatorname{div} u_m \to \operatorname{div} u \text{ in } L^2(\Omega) \text{ weakly} \\ \varepsilon_{ij}(u_m) \to \varepsilon_{ij}(u) \text{ in } M_1(\Omega) \text{ weakly.} \end{cases}$$

To deduce from this, by the classical methods of the calculus of variations, that u is a solution of $\mathscr{Q}$, one would have to

(6.32) be able to assert that $u \in \tilde{\mathscr{C}}_a$, i.e. that $u.\nu = u_0.\nu$ on Γ_0 ,

(6.33) know that $J(v) - L(v)$ is l.s.c. for the weak topology of $U(\Omega)$ (i.e. the topology corresponding to the convergence defined by (6.31), see (3.35) (3.36)).

The assertion (6.32) will be one of the results in Section 7, and the point raised in (6.33) will be established in Section 8. Let us note in this connection that $\int_{\Gamma_1} g\, v\, d\Gamma$ and $\int_{\Gamma_0} \psi_\infty(\mathscr{T}^D(\gamma_\tau(v-u_0)))d\Gamma$ *are not l.s.c.* for this topology.

Remark 6.3. To end this section we describe a special case where the difficulties (6.32) and (6.33) are considerably lessened: this is the one where

(6.34) $$\Gamma_0 = \emptyset \ , \ \Gamma_1 = \Gamma \ , \quad g = 0 \ .$$

In this case the difficulty (6.32) disappears, because $\tilde{\mathscr{C}}_a = U(\Omega)$. As regards the lower semi-continuity of $J(v) - L(v)$, we note that L is continuous because

$$L(v) = \int_\Omega fv \, dx \ .$$

On the other hand the functional $J(v)$ reduces to

$$\frac{\kappa}{2}\int_\Omega (\text{div } v)^2 \, dx + \int_\Omega \psi^D(\varepsilon^D(v)) \ .$$

The lower semi-continuity of the first term is a standard result : that of the second summand has already been mentioned in Section 5.

Finally, if $\inf \mathscr{Q} > -\infty$, and more precisely if the inequality in (6.22) is strict, we obtain the existence of a solution to $\mathscr{Q}$ in $U(\Omega)/\mathscr{R}$ and hence in $U(\Omega)$, in the case (6.34).

7. Duality between the generalised stresses and strains

If σ is a field of stresses in $L^2(\Omega;E)$ and u a field of strains in $BD(\Omega)$, the product of σ by $\varepsilon(u)$ has no meaning in general; we have no generalised Green's formula I.(1.49), I.(1.51) between σ and $\varepsilon(u)$, and there is *a priori*, no immediate means of setting up a duality between the generalised strain problem introduced in Section 6, i.e. the problem (6.18), and the stress problem I.(4.29), I.(4.30), which we recall is:

$$\text{(7.1)} \qquad \operatorname*{Sup}_{\sigma} \{ -\frac{1}{2}\mathscr{A}(\sigma,\sigma) + \int_{\Gamma_0} (\sigma.\nu).u_0 \, d\Gamma \}$$

where σ is constrained to lie within $\mathscr{S}_a \cap \mathscr{K}_a$ and where

$$\text{(7.2)} \qquad \mathscr{S}_a = \{\sigma \in L^2(\Omega;E),\ \operatorname{div} \sigma + f = 0 \ \text{ in } \ \Omega,\ \sigma.\nu = g \ \text{ on } \ \Gamma_1\}$$

$$\text{(7.3)} \qquad \mathscr{K}_a = \{\sigma \in L^2(\Omega;E),\ \sigma^D(x) \in K^D \ \text{p.p} \ x \in \Omega\} \ .$$

and

$$\text{(7.4)} \qquad \mathscr{A}(\sigma,\tau) = \frac{1}{n^2\kappa} \int_\Omega \operatorname{tr} \sigma . \operatorname{tr} \tau \, dx + \frac{1}{2\mu} \int_\Omega \sigma^D . \tau^D \, dx \ .$$

Our object in this Section is to give a meaning to the product $\sigma.\varepsilon(u)$, to generalise Green's formula I.(1.51), and to establish a duality between (6.18) and (7.1). We shall do all this by making fuller use of the information which we have when u is admissible for (6.18)

$(u \in \tilde{\mathscr{C}}_a)$ and σ is admissible for (7.1) $(\sigma \in \mathscr{S}_a \cap \mathscr{K}_a)$. We shall follow essentially R. Kohn - R. Temam [1] [2]. We refer the reader to O. Debordes [2] for a different approach to the problem, and to G. Anzelotti [1] for certain results which overlap ours.

In Section 7.1 we give a few complementary results on stress tensors, and in particular a technical result on the approximation of a σ, admissible for (7.1), by more regular functions and also a regularity result for such tensors which generalises the Sobolev embedding theorem (I.(1.20)). In Section 7.2 we generalise the trace theorem given in (1.41) Ch.I and show that it is possible to define $u.\nu_{|\Gamma}$, when $u \in L^1(\Omega)^n$ and $\operatorname{div} u \in M_1(\Omega)$. In Section 7.3 we go on to give a meaning to the product $\sigma.\varepsilon(u)$ when $\sigma \in \mathscr{S}_a \cap \mathscr{K}_a$ and $u \in \tilde{\mathscr{C}}_a$; $\sigma.\varepsilon(u)$ is defined as a bounded measure on Ω (and not as a function). Finally in Section 7.4 we present the new duality between (6.18) and (7.1).

7.1. Further results on the stress tensor

We introduce the space (1)

$$(7.5) \qquad \Sigma(\Omega) = \{\sigma \in L^2(\Omega;E)\ ,\ \operatorname{div}\sigma \in L^{n+1}(\Omega)^n\ ,\ \sigma^D \in L^\infty(\Omega;E)\}\ .$$

It is clear that if σ is admissible for (7.1), then $\sigma \in \Sigma(\Omega)$ provided that

$$f \in L^{n+1}(\Omega)^n\ .$$

$\Sigma(\Omega)$ is a Banach space for the natural norm:

(1) We can replace the condition $\operatorname{div}\sigma \in L^{n+1}$ by the condition $\operatorname{div}\sigma \in L^{n+\alpha}$, with $\alpha>0$ and arbitrarily small; we can even allow $\alpha = 0$ at the cost of some purely technical modifications.

$$\|\sigma\|_{\Sigma(\Omega)} = |\sigma|_{L^2(\Omega;E)} + \|\text{div}\,\sigma\|_{L^{n+1}(\Omega)^n} + \|\sigma^D\|_{L^\infty(\Omega;E^D)} .$$

The space $\mathscr{C}^\infty(\bar{\Omega};E)$ is not dense in $\Sigma(\Omega)$ for the above norm, but we shall see in what sense an element σ of $\Sigma(\Omega)$ can be approximated by regular functions.

LEMMA 7.1

Let Ω be a bounded open set in $\mathbb{R}^n$, such that $\complement\bar{\Omega}$ satisfies the cone condition (cf. I(1.16)). Let σ be given in $L^p(\Omega;E)$, $1\leq p<\infty$ with $\sigma^D \in L^\infty(\Omega;E^D)$ and $div\ \sigma \in L^r(\Omega)^n$, $1\leq r\leq p$. Then there exists a sequence σ_m of elements of $\mathscr{C}^\infty(\bar{\Omega};E)$ which satisfy:

(7.6) $$\|\sigma_m\|_{L^\infty(\Omega;E^D)} \leq c(\Omega)\ \|\sigma^D\|_{L^\infty(\Omega;E^D)}$$

where the constant $c(\Omega)$ depends only on Ω, and, as $m \to \infty$,

(7.7) $$\sigma_m \to \sigma \quad in \quad L^p(\Omega;E)$$

(7.8) $$div\ \sigma_m \to div\ \sigma \quad in \quad L^r(\Omega)^n .$$

Proof

The proof is essentially that of Proposition 1.3, Ch.I: we take $m = n^2$ and $\alpha_0 = p$ in I.(1.74); then $\mathscr{A}_1\sigma = \text{div}\,\sigma$, $\ell_1 = 3$, $\alpha_1 = r$ and $\mathscr{A}_2\sigma = \sigma^D$, $\ell_2 = n^2$, $\alpha_2 = \infty$; the hypothesis $r\leq p$ ensures that the space is of local type. The construction used in Proposition I.1.3 consists of writing

(7.9) $$\sigma = \phi\sigma + \sum_{i\in I} \phi_i\sigma ,$$

where $\phi \in \mathscr{C}_0^\infty(\Omega)$, $0\leq\phi\leq1$, $\phi_i \in \mathscr{C}_0^\infty(\mathscr{O}_i)$, $0\leq\phi_i\leq1$, the $\mathscr{O}_i$ being balls centred on points of Γ whose radius is small enough to ensure that the condition I.(1.16) (for $\complement\bar{\Omega}$) is satisfied; since the boundary Γ is compact, it can be covered by a finite number of these $\mathscr{O}_i$, and the partition of unity,

(7.10) $$1 = \phi + \sum_{i\in I} \phi_i$$

subordinated to this covering of $\bar{\Omega}$ by Ω and the $\mathcal{O}_i$ is therefore finite.

The functions $\rho_\eta(\sigma\phi)$ constructed in para. ii) of the proof of Proposition 1.3 satisfy

$$\rho_\eta * (\sigma\phi) \to \sigma\phi \quad \text{in} \quad L^p(\Omega;E)$$

$$\text{div}\, \rho_\eta * (\sigma\phi) \to \text{div}\,(\sigma\phi) \quad \text{in} \quad L^r(\Omega)^n$$

as $\eta \to 0$, and of course

(7.11) $$\|\rho_\eta*(\sigma^D\phi)\|_{L^\infty} \leq \|\sigma^D\phi\|_{L^\infty} \leq \|\sigma^D\|_{L^\infty} .$$

Similar remarks apply to the restrictions to Ω of the regularised functions obtained from the functions $\sigma\phi_i$. By addition (with $\eta = \frac{1}{m}$) we obtain (7.7) (7.8) and also (7.6) since the sum (7.10) is finite.

Remark 7.1. i) Let $x^i \in \Gamma$ be the centre of the ball $\mathcal{O}_i$ mentioned in the proof of Lemma 7.1. The above functions ϕ and ϕ_j with $j \neq i$, vanish in a neighbourhood of x^i , and so, when m is large enough, σ_m coincide with the regularised function derived from $\phi_i\sigma$ throughout a sufficiently small neighbourhood of x^i . Hence we have the inequality, corresponding to (7.11),

(7.12) $$\|\sigma_m^D\|_{L^\infty(\mathcal{O}_i'\cap\Omega;E)} \leq \|\sigma^D\|_{L^\infty(\Omega;E)} ,$$

where $\mathcal{O}_i' \subset \mathcal{O}_i$ is a suitable ball of centre x^i .

Lastly x^i can be chosen arbitrarily on Γ and we can make Lemma 7.1 more complete by adding the following statement:

(7.13) The sequence σ_m given by Lemma 7.1 can be chosen so that, at any fixed point $x \in \Gamma$:

$$\|\sigma_m^D\|_{L^\infty(\mathcal{O}\cap\Omega;E)} \leq \|\sigma^D\|_{L^\infty(\Omega;E)}$$

where $\mathcal{O}$ is a ball of centre x, whose radius is sufficiently small, but is independent of m.

ii) If, to be more precise, $\sigma^D(x)$ is in K^D for almost all x in Ω, then $\phi_i(x)\ \sigma^D(x) \in K^D$ p.p. (since $0 \in K^D$ and $0 \leq \phi_i(x) \leq 1$), and the above argument shows that $\sigma_m^D(x) \in K^D$ for almost all x of $\mathcal{O}\cap\Omega$, where $\mathcal{O}$ is defined as above.

iii) By replacing ϕ_i by the function ϕ appearing in (7.10) we see that a result identical to (7.13), or to the one in para. ii) above, holds for any open $\mathcal{O}$ which is a relatively compact subset of Ω, with $\mathcal{O} \subset \bar{\mathcal{O}} \subset \Omega$; this is a consequence of (7.11).

□

We have commented on, and used on several occasions, the fact that if $\sigma \in L^2(\Omega;E)$ with div $\sigma \in L^2(\Omega)^n$, then the trace of $\sigma.\nu$ on Γ can be defined with $\gamma_\nu\sigma = \sigma.\nu_{|\Gamma} \in H^{-\frac{1}{2}}(\Gamma)^n$ (cf. I.(1.47)). We can improve on this result when $\sigma \in \Sigma(\Omega)$.

We shall suppose from now on that

(7.14) Ω is a bounded open set of class $\mathscr{C}^2$ in $\mathbb{R}^n$.

LEMMA 7.2.

If $\sigma \in \Sigma(\Omega)$ *and* Ω *satisfies (7.14), then* $(\sigma.\nu)_\tau \in L^\infty(\Gamma)^n$ *and*

(7.15) $$\|(\sigma.\nu)_\tau\|_{L^\infty(\Gamma)^n} \leq \frac{1}{\sqrt{2}}\|\sigma^D\|_{L^\infty(\Omega;E)}$$

where $(\sigma.\nu)_\tau$ *is the tangential component on* Γ *of* $\sigma.\nu$ *:*

(7.16) $$(\sigma.\nu)_\tau = \sigma.\nu - [(\sigma.\nu).\nu]\nu .$$

<u>Proof</u>

Let σ_m be the sequence, given by Lemma 7.1, and which approximates

σ. By (7.7) and (7.8) (with $p = r = 2$) and the trace theorem I.(1.47),

(7.17) $$\sigma_m.\nu_{|\Gamma} \to \sigma.\nu_{|\Gamma} \quad \text{in} \quad H^{-1/2}(\Gamma)^n ,$$

as $n \to \infty$. The result will therefore be established if we show that (7.15) holds for the functions σ_m ; it will even suffice, after Remark 7.1 and (7.13), to prove that

$$\|(\sigma_m.\nu)_\tau\|_{L^\infty(\mathcal{O}\cap\Gamma)^n} \leqslant \frac{1}{\sqrt{2}} \|\sigma_m^D\|_{L^\infty(\Omega;E)}$$

where $\mathcal{O}$ is a small enough ball centred on any arbitrary point x of Γ.

In fact for a regular function the property we are looking for is a local algebraic property: at each point $u \in \Gamma \cap \mathcal{O}$, we choose an orthonormal frame of reference, one of whose axes is parallel to the normal $\nu(y)$ at y, and we find by an elementary calculation that:

(7.18) $$|(\sigma_m(y).\nu(y))_\tau|^2 = \sum_{j=1}^{n-1} (\xi_{jn})^2 \leqslant \frac{1}{2} \sum_{\substack{i,j=1 \\ i\neq j}}^{n-1} (\xi_{ij})^2 \leqslant \frac{1}{2} |\sigma_m^D(y)|^2 ;$$

we have written ξ_{ij} for the components of $\sigma_m(y)$ with respect to this frame of reference and chosen ν to be parallel to the last axis, $\nu = (0,\ldots,0,1)$.

Remark 7.2. The proof of Lemma 7.2, in conjunction with (7.17) shows that the sequence σ_m given by Lemma 7.1 also satisfies

(7.19) $$(\sigma_m.\nu)_\tau \to (\sigma.\nu)_\tau \quad \text{in} \quad L^\infty(\Gamma)^n \text{ weak-star,}$$

when $m \to \infty$ (in addition to (7.17)).

□

We now state a regularity result for the functions of $\Sigma(\Omega)$ (inclusion in L^p spaces).

PROPOSITION 7.1.

Let Ω be a bounded open set in $\mathbb{R}^n$ of class $\mathscr{C}^2$.

i) Let $\sigma \in L^1(\Omega;E)$ *with* $\sigma^D \in L^\infty(\Omega;E)$ *and* $\operatorname{div} \sigma \in L^p(\Omega)^n$, $1 \leq p < n$. *Then*

$$\sigma \in L^{p^*}(\Omega;E) \quad , \quad p^* = \frac{np}{n-p} \tag{7.20}$$

and there exists a constant $c = c(\Omega,p)$ *depending only on* Ω *and* p, *and such that*

$$\|tr\, \sigma\|_{L^{p^*}(\Omega)} \leq c(\Omega,p).\{\|tr\, \sigma\|_{L^1(\Omega)} + \|\sigma^D\|_{L^\infty(\Omega;E)} + \|\operatorname{div} \sigma\|_{L^p(\Omega)^n}\} \tag{7.21}$$

ii) In particular, if $\sigma \in \Sigma(\Omega)$, *then* $\sigma \in L^p(\Omega)$, *for all finite* p *with* $1 \leq p < \infty$, *and (7.21) gives*

$$\|\sigma\|_{L^p(\Omega;E)} \leq c'(\Omega,p)\ \|\sigma\|_{\Sigma(\Omega)} . \tag{7.22}$$

Proof

We need to prove i) only, since ii) is an immediate consequence. We first assume that σ also satisfies $\sigma \in L^p(\Omega;E)$. We have to show that $tr\,\sigma \in L^{p^*}(\Omega)$ and that (7.21) holds. To do this it is sufficient to show that, for any ψ in $\mathscr{C}^\infty(\bar{\Omega})$,

$$\left|\int_\Omega \psi\ tr\,\sigma\ dx\right| \leq C(\Omega,p,\sigma)\ \|\psi\|_{L^q(\Omega)} , \tag{7.23}$$

where q is the exponent conjugate to p^*, that is $q = p^*/(p^*-1)$ and $C(\Omega,p,\sigma)$ is the expression on the right of the inequality (7.21).

We first examine the case where

$$\int_\Omega \psi\ dx = 0 . \tag{7.24}$$

In this case we write Ψ for the solution of Neumann's problem

(7.25)
$$\begin{cases} \Delta\Psi = \psi & \text{in} \quad \Omega \\ \dfrac{\partial\Psi}{\partial\nu} = 0 & \text{on} \quad \Gamma , \end{cases}$$

and put $u = \nabla\Psi$. As Ω is of class $\mathscr{C}^2$, the usual regularity results for Neumann's problem (cf. for example J.L. Lions - E. Magenes [1]) imply that $\Psi \in W^{2,q}(\Omega)$, and that

(7.26)
$$||\Psi||_{W^{2,q}(\Omega)} \leq c_1' ||\psi||_{L^q(\Omega)}$$

where c_1' depends only on Ω and q (and therefore p) .

Now consider the sequence of functions σ_m which approximate to σ under the conditions specified in Lemma 7.1 (with p and r replaced by p). For each m we apply the usual Green's formula I.(1.49) with σ_m and u . By noting that

(7.27)
$$\sigma.\varepsilon(u) = \frac{1}{n} \text{tr } \sigma.\text{div } u + \sigma^D.\varepsilon^D(u) ,$$

we obtain

(7.28)
$$\frac{1}{n} \int_\Omega (\text{tr } \sigma_m)(\text{div } u)dx =$$
$$= \int_\Gamma [(\sigma_m.\nu).u]d\Gamma - \int_\Omega (\text{div } \sigma_m).u \ dx - \int_\Omega \sigma_m^D.\varepsilon^D(u) \ dx .$$

Since $u \in W^{1,q}(\Omega)$, the functions $\text{div } u$, $\varepsilon_{ij}^D(u)$ and u are L-integrable on Ω and the trace of u on Γ is in $L^1(\Gamma)$ (by I.(1.32)). Also $u.\nu = \frac{\partial\Psi}{\partial\nu} = 0$ on Γ , so that

(7.29)
$$\int_\Gamma [(\sigma_m.\nu).u]d\Gamma = \int_\Gamma (\sigma_m.\nu)_\tau.u_\tau \ d\Gamma \quad .$$

Under these conditions the convergence relations (7.7) (7.8) (with $r = p$) and (7.19) suffice to justify the passage to the limit $m \to \infty$ in (7.28). We thus obtain

(7.30) $\frac{1}{n}\int_\Omega (\text{tr}\ \sigma)(\text{div}\ u)\ dx =$

$$= \int_\Gamma (\sigma.\nu)_\tau . u_\tau\ d\Gamma - \int_\Omega (\text{div}\ \sigma) u\ dx - \int_\Omega \sigma^D . \varepsilon^D(u)\ dx\ ,$$

and

(7.31) $|\int_\Omega (\text{tr}\ \sigma)\psi\ dx| \leqslant n\int_\Gamma |(\sigma.\nu)_\tau . u_\tau| d\Gamma + n\int_\Omega |(\text{div}\ \sigma).u| dx$

$$+ n\int_\Omega |\sigma^D . \varepsilon^D(u)| dx\ .$$

We now majorise the right-hand side of (7.31). Thanks to (7.15), the first integral appearing on the right of (7.31) is majorised by

$$n\|(\sigma.\nu)_\tau\|_{L^\infty(\Gamma)} \int_\Gamma |u| d\Gamma \leqslant n\|\sigma^D\|_{L^\infty(\Omega)} . \int_\Gamma |u|\ d\Gamma\ .$$

and with the help of Gagliardo's trace theorem (I.(1.32)) and (7.26) we have

$$\int_\Gamma |u| d\Gamma \leqslant c_2'(\Omega)\ \|u\|_{W^{1,1}(\Omega)^n} \leqslant c_3'\|\Psi\|_{W^{2,1}(\Omega)} \leqslant c_4'|\psi|_{L^q(\Omega)}\ .$$

Using Hölder's inequality we can majorise the second integral on the right of (7.31) by

$$n\|\text{div}\ \sigma\|_{L^p(\Omega)^n}\|u\|_{L^{p'}(\Omega)^n}$$

where $p' = p/(p-1)$. But $u = \text{grad}\ \Psi$ is in $W^{1,q}(\Omega)^n$ and so by Sobolev's injection theorem (cf. I.(1.28)), this space is precisely contained in $L^{p'}(\Omega)^n$ $(\frac{1}{p'} = \frac{1}{q} - \frac{1}{n})$. The end result is that the second term of the right-hand side of (7.31) is majorised by

$$c_5'\|\text{div}\ \sigma\|_{L^p(\Omega)^n}\|\Psi\|_{W^{2,q}(\Omega)} \leqslant$$

$$\leqslant c_1'\ c_5'\|\text{div}\ \sigma\|_{L^p(\Omega)^n}\|\psi\|_{L^q(\Omega)}\ .$$

The third term of the right-hand side of (7.31) is similarly majorised by

$$n\|\sigma^D\|_{L^\infty(\Omega;E)} \cdot \|\varepsilon^D(u)\|_{L^1(\Omega;E)} \leq$$

$$\leq c_6'\|\sigma^D\|_{L^\infty(\Omega;E)} \|u\|_{W^{1,1}(\Omega)^n}$$

$$\leq c_7'\|\sigma^D\|_{L^\infty(\Omega;E)} \|\psi\|_{L^q(\Omega)} .$$

By rearranging these various upper bounds, we see that (7.31) implies (7.23) when ψ satisfies (7.24).

To conclude the proof it remains only to establish an inequality of the type of (7.23) when $\psi \in \mathscr{C}^\infty(\bar{\Omega})$ does not satisfy (7.24). In this case we write, as we may, (7.23) with ψ replaced by

$$\psi - \frac{1}{|\Omega|}\int_\Omega \psi \, dx \;, \tag{7.32}$$

where $|\Omega|$ is the measure of Ω. It then follows easily that

$$|\int_\Omega \psi \operatorname{tr}\sigma \, dx| \leq \frac{1}{|\Omega|}\left(\int_\Omega |\psi| dx\right).\left(\int_\Omega |\operatorname{tr}\sigma| dx\right)$$
$$+ C(\Omega,p,\sigma)\left(\|\psi\|_{L^q(\Omega)} + c''(\Omega,q)\int_\Omega |\psi| dx\right) .$$

This inequality clearly implies an inequality of the same type as (7.23) and the Proposition is therefore proved when $\sigma \in L^p(\Omega;E)$. If we assume only that $\sigma \in L^1(\Omega;E)$, we apply the Proposition which has just been proved with $p = 1$ and obtain the result that $\sigma \in L^{1^*}(\Omega;E)$, $1^* = n/(n-1)$. We begin again with $p = 1^*$ to obtain $\sigma \in L^{(1^*)^*}(\Omega;E)$ and then $p = (1^*)^*$ and so on. At a certain stage we obtain $\sigma \in L^p(\Omega;E)$ and we are then in the situation envisaged at the outset.

Remark 7.3. When σ is a spherical tensor, $\sigma_{ij} = v\delta_{ij}$, the hypotheses on σ in Proposition 7.1 i) mean that $v \in W^{1,p}(\Omega)$, $1<p<n$ and then the conclusion (7.20) (7.21) is identical to that of the Sobolev embedding theorem for $W^{1,p}(\Omega)$ (that is to say I.(1.28) with $m = 1$). □

7.2 A trace theorem

We give here a variant of the trace theorem I.(1.41) which gives a meaning to $v.\nu_{|\Gamma}$ when $v \in L^2(\Omega)^n$ and $\operatorname{div} v \in L^2(\Omega)$ and which allows us to write the generalised Green's formula I.(1.42), (cf. J.L. Lions - E. Magenes [1] or R. Temam [6], Theorem 1.2, Ch.I).

We define the space

$$Z(\Omega) = \{v \in L^1(\Omega)^n \ , \ \operatorname{div} v \in M_1(\Omega)\} \ ; \tag{7.33}$$

This is a Banach space for the natural norm

$$||v||_{Z(\Omega)} = ||v||_{L^1(\Omega)^n} + ||\operatorname{div} v||_{M_1(\Omega)} \ .$$

We have

PROPOSITION 7.2.

Let Ω be a bounded open set in $\mathbb{R}^n$ of class $\mathscr{C}^2$. Then there exists a continuous linear mapping γ_ν of $Z(\Omega)$ into $\mathscr{C}^1(\Gamma)'$ such that

$$\gamma_\nu(v) = v.\nu_{|\Gamma} \ , \ \textit{for all} \ \ v \in \mathscr{C}^1(\bar{\Omega})^n \ , \tag{7.34}$$

and

$$\langle\gamma_\nu(v),\phi\rangle = \int_\Omega v.\operatorname{grad} \Phi \ dx + \int_\Omega \Phi.\operatorname{div} v \ , \tag{7.35}$$

for all $v \in Z(\Omega)$, for all $\phi \in \mathscr{C}^1(\Gamma)$ and all $\Phi \in \mathscr{C}^1(\bar{\Omega})$ such that $\Phi_{|\Gamma} = \phi$.

<u>Proof</u>

i) We write $\Lambda(v,\Phi)$ for the right-hand side of (7.35); we shall show that it depends only on $\Phi_{|\Gamma} = \phi$, and that the mapping $\phi \mapsto \Lambda(v,\Phi)$ is linear and continuous on $\mathscr{C}^1(\Gamma)$.

Proving that $\Lambda(v,\Phi)$ depends only on $\Phi_{|\Gamma}$ amounts to the same thing as showing that $\Lambda(v,\Phi) = 0$ whenever $\Phi \in \mathscr{C}^1(\bar{\Omega})$ and $\Phi_{|\Gamma} = 0$.

If $\Phi \in \mathscr{C}^1(\bar{\Omega})$ and $\Phi_{|\Gamma} = 0$, we consider the sequence of functions Φ_m defined by:

$$\Phi_m(x) = \Phi(x)\,\theta(md(x)) \,, \tag{7.36}$$

where $m \in \mathbb{N}$, $d(x)$ is the distance from x to Γ and θ is a function of $\mathscr{C}^\infty$ mapping $(0,\infty)$ into $[0,1]$, such that $\theta(s) = 0$ for $0<s\leq 1$ and $\theta(s) \equiv 1$ for $s\geq 2$.

It is easily verified, by regularising Φ_m and proceeding to the limit, that

$$\Lambda(v,\Phi_m) = 0 \,,\ \forall\, m \,. \tag{7.37}$$

To deduce that $\Lambda(v,\Phi) = 0$, it is thus sufficient to show that

$$\Lambda(v,\Phi) = \lim_{m\to\infty} \Lambda(v,\Phi_m) \,. \tag{7.38}$$

But

$$\sup_{x\in\bar{\Omega}} |\Phi_m(x)-\Phi(x)| \leq \sup_{d(x)\leq 2/m} |\Phi(x)|$$

and the latter tends to 0 as $m \to \infty$, because Φ is uniformly continuous in $\bar{\Omega}$ and $\Phi = 0$ on Γ ; it follows therefore that

$$\int_\Omega \Phi_m.\mathrm{div}\, v \;\to\; \int_\Omega \Phi.\mathrm{div}\, v$$

as $m \to \infty$.

Also

$$\left|\int_\Omega v.\mathrm{grad}\,\Phi\, dx - \int_\Omega v.\mathrm{grad}\,\Phi_m\, dx\right| \leq$$

$$\leq \int_\Omega |v(x)|\,|\mathrm{grad}\,\Phi(x)|\,\theta(md(x))dx \,+$$

$$+ \int_\Omega m|\Phi(x)|\,|\theta'(md(x))|\,|\mathrm{grad}\,d(x)|dx$$

$$\leq \sup_{\Omega} |\text{grad}\ \phi| . \int_{d(x)\leq 2/m} |v(x)|dx + m \sup_{s\geq 0} |\theta'(s)| . \int_{d(x)\leq 2/m} |\phi(x)|dx$$

(since $|\text{grad}\ d(x)| \leq 1$).

Since the measure of the set $\{x \in \Omega\ ,\ d(x) \leq \frac{2}{m}\}$ is less than $c\ m$ for some approprriate constant c and since $\Phi = 0$ on Γ , the above expression tends to 0 as $m \to \infty$ and thus (7.38) is proved.

ii) To show that the mapping

$$\phi \in \mathscr{C}^1(\Gamma) \to \Lambda(v,\phi) \tag{7.39}$$

is linear and continuous, it now suffices to show that for any $\phi \in \mathscr{C}^1(\Gamma)$ there exists a $\Phi \in \mathscr{C}^1(\bar{\Omega})$ such that $\Phi_{|\Gamma} = \phi$, the mapping $\phi \to \Phi$ being moreover linear and continuous. This is standard; the open set being of class $\mathscr{C}^1$ one can take, for example,

$$\Phi(x) = [1 - \theta(md(x))]\ \phi(\pi(x)) \quad ,$$

where θ is as above, m large enough, and $\pi(x)$ is the projection of x on Γ which is defined and of class $\mathscr{C}^1$ when $d(x) \leq \eta_0$, and $\eta_0 > 0$ is small enough (and thus $m \geq 2/\eta_0$) .

We denote by $\gamma_\nu(v)$ the linear form defined by (7.39) thus proving (7.35).

To establish (7.34) we note that if $v \in \mathscr{C}^1(\bar{\Omega})^n$, then by comparing (7.35) with the classical Green's formula, we obtain

$$\langle \gamma_\nu(v) - v.\nu_{|\Gamma}, \phi \rangle = 0\ ,\quad \forall\ \phi \in \mathscr{C}^1(\Gamma)\ .$$

This implies that $\gamma_\nu(v) = v.\nu_{|\Gamma}$ in $\mathscr{C}^1(\Gamma)'$ (and hence also in the sense of distributions on Γ) .

Remark 7.4. It follows from (7.34) and (7.35) that γ_ν is the *one and only* extension of the mapping

$$v \in \mathscr{C}^1(\bar{\Omega})^n \to v.\nu_{|\Gamma} \in \mathscr{C}(\Gamma)$$

into a continuous linear mapping of $Z(\Omega)$ into $\mathscr{C}^1(\Gamma)'$. The condition (7.34) alone is not sufficient to ensure the uniqueness of γ_ν because $\mathscr{C}^1(\bar{\Omega})^n$ is not dense in $Z(\Omega)$; a similar situation arose in connection with the trace theorem for $BD(\Omega)$ (Theorem 2.1).

7.3. Definition of $\sigma.\varepsilon(u)$ as a measure

We shall show that it is possible to define $\sigma.\varepsilon(u)$ as a bounded measure on Ω when $u \in U(\Omega)$ and $\sigma \in \Sigma(\Omega)$.

It is easy to define $\sigma.\varepsilon(u)$ as a distribution on Ω. If σ and u are sufficiently regular and Green's formula I.(1.51) is written with u replaced by ϕu where $\phi \in \mathscr{C}_0^\infty(\Omega)$, we obtain

$$(7.40) \qquad \int_\Omega \sigma.\varepsilon(u)\phi \, dx = - \int_\Omega (\text{div } \sigma).u \, \phi \, dx - \int_\Omega \sigma.(u \otimes \text{grad } \phi) dx \, .$$

When $\sigma \in \Sigma(\Omega)$, $u \in U(\Omega)$ and $\phi \in \mathscr{C}_0^\infty(\Omega)$, the right-hand side of (7.40) has a meaning thanks to Theorem 2.2 ($u \in L^{n^*}(\Omega)^n$, $n^* = n/(n-1)$) and Proposition 7.1 ($\sigma \in L^n(\Omega;E)$ in particular); also the right-hand side of (7.40) depends linearly on ϕ, and continuously, for the topology of uniform convergence of ϕ and its first derivatives. It therefore defines a distribution which we denote by $\sigma.\varepsilon(u)$.

By writing the usual algebraic relation

$$(7.41) \qquad \sigma.\varepsilon(u) = \frac{1}{n} (\text{div } u)(\text{tr } \sigma) + \sigma^D.\varepsilon^D(u) \quad ,$$

and observing that $(\text{div } u)(\text{tr } \sigma)$ is a summable function, we define also a distribution $\sigma^D.\varepsilon^D(u)$:

$$(7.42) \qquad \langle \sigma.\varepsilon(u),\phi \rangle = - \int_\Omega (\text{div } \sigma)u \, \phi \, dx - \int_\Omega \sigma.(u \otimes \text{grad } \phi) dx \quad ,$$

$$(7.43) \quad \langle \sigma^D.\varepsilon^D(u),\phi \rangle = - \frac{1}{n} \int_\Omega (\text{div } u)(\text{tr } \sigma)\phi \, dx - \int_\Omega (\text{div } \sigma)u \, \phi \, dx - \int_\Omega \sigma.(u \otimes \text{grad } \phi) dx \, ,$$

for all ϕ of the class $\mathscr{C}_0^\infty(\Omega)$.

In fact we have

LEMMA 7.3

The distributions $\sigma.\varepsilon(u)$ *and* $\sigma^D.\varepsilon^D(u)$ *defined by (7.42) (7.43) are bounded measures on* Ω ; $\sigma^D.\varepsilon^D(u)$ *is absolutely continuous with respect to* $|\varepsilon^D(u)|$ *and*

$$(7.44)\qquad \left|\int_\Omega \phi\, \sigma^D.\varepsilon^D(u)\right| \leq \|\sigma^D\|_{L^\infty(\Omega;E_n)} \cdot \int_\Omega |\phi||\varepsilon^D(u)|$$

for all ϕ *of* $\mathscr{C}_o(\Omega)$.

Proof

It suffices to prove the result for $\sigma^D.\varepsilon^D(u)$. We consider the sequence u_m approximating u which is given by Theorem 3.4 and Remark 3.5 : the sequence u_m converges to u in the sense of (3.37) (3.39) (3.40) and (3.55) : when $m \to \infty$,

$$(7.45)\qquad \begin{cases} u_m \to u \quad \text{in } L^p(\Omega)^n \text{ strongly, } \forall\, p<n^* \\ \operatorname{div} u_m \to \operatorname{div} u \quad \text{in } L^2(\Omega) \text{ strongly} \\ \varepsilon_{ij}(u_m) \to \varepsilon_{ij}(u) \text{ weakly, } \forall\, i,j \\ \int_\Omega |\varepsilon^D(u_m)| \to \int_\Omega |\varepsilon^D(u)| \ ; \end{cases}$$

for the last convergence we have written (5.42) with $f(\xi^D) = |\xi^D|$ (1).
By (2.15) we have therefore, for all $\phi \in \mathscr{C}(\bar{\Omega})$

$$(7.46)\qquad \int_\Omega \phi|\varepsilon^D(u_m)| \to \int_\Omega \phi|\varepsilon^D(u)| \ .$$

(1) See also Remarks 5.3 and 5.4.

Also it is clear from (7.43) and (7.45) that $\sigma^D.\varepsilon^D(u_m)$ converges to $\sigma^D.\varepsilon^D(u)$ in the distribution sense,

$$\langle\sigma^D.\varepsilon^D(u_m),\phi\rangle \rightarrow \langle\sigma^D.\varepsilon^D(u),\phi\rangle \,, \tag{7.47}$$

as $m \rightarrow \infty$, for all $\phi \in \mathscr{C}_0^\infty(\Omega)$. The inequality

$$\left|\int_\Omega (\sigma^D.\varepsilon^D)(u_m))\phi \; dx\right| \leq \|\sigma^D\|_{L^\infty(\Omega;E)} \int_\Omega |\phi|\,|\varepsilon^D(u_m)|$$

gives, in the limit, with (7.46) (applied to $|\phi|$) and (7.47) :

$$|\langle\sigma^D.\varepsilon^D(u),\phi\rangle| \leq \|\sigma^D\|_{L^\infty(\Omega;E)} \int_\Omega |\phi|\,|\varepsilon^D(u)|$$

for all $\phi \in \mathscr{C}_0^\infty(\Omega)$. It follows from this by classical arguments (cf. N. Bourbaki [3], L. Schwartz [1]) that $\sigma^D.\varepsilon^D(u)$ is a bounded measure satisfying (7.44) (with $\phi \in \mathscr{C}_0(\Omega)$) and that it is absolutely continuous with respect to $|\varepsilon^D(u)|$.

□

The measures $\sigma.\varepsilon(u)$ and $\sigma^D.\varepsilon^D(u)$ are of course bilinear functions of σ and u. They also depend continuously on σ and u under certain topologies. As regards u we have :

LEMMA 7.4.

Let $\sigma \in \Sigma(\Omega)$ and $u \in U(\Omega)$; the measures $\sigma.\varepsilon(u)$ and $\sigma^D.\varepsilon^D(u)$ depend continuously on u in the following sense : if u_m converges to u in the sense of (7.45) then, when $m \rightarrow \infty$,

$$\int_\Omega \phi(\sigma^D.\varepsilon^D(u_m)) \rightarrow \int_\Omega \phi(\sigma^D.\varepsilon^D(u)) \tag{7.48}$$

$$\int_\Omega \phi(\sigma.\varepsilon(u_m)) \rightarrow \int_\Omega \phi(\sigma.\varepsilon(u)) \tag{7.49}$$

for all $\phi \in \mathscr{C}(\bar{\Omega})$.

Proof

It is sufficient to prove (7.48). For all $\delta>0$, a compact $C \subset \Omega$ can be found such that

$$\int_{\Omega\setminus C} |\varepsilon^D(u)| \leq \delta \tag{7.50}$$

and

$$\int_{\partial C} |\varepsilon^D(u)| = 0 . \tag{7.51}$$

Thus let $\psi \in \mathscr{C}_0^\infty(\Omega)$ with $0\leq\psi\leq 1$ and $\psi = 1$ on C. We write

$$\begin{aligned}
&\left|\int_\Omega \phi(\sigma^D.\varepsilon^D(u_m)) - \int_\Omega \phi(\sigma^D.\varepsilon^D(u))\right| \leq \\
&\quad \leq \left|\int_\Omega \phi\psi(\sigma^D.\varepsilon^D(u_m)) - \int_\Omega \phi\psi(\sigma^D.\varepsilon^D(u))\right| \\
&\qquad + \left|\int_\Omega \phi(1-\psi)(\sigma^D.\varepsilon^D(u_m))\right| + \left|\int_\Omega \phi(1-\psi)(\sigma^D.\varepsilon^D(u))\right| \\
&\quad \leq \left|\int_\Omega \phi\psi(\sigma^D.\varepsilon^D(u_m)) - \int_\Omega \phi\psi(\sigma^D.\varepsilon^D(u))\right| \\
&\qquad + |\phi|_{\mathscr{C}(\bar\Omega)} \|\sigma^D\|_{L^\infty(\Omega;E)} \left\{\int_{\Omega\setminus C} |\varepsilon^D(u_m)| + \int_{\Omega\setminus C} |\varepsilon^D(u)|\right\}
\end{aligned}$$

(by (7.44)).

By Lemma 2.1, (7.45) and (7.51),

$$\lim_{m\to\infty} \int_{\Omega\setminus C} |\varepsilon^D(u_m)| = \int_{\Omega\setminus C} |\varepsilon^D(u)| \ ;$$

so that Lemma 7.3 and (7.50) imply

$$\limsup_{m\to\infty} \left|\int_\Omega \phi(\sigma^D.\varepsilon^D(u_m)) - \int_\Omega \phi(\sigma^D.\varepsilon^D(u))\right| \leq 2\delta |\phi|_{\mathscr{C}(\bar\Omega)} \|\sigma^D\|_{L^\infty(\Omega;E)} .$$

Since $\delta>0$ is arbitrarily small, the result is proved.

□

As regard continuity with respect to σ we consider a sequence $\sigma_m \in \Sigma(\Omega)$ such that, as $m \to \infty$,

(7.52) $$\sigma_m \to \sigma \text{ in } L^2(\Omega;E) \text{ strongly}$$

(7.53) $$\operatorname{div} \sigma_m \to \operatorname{div} \sigma \text{ in } L^{n+1}(\Omega)^n \text{ strongly,}$$

(7.54) $$\sigma_m^D \to \sigma^D \text{ in } L^\infty(\Omega;E^D) \text{ weak-star convergence}$$

Note that if σ_m satisfies (5.52) (7.53) then (7.54) holds if and only if σ^D is bounded in L^∞,

(7.55) $$\|\sigma_m^D\|_{L^\infty(\Omega;E^D)} \leq \text{const.}$$

By Lebesgue's theorem it follows easily from (7.52) and (7.54) that σ_m^D then converges to σ^D in $L^p(\Omega;E^D)$ for all finite $p \geq 1$. We next see, by Proposition 7.1, that

(7.56) $\sigma_m \to \sigma$ in $L^p(\Omega;E)$ strongly for all p such that $1 \leq p < \infty$.

We have

LEMMA 7.5

The measures $\sigma.\varepsilon(u)$ *and* $\sigma^D.\varepsilon^D(u)$ *depend continuously on* σ *in the following sense: if* $u \in U(\Omega)$, $\sigma \in \Sigma(\Omega)$ *and* σ_m *converge to* σ *in the sense of (7.52)-(7.54), then, as* $m \to \infty$,

(7.57) $$\int_\Omega \phi(\sigma_m^D.\varepsilon^D(u)) \to \int_\Omega \phi(\sigma^D.\varepsilon^D(u))$$

(7.58) $$\int_\Omega \phi(\sigma_m.\varepsilon(u)) \to \int_\Omega \phi(\sigma.\varepsilon(u)),$$

for all $\phi \in \mathscr{C}(\bar{\Omega})$.

Proof

It is sufficient to prove (7.57). If $\phi \in \mathscr{C}_0^\infty(\Omega)$ then (7.57) follows easily from the definition (7.43). If $\phi \in \mathscr{C}(\bar{\Omega})$ then for an arbitrary $\delta>0$, there exists a $\psi \in \mathscr{C}_0^\infty(\Omega)$ such that $0\leq\psi\leq 1$ and

$$\int_\Omega (1-\psi)|\varepsilon^D(u)| \leq \delta .$$

By (7.44)

$$|\int_\Omega \phi(1-\psi)(\sigma^D.\varepsilon^D(u))| \leq \|\sigma^D\|_{L^\infty(\Omega;E)} \int_\Omega (1-\psi)|\varepsilon^D| \leq \delta\|\sigma^D\|_{L^\infty(\Omega;E)} ,$$

and similarly for all m, using (7.55),

$$|\int_\Omega \phi(1-\psi)(\sigma_m^D.\varepsilon^D(u))| \leq c\delta .$$

It follows that

$$|\int_\Omega \phi(\sigma_m^D.\varepsilon^D(u)) - \int_\Omega \phi(\sigma^D.\varepsilon^D(u))| \leq |\int_\Omega \phi\psi(\sigma_m^D.\varepsilon^D(u)) - \int_\Omega \phi\psi(\sigma^D.\varepsilon^D(u))|$$
$$+ \delta(c+\|\sigma^D\|_{L^\infty}) .$$

The upper limit of this quantity as $m \to \infty$ is thus less than $\delta(c+\|\sigma^D\|_{L^\infty})$ and since $\delta>0$ is arbitrarily small, this upper limit is zero which proves (7.57). □

The following remark concerns the trace on Γ of $(\sigma.\nu).u$. We have

LEMMA 7.6

Assuming Ω to be an open set of class $\mathscr{C}^2$, there exists an unique distribution on Γ, denoted by $(\sigma.\nu).u$, which belongs to $\mathscr{C}^1(\Gamma)'$ and which satisfies

$$(7.59)\quad \langle(\sigma.\nu).u,\phi\rangle = \int_\Omega \Phi(\sigma^D.\varepsilon^D(u)) + \frac{1}{n}\int_\Omega (\mathrm{div}\ u)(\mathrm{tr}\ \sigma)\Phi\ dx$$
$$+ \int_\Omega (\mathrm{div}\ \sigma).u\ \Phi\ dx$$
$$+ \int_\Omega \sigma.(u \otimes \mathrm{grad}\ \Phi)\ dx$$

for all $\phi \in \mathscr{C}^1(\Gamma)$ *and all* $\Phi \in \mathscr{C}^1(\bar{\Omega})$ *such that* $\Phi_{|\Gamma} = \phi$.

Proof

Since $u \in L^{n/(n-1)}(\Omega)^n$, it follows from Proposition 7.1 that $\sigma.u \in L^1(\Omega)^n$. Also the formula (7.42) means precisely

$$\operatorname{div}(\sigma.u) = \sigma.\varepsilon(u) + (\operatorname{div}\sigma).u$$

$$= \sigma^D.\varepsilon^D(u) + \frac{1}{n}(\operatorname{div} u)(\operatorname{tr}\sigma) + (\operatorname{div}\sigma)u .$$

By the preceding results, $\operatorname{div}(\sigma.u)$ is a bounded measure on Ω , and we can apply Proposition 7.2 with $v = \sigma.u$; this proposition allows us to define $\gamma_\nu(v) = \gamma_\nu(\sigma.u)$, which satisfies Green's formula (7.35). If we agree, by a conventional abuse of notation, to write $(\sigma.\nu).u$ or $(\sigma.u).\nu$ for the distribution $\gamma_\nu(\sigma.u)$ then (7.35) becomes identical to (7.59) and the lemma is proved.

□

We can be slightly more precise about the definition of $(\sigma.\nu).u$. When σ and u are sufficiently regular, we can write:

(7.60) $$(\sigma.\nu).u = (\sigma.\nu)_\tau.u_\tau + (\sigma.\nu)_\nu.u_\nu .$$

In the present case $(\sigma.\nu)_\tau = (\sigma^D.\nu)_\tau$ by Lemma 7.2 and this belongs to $L^\infty(\Gamma)^n$; on the other hand $u_\tau \in L^1(\Gamma)^n$ by the trace theorem for $BD(\Omega)$; the scalar product $(\sigma.\nu)_\tau.u_\tau$ is well-defined and thus belongs to $L^1(\Gamma)$; the product $(\sigma.\nu)_\nu.u_\nu$ is not defined, on the other hand, when $\sigma \in \Sigma(\Omega)$ and $u \in U(\Omega)$. We shall however establish

LEMMA 7.7

If $\sigma \in \Sigma(\Omega)$, $u \in U(\Omega)$ *and*

(7.61) $$u_\nu \in H^{1/2}(\Gamma)$$

or

(7.62) $$(\sigma.\nu)_\nu \in L^\infty(\Gamma) ,$$

then the relation (7.60) holds.

Proof

We first note that each term in (7.60) does have a meaning: $(\sigma.\nu).u$ by Lemma 7.6, $(\sigma.\nu)_\tau.u_\tau$ for the reason described above; and finally, with regard to $(\sigma.\nu)_\nu.u_\nu$, if (7.61) holds then $u_\nu \in H^{1/2}(\Gamma)$, $(\sigma.\nu)_\nu \in H^{-1/2}(\Gamma)$ by I.(1.47) so that $(\sigma.\nu)_\nu.u_\nu$ is well-defined as a distribution on $\Gamma(\in \mathscr{C}^1(\Gamma)'$, since the functions in $\mathscr{C}^1(\Gamma)$ operate multiplicatively in $H^{1/2}(\Gamma))$. If (7.62) holds then we have, more simply $(\sigma.\nu)_\nu \in L^\infty(\Gamma)^n$, $u_\nu \in L^1(\Gamma)^n$ by the trace theorem for $BD(\Omega)$ and the product $(\sigma.\nu)_\nu.u_\nu$ is an L-integrable function on Γ.

To prove (7.60) we have to establish that

$$(7.63)\qquad \langle(\sigma.\nu).u,\phi\rangle = \langle(\sigma.\nu)_\tau.u_\tau,\phi\rangle + \langle(\sigma.\nu)_\nu.u_\nu,\phi\rangle ,$$

for all $\phi \in \mathscr{C}^1(\bar{\Omega})$.

In the case (7.61), we approximate σ by a sequence of regular functions σ_m which by Lemma 7.1, satisfy (7.52)-(7.54). For each m Green's formula (7.59), with σ replaced by σ_m, holds true and it is clear that in this case

$$(7.64)\qquad \langle(\sigma_m.\nu).u,\phi\rangle = \int_\Gamma (\sigma_m.\nu).u\ \phi\ d\Gamma =$$

$$= \int_\Gamma (\sigma_m.\nu)_\tau.u_\tau\ \phi\ d\Gamma + \int_\Gamma (\sigma_m.\nu)_\nu.u_\nu\ \phi\ d\Gamma .$$

The passage to the limit in formulae (7.59) and (7.64) is legitimate, yielding (7.63); to pass to the limit in (7.64) we note that by (7.52)-(7.54) and I.(1.47),

$$(7.65)\qquad (\sigma_m.\nu)_\nu \to (\sigma.\nu)_\nu \text{ in } H^{-1/2}(\Gamma) .$$

Since $u \in H^{1/2}(\Gamma)$, we have indeed

$$(7.66)\qquad \int_\Gamma (\sigma_m.\nu)_\nu\ u_\nu\phi\ d\Gamma = \langle(\sigma_m.\nu)_\nu,u_\nu\phi\rangle \to \langle(\sigma.\nu)_\nu,u_\nu\phi\rangle$$

as $m \to \infty$.

The argument is similar in the case (7.62): one then approximates u by a sequence u_m satisfying (7.45).

We summarise all these lemmas in

THEOREM 7.1.

For $u \in U(\Omega)$ and $\sigma \in \Sigma(\Omega)$, it is possible to define bounded measures on Ω, $\sigma.\varepsilon(u)$, $\sigma^D.\varepsilon^D(u)$ defined by (7.42) and (7.43). These measures depend linearly and continuously on σ and u (separately), under the topologies indicated in Lemmas 7.4 and 7.5.

Finally, one can define a distribution $(\sigma.\nu).u \in \mathscr{C}^1(\Gamma)'$ such that the generalised Green's formula (7.59) is satisfied.

Remark 7.5. As an adjunct to Lemma 7.7 let us note the following : if $\sigma \in \mathscr{S}_a \cap \mathscr{K}_a$ (cf. (7.2) (7.3)) and $u \in \tilde{\mathscr{C}}_a$ (cf. (6.17)), then $u_{\nu|\Gamma_0} = u_{0\nu|\Gamma_0} \in H^{1/2}(\Gamma_0)$ and $\sigma.\nu_{|\Gamma_1} = g_{|\Gamma_1}$, so that $(\sigma.\nu)_{\nu|\Gamma_1} = g_{\nu|\Gamma_1} \in L^\infty(\Gamma_1)$ by hypothesis (6.14).

We thus have a combination of (7.61) and (7.62). It can be shown, in this case, that

$$(7.67) \qquad \langle(\sigma.\nu).u,\phi\rangle = \int_\Gamma (\sigma.\nu)_\tau \, u_\tau \, \phi \, d\Gamma + \int_{\Gamma_1} (\sigma.\nu)_\nu \, u_\nu \phi \, d\Gamma + \langle(\sigma.\nu)_\nu, u_{0\nu} \, \phi\rangle ,$$

for all $\phi \in \mathscr{C}^1(\Gamma)$. The meaning of each of the terms in (7.67) is clear enough, except for the last which represents, let us recall, the scalar product in the duality pairing between $(\sigma.\nu)_\nu \in H^{-1/2}(\Gamma)$ and $u_{0\nu} \phi \in H^{1/2}(\Gamma)$, the function u_0 being supposed, by hypothesis, to vanish on Γ_1 (cf. I.(4.2)). We refer the reader to R. Kohn - R. Temam [2] for the somewhat technical details of this result.

□

We now add a few remarks complementing the preceding results.

Remark 7.6. The measures $\sigma.\varepsilon(u)$ and $\sigma^D.\varepsilon^D(u)$ are equal to the ordinary scalar product of σ and $\varepsilon(u)$ (respectively σ^D and $\varepsilon^D(u)$) in certain cases where this product has a meaning; for example, if, in addition to the preceding condition ($\sigma \in \Sigma(\Omega)$, $u \in U(\Omega)$) we have :

$$\sigma^D \in \mathscr{C}(\bar{\Omega};E^D) \tag{7.68}$$

or if

$$\varepsilon^D(u) \in L^1(\Omega;E^D) . \tag{7.69}$$

To see this it is sufficient, in either case to verify that (compare with (7.40) (7.42) (7.43))

$$\langle\sigma^D.\varepsilon^D(u),\phi\rangle = \int_\Omega \phi\ \sigma^D.\varepsilon^D(u) = -\frac{1}{n}\int_\Omega (\text{div } u)(\text{tr } \sigma)\phi\ dx - \int_\Omega (\text{div } \sigma)u\ \phi\ dx - \int_\Omega \sigma.(u \otimes \text{grad } \phi)dx , \tag{7.70}$$

for all $\phi \in \mathscr{C}_0^\infty(\Omega)$; in the integral $\int_\Omega \phi\ \sigma^D.\varepsilon^D(u)$, $\sigma^D.\varepsilon^D(u)$ represents the product of the continuous function σ^D by the measure $\varepsilon^D(u)$ in the case (7.68), and the product of the function σ^D (of class L^∞) by the function $\varepsilon^D(u)$ (of class L^1) in the case (7.69). In view of Green's formula I.(1.51), the relation (7.70) is obvious if $\sigma \in \Sigma(\Omega)$ and u is regular enough, for example $u \in H^1(\Omega)^n$. In the case (7.68), we can obtain (7.70) by approximating u by a sequence u_m satisfying (7.45) and proceeding to the limit; in the case (7.69) we approximate u by a sequence $u_m \in \mathscr{C}^\infty(\bar{\Omega})^n$ which converges to u in $LD(\Omega) \cap U(\Omega)$. ($\mathscr{C}^\infty(\bar{\Omega})^n$ is dense in $LD(\Omega) \cap U(\Omega)$, cf. Remark 3.4, ii)).

Remark 7.7. i) In the general case ($\sigma \in \Sigma(\Omega)$ and $u \in U(\Omega)$) , the products $\sigma.\varepsilon(u)$ and $\sigma^D.\varepsilon^D(u)$ are *local*, that is to say

$$(\theta\sigma).\varepsilon(u) = \theta(\sigma.\varepsilon(u)) \tag{7.71}$$

and

$$\sigma.\varepsilon(\theta u) = (\sigma.\varepsilon(u))\theta + \sigma.(u \otimes \text{grad } \theta) , \tag{7.72}$$

for all $\varepsilon \in \mathscr{C}^1(\bar{\Omega})$; this follows from (7.42) (7.43). Consequently if $\mathcal{O}$ is an open subset of Ω, then

$$(7.73) \qquad \begin{cases} \sigma_1.\varepsilon(u) = \sigma_2.\varepsilon(u) \quad \text{on } \mathcal{O}, \text{ if} \\ \sigma_1, \sigma_2 \in \Sigma(\Omega) \ , \ \sigma_1 = \sigma_2 \quad \text{on } \mathcal{O} \text{ and } u \in U(\Omega) \ , \end{cases}$$

$$(7.74) \qquad \begin{cases} \sigma.\varepsilon(u_1) = \sigma.\varepsilon(u_2) \quad \text{on } \mathcal{O} \ , \text{ if} \\ \sigma \in \Sigma(\Omega) \text{ and } u_1, u_2 \in U(\Omega) \ , \ u_1 = u_2 \quad \text{on } \mathcal{O} \ . \end{cases}$$

ii) If $u \in U(\Omega)$, then $\varepsilon^D(u) \in M_1(\Omega;E^D)$ and its Lebesgue decomposition can be written

$$(7.75) \qquad \varepsilon^D(u) = h\ dx + \varepsilon^D(u)^S \ ,$$

where $h \in L^1(\Omega;E^D)$ and $\varepsilon^D(u)^S \in M_1(\Omega;E^D)$ is singular for the Lebesgue measure dx . If now $\sigma \in \Sigma(\Omega)$, we can define the measure $\sigma^D.\varepsilon^D(u)^S$ by putting

$$(7.76) \qquad \sigma^D.\varepsilon^D(u)^S = \sigma^D.\varepsilon^D(u) - \sigma^D.h\ dx \ ,$$

where $\sigma^D.\varepsilon^D(u) \in M_1(\Omega)$ has been defined earlier and $\sigma^D.h \in L^1(\Omega)$. We have

(7.77) $\sigma^D.\varepsilon^D(u)^S$ is singular with respect to the Lebesgue measure dx , it is also absolutely continuous with respect to $|\varepsilon^D(u)|$ and

$$\left|\int_\Omega \phi\ \sigma^D.\varepsilon^D(u)^S\right| \leq \|\sigma^D\|_{L^\infty(\Omega;E^D)} \int_\Omega |\phi|\,|\varepsilon^D(u)^S| \ ,$$

for all $\phi \in \mathscr{C}_0(\Omega)$.

To check this we have only to approximate σ by a sequence of functions $\sigma_m \in \mathscr{C}^\infty(\bar{\Omega};E)$ satisfying (7.52)-(7.55), noting that, after Remark 7.1, ii), we can always arrange that

(7.78) $$\|\sigma_m^D\|_{L^\infty(\mathcal{O};E)} \leq \|\sigma^D\|_{L^\infty(\Omega;E)} ,$$

where $\mathcal{O}$ is an open neighbourhood of the support of ϕ . Since $\sigma_m^D.\varepsilon^D(u)$ converges weakly to $\sigma^D.\varepsilon^D(u)$ and $\sigma_m^D.h$ converges to $\sigma^D.h$ in $L^1(\Omega)$, by Lebesgue's Theorem, it follows that $\sigma_m^D.\varepsilon^D(u)^S$ converges weakly to $\sigma^D.\varepsilon^D(u)^S$; moreover by Remark 7.6 and (7.75)

$$\sigma_m^D.\varepsilon^D(u) = \sigma_m^D.h\ dx + \sigma_m^D.\varepsilon^D(u)^S ,$$

and the inequality

$$\left|\int_\Omega \phi.\sigma_m^D\ \varepsilon^D(u)^S\right| \leq \|\sigma_m^D\|_{L^\infty(\mathcal{O};E)}\int_\Omega |\phi|\,|\varepsilon^D(u)^S|$$
$$\leq \|\sigma^D\|_{L^\infty(\Omega;E)}\int_\Omega |\phi|\,|\varepsilon^D(u)^S|$$

implies, in the limit, the inequality (7.77). It follows by the classical argument that $\sigma^D.\varepsilon^D(u)^S$ is absolutely continuous with respect to $|\varepsilon^D(u)^S|$ and $|\varepsilon^D(u)|$.

7.4. Generalised duality

We begin by observing the following

LEMMA 7.8

For all $u \in \tilde{\mathscr{C}}_a$ *and all* $\sigma \in \mathscr{S}_a \cap \mathscr{K}_a$

(7.79) $$\frac{\kappa}{2}\,(div\ u)^2 \geq \frac{1}{n}\,(div\ u)(tr\ \sigma) - \frac{1}{2n^2\kappa}\,(tr\ \sigma)^2 \quad in\ \ \Omega ,$$

(7.80) $$\psi(\varepsilon^D(u)) \geq \sigma^D.\varepsilon^D(u) - \frac{1}{4\mu}\,|\sigma^D|^2 \quad in\ \ \Omega$$ (1)

(7.81) $$\psi_\infty(\mathscr{T}^D(u_{0\tau}-u_\tau)) \geq (u_{0\tau}-u_\tau).(\sigma.\nu)_\tau \quad on\ \ \Gamma_0$$ (2) .

---

(1) k and μ are positive coefficients in I.(2.65), I.(2.66); μ is not a measure here!

(2) In this form, the relation is true on the whole of Γ , but we shall use it only on Γ_0 .

Proof

The relation (7.79) which holds almost everywhere in Ω is obvious. To prove (7.80) we approximate u by a sequence u_m satisfying (7.45). By the definition of ψ^* and since $\sigma^D(x) \in K^D$ p.p., we have, for all m, the relation

$$\psi(\varepsilon^D(u_m)(x)) \geq \sigma^D(x).\varepsilon^D(u_m)(x) - \frac{1}{4\mu}|\sigma^D(x)|^2$$

which holds for almost all x in Ω ; hence, in the distribution sense

$$\psi(\varepsilon^D(u_m)) \geq \sigma^D.\varepsilon^D(u_m) - \frac{1}{4\mu}|\sigma^D|^2 . \tag{7.82}$$

When $m \to \infty$, $\sigma^D.\varepsilon^D(u_m)$ converges weakly to $\sigma^D.\varepsilon^D(u)$ by Lemma 7.4. By Theorem 5.3 and Remark 5.3, the sequence u_m can be chosen in such a way that $\psi(\varepsilon^D(u_m))$ also converges weakly to $\psi(\varepsilon^D(u))$; (7.82) then yields (7.80) in the limit.

If $\sigma \in \mathscr{S}_a \cap \mathscr{K}_a \cap \mathscr{C}(\bar{\Omega};E^D)$, we see by continuity that $\sigma^D(x) \in K^D$, $\forall\, x \in \Gamma$; this implies, by definition of ψ_∞ , that, at every point of Γ_0 ,

$$\psi_\infty(\mathscr{T}^D(u_{0\tau}-u_\tau)) \geq \sigma^D.\mathscr{T}^D(u_{0\tau}-u_\tau) \ ;$$

by a simple algebraic calculation, detailed in I.(6.26), the right-hand side of this inequality is equal to $(u_\tau-u_{0\tau}).(\sigma.\nu)_\tau$; thus proving (7.81) in this case. When σ is only in $\mathscr{S}_a \cap \mathscr{K}_a$ we approximate it by a sequence σ_m which satisfies (7.52)-(7.54) and $\sigma_m^D(x) \in K^D$ in the neighbourhood $\mathscr{O}_{x_0} \cap \Omega$ of a point x_0 on Γ (cf. Remark 7.1, ii)). By the above argument the inequality

$$\psi_\infty(\mathscr{T}^D(u_{0\tau}-u_\tau)) \geq (u_{0\tau}-u_\tau).(\sigma_m.\nu)_\tau$$

is true $d\Gamma$ - p.p. on $\mathscr{O}_{x_0} \cap \Gamma_0$, for all m . When $m \to \infty$, $(\sigma_m^D.\nu)_\tau$ converges to $(\sigma^D.\nu)_\tau$ in $L^\infty(\Gamma)^n$ weak-star, thanks to Lemma 7.2;

it follows that the inequality (7.81) holds on $\mathcal{O}_{x_0} \cap \Gamma_0$ since x_0 is an arbitrary point of Γ_0.

□

Now by adding (7.79) and (7.80) and integrating over Ω, we see that

$$(7.83) \qquad \Psi(\varepsilon^D(u)) \geq \int_\Omega \sigma.\varepsilon(u) - \frac{1}{2}\mathcal{A}(\sigma,\sigma) ,$$

for all $u \in \tilde{\mathscr{C}}_a$ and all $\sigma \in \mathscr{S}_a$ (cf. (6.12) and (7.3)). By integrating (7.81) over Γ_0 and adding (7.83), we obtain

$$\Psi(\varepsilon^D(u)) + \int_{\Gamma_0} \psi_\infty(\mathscr{T}^D(u_{0\tau}-u_\tau)d\Gamma \geq$$

$$\geq \int_\Omega \sigma.\varepsilon(u) + \int_{\Gamma_0} (u_{0\tau}-u_\tau).(\sigma.\nu)_\tau - \frac{1}{2}\mathcal{A}(\sigma,\sigma) .$$

By Green's formula (7.59) with $\Phi \equiv 1$, (7.60) and (7.67) (with $\Phi \equiv 1$), the right-hand side of (7.84) is simply

$$L(u) + \int_{\Gamma_0} (\sigma.\nu).u_0 \, d\Gamma - \frac{1}{2}\mathcal{A}(\sigma,\sigma) .$$

Defining J as in (6.25), namely

$$(7.85) \qquad J(u) = \Psi(\varepsilon(u)) + \int_{\Gamma_0} \psi_\infty(\mathscr{T}^D(u_{0\tau}-u_\tau))d\Gamma ,$$

we obtain

$$(7.86) \qquad J(u) - L(u) \geq \int_\Gamma (\sigma.\nu)u_0 \, d\Gamma - \frac{1}{2}\mathcal{A}(\sigma,\sigma) ,$$

for all $u \in \tilde{\mathscr{C}}_a$ and all $\sigma \in (\mathscr{S}_a \cap \mathscr{K}_a)$. Remembering that

$$\operatorname*{Sup}_{\sigma \in \mathscr{S}_a \cap \mathscr{K}_a} \{\int_{\Gamma_0} (\sigma.\nu)u_0 \, d\Gamma - \frac{1}{2}\mathcal{A}(\sigma,\sigma)\} = \text{Sup } (7.2)$$

$$\sup_{\sigma \in \mathscr{S}_a \cap \mathscr{K}_a} \{\int_{\Gamma_0} (\sigma.\nu)u_0 \, d\Gamma - \frac{1}{2}\mathscr{A}(\sigma,\sigma)\} = \text{Inf I.(6.10) by Theorem 6.1}$$

$$\geq \text{Inf (6.18)}$$

since (6.18) is an extension of I.(6.10). We find ourselves back to the result of Theorem 6.1.

(7.87) Inf I.(4.16) = Inf I.(6.4) = Inf (6.18) = Sup (7.2).

Thus

PROPOSITION 7.2.

On the hypotheses (6.14) and (7.14), the inequality (7.86) holds for all $u \in \tilde{\mathscr{C}}_a$ *and all* $\sigma \in (\mathscr{S}_a \cap \mathscr{K}_a)$ *; it implies in particular (7.87), that is*

(7.88) $$\mathit{Inf}\ \mathscr{Q} = \mathit{Sup}\ \mathscr{P}^*$$

In order to pinpoint the duality between (6.18) and (7.2), we could use the general framework of § 2, Ch.I, but this is ill-adapted to the present situation *by reason of the complexity of the duality between* $\Sigma(\Omega)$ and $U(\Omega)$. We prefer to introduce the Lagrangian function of the problem, leading to duality through the Lagrangians (cf. for example, Ekeland - Temam [1], Ch.6) : the function $\mathscr{L}$ is defined on $\tilde{\mathscr{C}}_a \times (\mathscr{S}_a \cap \mathscr{K}_a)$ by

(7.89) $$\mathscr{L}(u,\sigma) = -L(u) - \frac{1}{2}\mathscr{A}(\sigma,\sigma) + \int_\Omega \sigma.\varepsilon(u) + $$

$$+ \int_{\Gamma_0} (u_{0\tau} - u_\tau).(\sigma.\nu)_\tau \, d\Gamma .$$

The primal and dual problems then take the form (cf. Ekeland - Temam, *loc. cit.*)

$$\text{(7.90)} \qquad \operatorname*{Inf}_{u \in \tilde{\mathscr{C}}_a} \{ \operatorname*{Sup}_{\sigma \in \mathscr{S}_a \cap \mathscr{K}_a} \mathscr{L}(u,\sigma) \}$$

and

$$\text{(7.91)} \qquad \operatorname*{Sup}_{\sigma \in \mathscr{S}_a \cap \mathscr{K}_a} \{ \operatorname*{Inf}_{u \in \tilde{\mathscr{C}}_a} \mathscr{L}(u,\sigma) \} .$$

By integrating (7.81) over Γ_0 we find

$$\text{(7.92)} \qquad \int_{\Gamma_0} \psi_\infty(\mathscr{T}^D(u_{0\tau}-u_\tau))d\Gamma \geq \int_{\Gamma_0} (u_{0\tau}-u_\tau).(\sigma.\nu)_\tau \, d\Gamma$$

and

$$\begin{aligned} \text{(7.93)} \qquad \mathscr{L}(u,\sigma) &\leq - L(u) + \int_{\Gamma_0} \psi_\infty(\mathscr{T}^D(u_{0\tau}-u_\tau)) \, d\Gamma + \\ &\quad + \int_\Omega \sigma.\varepsilon(u) - \frac{1}{2} \mathscr{A}(\sigma,\sigma) \\ &\leq - L(u) + \int_\Gamma \psi_\infty(\mathscr{T}^D(u_{0\tau}-u_\tau))d\Gamma + \Psi(\varepsilon(u)) \\ &\leq J(u) - L(u) . \end{aligned}$$

Similarly using (7.59) with $\Phi \equiv 1$, (7.60) and (7.67) (with $\phi \equiv 1$) :

$$\begin{aligned} \text{(7.94)} \quad \mathscr{L}(u,\sigma) &= - \frac{1}{2}\mathscr{A}(\sigma,\sigma) - \int_{\Gamma_1} g \, u \, d\Gamma + \int_\Gamma (\sigma.\nu).u \, d\Gamma + \\ &\quad + \int_{\Gamma_0} (u_{0\tau}-u_\tau).(\sigma.\nu)_\tau \, d\Gamma \\ &= - \frac{1}{2}\mathscr{A}(\sigma,\sigma) + \int_{\Gamma_0} (\sigma.\nu).u_0 \, d\Gamma \quad (\text{since } \sigma.\nu = g \text{ on } \Gamma_1 \\ &\qquad\qquad \text{and } u.\nu = u_0.\nu \text{ on } \Gamma_0) \end{aligned}$$

It follows from the foregoing that the infimum of problem (7.90) and the supremum of problem (7.91) are equal to the number (7.88).

Remark 7.8. We can also consider $\mathscr{L}(u,\sigma)$ on $\tilde{\mathscr{C}}_a \times \mathscr{E}_a$,

$$\mathscr{E}_a = \{\sigma \in L^2(\Omega;E),\ \operatorname{div}\sigma \in L^{n+1}(\Omega)^n,\ \sigma^D(x) \in K^D \text{ p.p. } x \in \Omega\} .$$

Then, instead of (7.93)

$$\operatorname*{Sup}_{\sigma \in \mathscr{E}_a} \mathscr{L}(u,\sigma) = J(u) - L(u) ,$$

and equality in (7.94) occurs only if $\sigma \in \mathscr{S}_a \cap \mathscr{E}_a = \mathscr{S}_a \cap \mathscr{K}_a$:

$$\operatorname*{Inf}_{u \in \tilde{\mathscr{C}}_a} \mathscr{L}(u,\sigma) = \begin{cases} -\frac{1}{2}\mathscr{A}(\sigma,\sigma) + \int_\Gamma (\sigma.\nu).u_0 \, d\Gamma & \text{if } \sigma \in \mathscr{S}_a \cap \mathscr{K}_a \\ -\infty & \text{if not.} \end{cases}$$

Under these conditions problems (7.90) and (7.91), with $\mathscr{S}_a \cap \mathscr{K}_a$ replaced by $\mathscr{E}_a$, are identical to the problems $\mathscr{Q}$ and $\mathscr{P}^*$, i.e. (6.18) and (7.2).

8. Existence of solutions to the generalised strain problem

We now reach the ultimate goal of Chapter II, the existence of a solution to the generalised strain problem, the problem $\mathcal{Q}$ introduced in Section 6 (cf. (6.18)). This result was proved, on very special assumptions, in Section 6.3; we shall prove it here under the most general conditions. Section 8.1 contains some technical preliminaries; in Section 8.2 we give the two main results, the existence of a solution and the conditions for optimality The results given follow those of F. Demangel and R. Temam [1], of R. Kohn and R. Temam [1][2], and as far as existence is concerned, generalise the partial results of G. Anzelotti and M. Giaquinta [1], J. Naumann [1][2], R. Temam [10].

8.1 A result on lower semicontinuity

We adhere to the notation of the previous Sections, in particular Sections 6 and 7.

We extend the function u_0 into a function $\tilde{u}_0$ from $\mathbb{R}^n$ into $\mathbb{R}^n$, with

(8.1) $$\tilde{u}_0 \in H^1(\mathbb{R}^n)^n \ , \ \tilde{u}_{0|\Omega} = u_0 \ .$$

To simplify we shall suppose, without loss of generality, that $\tilde{u}_0$ has compact support in $\mathbb{R}^n$. For each function from Ω into $\mathbb{R}^n$, we similarly define the function $\tilde{u}$ from $\mathbb{R}^n$ into itself, such that

$$\tilde{u} = u \ \text{ on } \ \Omega \ , \ \tilde{u} = \tilde{u}_0 \ \text{ on } \ \mathbb{R}^n \setminus \bar{\Omega} \ .$$

It can be verified, exactly as in Lemma 2.2, that $\varepsilon_{ij}(\tilde{u})$ is equal in $\mathbb{R}^n$ to the sum of three measures

$$(8.2) \qquad \varepsilon_{ij}(\tilde{u}) = \theta_\Omega \, \varepsilon_{ij}(u) + \theta_{\mathbb{R}^n \setminus \bar{\Omega}} \, \varepsilon_{ij}(\tilde{u}_0) + \mathcal{T}_{ij}(u_0 - \gamma_0(u)) \, \delta_\Gamma ,$$

where θ_B is the characteristic function of a set B and $\gamma_0(u)$ is the trace of u on Γ (the internal trace of $u \in BD(\Omega)$, whereas the external trace of $\tilde{u}$ on Γ is that of $\tilde{u}_0 \in BD(\mathbb{R}^n \setminus \bar{\Omega})$, that is to say u_0). It follows clearly from (8.2), and the compactness of the support of $\tilde{u}_0$, that if $u \in U(\Omega)$, then $\tilde{u} \in U(\mathbb{R}^n)$, i.e. $\tilde{u} \in L^1(\mathbb{R}^n)$, $\varepsilon_{ij}(\tilde{u}) \in M_1(\mathbb{R}^n) \ \forall \ i,j$, $\operatorname{div} \tilde{u} \in L^2(\mathbb{R}^n)$. Of course $\mathbb{R}^n$ can be replaced here by any ball containing $\bar{\Omega}$.

The result we wish to establish here is one on lower semi-continuity. Consider a sequence u_m in $U(\Omega)$ which converges weakly to u, i.e.

$$(8.3) \quad \begin{cases} u_m \to u \ \text{ in } \ L^1(\Omega)^n \ \text{strong and} \ L^{n/(n-1)}(\Omega)^n \ \text{weak} \\ \operatorname{div} u_m \to \operatorname{div} u \ \text{ in } \ L^2(\Omega) \ \text{ weak} \\ \varepsilon(u_m) \to \varepsilon(u) \ \text{ weakly in } \ M_1(\Omega;E) \end{cases}$$

If one now considers the functions $\tilde{u}_m$, defined by

$$\tilde{u}_m = u_m \text{ on } \Omega \ , \quad \tilde{u}_m = \tilde{u}_o \text{ on } \mathbb{R}^n \setminus \bar{\Omega} \ ,$$

then clearly, when $m \to \infty$

$$(8.4) \quad \begin{cases} \tilde{u}_m \to \tilde{u} \text{ in } L^1(\mathbb{R}^n)^n \text{ strong and } L^{n(n-1)}(\mathbb{R})^n \text{ weak} \\ \varepsilon(\tilde{u}_m) \to \varepsilon(\tilde{u}) \text{ weakly in } M_1(\mathbb{R}^n;E) \ . \end{cases}$$

Note that if $u_m.\nu = u_o.\nu$ on Γ_0 for all m , then by the trace theorem of Section 7.2 (cf. Proposition 7.2), we have $u.\nu = u_o.\nu$ on Γ_0 in the limit; if therefore the u_m is in $\tilde{\mathscr{C}}_a$, then so is $u \in \tilde{\mathscr{C}}_a$ as well.

LEMMA 8.1

Let u_m be a sequence in $\tilde{\mathscr{C}}_a$ which converges to $u \in \tilde{\mathscr{C}}_a$ in the sense of (8.3). If the set $\mathscr{S}_a \cap \mathscr{K}_a$ is not empty, then

$$(8.5) \qquad J(u) - L(u) \leq \liminf_{m\to\infty} \{J(u_m) - L(u_m)\} \ .$$

Proof

Let σ be an element of $\mathscr{S}_a \cap \mathscr{K}_a$. By definition of the sets $\mathscr{S}_a$, $\mathscr{K}_a$, $\tilde{\mathscr{C}}_a$ and Green's formula (7.59) (written with $\Phi \equiv 1$) , we have

$$J(v) - L(v) = \int_\Omega \{\frac{\kappa}{2}(\text{div } v)^2 - \frac{1}{n}(\text{div } v)(\text{tr } \sigma)\}dx +$$

$$+ \int_\Omega \psi(\varepsilon^D(v)) + \int_{\Gamma_0} \psi_\infty(\mathscr{T}^D(u_{0\tau} - v_\tau))d\Gamma$$

$$+ \int_\Gamma (\sigma.\nu).u_0 \, d\Gamma - \int_\Omega \sigma^D.\varepsilon^D(v) -$$

$$- \int_{\Gamma_0} (u_{0\tau} - v_\tau).(\sigma.\nu)_\tau \, d\Gamma .$$

As certain terms are obviously continuous or l.s.c. for the topology defined by (8.10), it will be sufficient to prove the lower semi-continuity of

(8.6) $$J_1(v) = \int_{\Omega\cup\Gamma_0} \psi(\varepsilon^D(\tilde{v})) - \int_{\Omega\cup\Gamma_0} \sigma^D.\varepsilon^D(\tilde{v}) .$$

In (8.6) we have written

(8.7) $$\int_{\Omega\cup\Gamma} \psi(\varepsilon^D(\tilde{v})) = \int_\Omega \psi(\varepsilon^D(v)) + \int_\Gamma \psi_\infty(\mathscr{T}^D(u_{0\tau} - v_\tau))d\Gamma$$

(8.8) $$\int_{\Omega\cup\Gamma_0} \sigma^D.\varepsilon^D(\tilde{v}) = \int_\Omega \sigma^D.\varepsilon^D(v) + \int_{\Gamma_0} (u_{0\tau} - v_\tau).(\sigma.\nu)_\tau d\Gamma ;$$

(8.7) is consistent with (8.2) and the definition of the measure $\psi(\mu)$ following from Section 4 while (8.8) should be regarded here merely as a convenient notation ([1]).

In order to prove the lower semi-continuity of J_1 let us consider for a given sufficiently small $\zeta>0$ a $\mathscr{C}^\infty$ function ϕ_ζ such that $0\leq\phi_\zeta\leq 1$, $\phi_\zeta = 1$ in a neighbourhood of Γ_1 and

$$\text{support } \phi_\zeta \subset \Omega_\zeta = \{x \in \mathbb{R}^n,\ \text{dist}(x, \Gamma_1) < \zeta\}\ .$$

Setting $g_\zeta = 1 - \phi_\zeta$ we write (7.59) with u and ϕ replaced by v and g_ζ ; taking into account that $g_\zeta = 0$ on Γ_1, and using the notation introduced in (8.8) we find :

$$\text{(8.9)} \quad \int_{\Omega\cup\Gamma} \sigma^D.\varepsilon^D(\tilde{v})g_\zeta = -\int_\Omega v.\text{div}\sigma\ g_\zeta - \int_\Omega \sigma.(v\otimes\text{grad}g_\zeta)dx$$
$$- \frac{1}{n}\int_\Omega (\text{div}v).(\text{tr}\sigma)g_\zeta dx + \int_{\Gamma_0} (\sigma.\nu).u_0 g\ d\Gamma.$$

([1]) If σ is properly extended to $\mathbb{R}^n$ as a tensor function $\tilde{\sigma}$, then $\tilde{\sigma}^D.\varepsilon^D(\tilde{v})$ would be, with the method of Sec.7, a well defined measure on $\mathbb{R}^n$ such that (cf. (8.2)) :

$$\tilde{\sigma}^D.\varepsilon^D(\tilde{v}) = \theta_\Omega\ \sigma^D.\varepsilon^D(v) + \theta_{\mathbb{R}^n\setminus\bar{\Omega}}\ \sigma^D.\varepsilon^D(\tilde{u}_0) + \sigma^D.\mathscr{T}^D(u_0-\gamma_0(v))\delta_\Gamma$$

and the left (and right) hand side of (8.8) could be identified with $\int_{\Omega\cup\Gamma_0} \tilde{\sigma}^D.\varepsilon^D(\tilde{v})$. There is however no need to develop this point here and we will simply set

$$\sigma^D.\varepsilon^D(\tilde{v})\big|_{\Omega\cup\Gamma} = \theta_\Omega\sigma^D.\varepsilon^D(v) + \sigma^D.\mathscr{T}^D(u_0-\gamma_0(v))$$

as a convenient notation.

It follows easily from (8.9) that

$$\lim_{m\to\infty} \int_{\Omega\cup\Gamma_0} \sigma^D.\varepsilon^D(\tilde{u}_m)g_\zeta = \int_{\Omega\cup\Gamma_0} \sigma^D.\varepsilon^D(\tilde{u})g_\zeta \quad . \tag{8.10}$$

Also since $g_\zeta \geq 0$ and ψ is convex, the integral $\int_{\mathcal{O}} \psi(\varepsilon^D(\tilde{v}))g_\zeta$ is lower semi-continuous for the topology (8.3) whenever $\mathcal{O}$ is open; we then conclude that

$$\int_{\Omega\cup\Gamma_0} \psi(\varepsilon^D(\tilde{v}))g_\zeta = \int_{\mathbf{R}^n} \psi(\varepsilon^D(\tilde{v}))g_\zeta - \int_{\mathbf{R}^n\setminus\bar{\Omega}} \psi(\varepsilon^D(u_0))g_\zeta \, dx \tag{8.11}$$

is lower semi-continuous for the topology (8.3) and thus

$$\int_{\Omega\cup\Gamma_0} \psi(\varepsilon^D(\tilde{u}))g_\zeta \leq \underline{\lim}_{m\to\infty} \int_{\Omega\cup\Gamma_0} \psi(\varepsilon^D(\tilde{u}_m))g_\zeta \quad . \tag{8.12}$$

We then write

$$\begin{aligned}
J_1(u) &= \int_{\Omega\cup\Gamma_0} \{\psi(\varepsilon^D(\tilde{u})) - \sigma^D.\varepsilon^D(\tilde{u})\} \\
&= \int_{\Omega\cup\Gamma_0} \{\psi(\varepsilon^D(\tilde{u})) - \sigma^D.\varepsilon^D(\tilde{u})\}(g_\zeta+\phi_\zeta) \\
&\leq \text{(by (8.10)(8.12))} \\
&\leq \int_{\Omega\cup\Gamma_0} \{\psi(\varepsilon^D(\tilde{u})) - \sigma^D.\varepsilon^D(\tilde{u})\}\phi_\zeta \\
&\quad + \underline{\lim}_{m\to\infty} \int_{\Omega\cup\Gamma_0} \{\psi(\varepsilon^D(\tilde{u}_m)) - \sigma^D.\varepsilon^D(\tilde{u}_m)\}g_\zeta \\
&\leq \underline{\lim}_{m\to\infty} J_1(u_m) + \int_{\Omega\cup\Gamma_0} \{\psi(\varepsilon^D(\tilde{u}) - \sigma^D.\varepsilon^D(\tilde{u})\}\phi_\zeta \\
&\quad + \overline{\lim}_{m\to\infty} \int_{\Omega\cup\Gamma_0} \{- \psi(\varepsilon^D(\tilde{u}_m)) + \sigma^D.\varepsilon^D(\tilde{u}_m)\}\phi_\zeta
\end{aligned}$$

Whence, with (7.80) (7.81)

(8.13) $$J_1(u) \leq \underline{\lim}_{m\to\infty} J_1(u_m) + E_\zeta$$

where

(8.14) $$E_\zeta = \int_{\Omega\cup\Gamma_0} \{\psi(\varepsilon^D(\tilde{u})) - \sigma^D.\varepsilon^D(\tilde{u})\}\phi_\zeta + \int_\Omega \psi^*(\sigma^D)\phi_\zeta \, dx$$

The desired conclusion

(8.15) $$J_1(u) \leq \underline{\lim}_{m\to\infty} J_1(u_m)$$

will follow from (8.13) if we show that E_ζ can be made arbitrarily small by choosing $\zeta>0$ sufficiently small. For both integrals appearing in (8.14) this is now a consequence of the Lebesgue dominated convergence theorem : whatever the choice made of ϕ_ζ, $\phi_\zeta(x) \to 0$, as $\zeta \to 0$, at <u>every point</u> x of $\Omega \cup \Gamma_0$ and thus almost everywhere for the Lebesgue measure dx and for the (bounded) measure $\psi(\varepsilon^D(\tilde{u})) - \sigma^D.\varepsilon^D(\tilde{u})$. Similarly $|\phi_\zeta(x)| \leq 1$ everywhere and thus a.e. for these measures. Whence by Lebesgue's theorem, $E_\zeta \to 0$ as $\zeta \to 0$ and the result is proved.

□

8.2 Existence of a solution and optimality conditions

We recall the hypotheses :

(8.16) Ω is an open set of class $\mathscr{C}^2$,

whose boundary Γ is partitioned into Γ_0, Γ_1, Γ_*. Γ_0 and Γ_1 are open in Γ while Γ_* the common boundary of Γ_0 and Γ_1 is closed in Γ and satisfies the condition I.(2.105). We are given functions f, g, u_0, with

(8.17) $$f \in L^\infty(\Omega)^n \quad (^1)$$

(8.18) $$g \in L^\infty(\Gamma_1)^n \quad ,$$

(8.19) $$u_0 \in H^1(\Omega)^n \ , \ \theta_{\Gamma_0}\gamma_0(u_0) = \gamma_0(u_0) \ ,$$

where θ_{Γ_0} is the characteristic function of Γ_0 in Γ and $\gamma_0(u_0)$ is the trace of u_0 on Γ, which for simplicity we also write u_0.

The strain problem is (6.18). It is dual to the stress problem (7.1) - (7.4), in the sense defined at the end of Section 7. Let us recall that the strain problem has a finite infimum

(8.20) $$\operatorname*{Inf}_{u \in \tilde{\mathscr{C}}_a} \{J(u) - L(u)\} > -\infty$$

if and only if one of the following equivalent condition is satisfied :

(^{1}) It would be sufficient to suppose that $f \in L^n(\Omega)^n$.

$$
(8.21) \quad \begin{cases} \operatorname{Inf} \Psi_\infty(\varepsilon^D(v)) \geq 1 , \\ \text{where the infimum is taken over all } v \in H^1(\Omega)^n \\ \text{such that } \operatorname{div} v = 0 , v = 0 \text{ on } \Gamma_0, L(v) = 1 \end{cases}
$$

or

$$
(8.22) \quad \begin{cases} \text{There exists a } \sigma \in \mathscr{S}_a , \text{ where} \\ \mathscr{S}_a = \{\sigma \in L^2(\Omega;E) , \operatorname{div} \sigma + f = 0 \text{ in } \Omega, \sigma.\nu = g \text{ on } \Gamma_1\}, \\ \text{and } \sigma^D(x) \in K^D \text{ p.p. } x \in \Omega . \end{cases}
$$

Either of the two following equivalent conditions

$$
(8.23) \quad \begin{cases} \operatorname{Inf}_v \Psi_\infty(\varepsilon^D(v)) > 1 , \text{ where the infimum is taken with} \\ \text{respect to the same set of } v \text{ as in (8.21).} \end{cases}
$$

$$
(8.24) \quad \begin{cases} \text{There exists a } \sigma \in \mathscr{S}_a \text{ such that } \sigma^D(x) \in \lambda K^D \text{ p.p.} \\ x \in \Omega \text{ where } 0 \leq \lambda < 1 \end{cases}
$$

is called the *safe load condition.*

By Proposition 6.1, if the *safe load condition* is satisfied then every minimising sequence for the strain problem is bounded in $U(\Omega)$ (respectively in $U(\Omega)/\mathcal{R}$ if $\Gamma_0 = \emptyset$) .

We now state

THEOREM 8.1

Under the hypotheses (8.16) - (8.19) , if $\Gamma_0 \neq \emptyset$ *and the safe load condition is satisfied, then the strain problem (6.18) has a solution in* $U(\Omega)$.

Under these same assumptions, if $u \in \tilde{\mathscr{C}}_a$ *and* $\sigma \in \mathscr{S}_a \cap \mathscr{K}_a$ *then* u *is a solution of (6.18) and* σ *a solution of (7.1) - (7.4) if and only if :*

(8.25) $$\psi(\varepsilon^D(u)) + \frac{1}{4\mu}\,|\sigma^D|^2 = \sigma^D.\varepsilon^D(u) \quad in \quad \Omega ,$$

(8.26) $$\kappa \; div \; u = \frac{1}{n}\, tr \quad \sigma \; in \quad \Omega ,$$

(8.27) $$\psi_\infty(\mathcal{T}^D(u_{o\tau} - u_\tau)) = (u_{o\tau} - u_\tau).(\sigma.\nu)_\tau \quad on \quad \Gamma_0 .$$

Proof

The existence of a solution can now easily be proved by the direct method of the calculus of variations. Let (u_m) be a minimising sequence of (6.18). It belongs to $\tilde{\mathscr{C}}_a$ and is bounded in $U(\Omega)$; it therefore contains a subsequence, which we shall still denote by (u_m) , which converges weakly to u , in the sense of (8.3) (cf. (3.4) and (3.37)). By Proposition 7.2, u is in $\tilde{\mathscr{C}}_a$ and by Lemma 8.1,

$$J(u) - L(u) \leq \liminf_{m\to\infty} \{J(u_m) - L(u_m)\} .$$

It follows that u is a solution of (6.18).

As in the proof of (7.86), if $u \in \tilde{\mathscr{C}}_a$ and $\sigma \in \mathscr{S}_a \cap \mathscr{K}_a$, there is equality in (7.86) if and only if there is equality in the relations (7.79) (7.81) which become identical to (8.25) - (8.27).

Remark 8.1. i) The existence proof in Theorem 8.1 consisted in showing that a minimising sequence for $\mathscr{Q}$ contained a subsequence converging weakly to a solution of $\mathscr{Q}$. More generally under the hypotheses of Theorem 8.1, *every minimising sequence for* $\mathscr{P}$, $\mathscr{P}\mathscr{R}$ *or* $\mathscr{Q}$ *contains points of accumulation for the weak topology of* $U(\Omega)$ *and every point of accumulation of a minimising sequence is a 'generalised solution', that is to say a solution to the generalised problem* $\mathscr{Q}$.

ii) When $\Gamma_0 = \emptyset$ and $\Gamma_1 = \Gamma$ and , if the condition

$$L(v) = 0 , \forall v \in \mathscr{R} \tag{8.28}$$

is satisfied, we prove the existence of a solution to the strain problem by considering a quotient problem as in Proposition 2.7, Ch.I.

Chapter III
Asymptotic problems and problems in the theory of plates

Introduction

The object of this chapter, which is shorter than its predecessors, is to study some problems allied to those in Chapters I and II. In Section 1 we study various asymptotic problems corresponding to imperfectly plastic models; in particular problems where the parameter characterising the degree of imperfection tends to zero. In Section 2, we consider a number of problems derived from the theory of plates, and make a study of these which parallels that made in three dimensions. More specifically we cover : the elastic problem : the existence of a solution and duality; the perfectly plastic problem : duality, study of the stress problem, introduction of a space analogous to the BD space for the strain problem (the space $HB(\Omega)$) , formulation of a generalised strain problem in $HB(\Omega)$ and existence of a solution to the latter.

1. Some asymptotic problems: problems of imperfectly plastic bodies

We shall study here various perturbed versions of the perfectly plastic model considered in Chapters I and II. The perturbations which we shall deal with, and which were mentioned in Chapter I, Section 3, are those associated with elasto-visco-plastic problems, Norton-Hoff type problems, and problems with hardening involved. These perturbations are considered in Sections 1.1, 1.2 and 1.3 respectively. In each case our study will lead us to consider certain salient aspects of the perturbed problem, namely the duality between the stress and strain problems, and the existence of a solution. We then go on to study the behaviour of the perturbed problem when the parameter characterising the extent of perturbation tends to zero; when this takes place we determine the convergence of the solutions of the perturbed problem to those of the perfectly plastic elastic problem.

1.1 Elasto-visco-plastic perturbations

We take up once more the assumptions and data of Section 4, Ch.I, and in particular (4.1)-(4.10). We shall denote by ψ_e and ψ_p the functions ψ in the elastic and perfectly-plastic cases respectively, and their conjugates by ψ_e^*, ψ_p^* ; that is to say (cf. (2.65), (2.68), (4.5), (4.9), Ch.I) we have :

(1.1) $$\psi_e(\xi) = \frac{\kappa}{2}(\mathrm{tr}\ \xi)^2 + \mu|\xi^D|^2 \quad , \quad \forall\ \xi \in E ,$$

(1.2) $$\psi_e^*(\eta) = \frac{1}{2}A\ \eta.\eta = \frac{(\mathrm{tr}\ \eta)^2}{18\kappa} + \frac{1}{4\mu}|\eta^D|^2 \quad , \forall\ \eta \in E ,$$

$$(1.3) \qquad \psi_p^*(\eta) = \psi_e^*(\eta) \ \text{ if } \ \eta \in K \ , +\infty \ \text{ if not}, \ \ \psi_p = (\psi_p^*)^* \ .$$

For any $\alpha>0$, we define a convex function ψ_α of E into $\mathbb{R}$, and its conjugate ψ^* ; we have (cf. I(3.6)) :

$$(1.4) \qquad \psi_\alpha^*(\eta) = \frac{1}{2} A\eta.\eta + \frac{1}{4\alpha} [\eta - P_K\eta]^2 \ ,$$

where P_K is the projection (operator) in E on K . By a result of E.H. Zarantonello [1], the function $\eta \to \frac{1}{2} [\eta - P_K\eta]^2$ is a convex, continuous, Gâteaux-differentiable function from E into $\mathbb{R}$, whose differential is the mapping $\eta \to \eta - P_K\eta$. It follows that ψ_α^* is a *strictly convex* continuous, Gâteaux differentiable function from E into $\mathbb{R}$, whose Gâteaux differential is the mapping

$$(1.5) \qquad \eta \ \to \ A\eta + \frac{1}{2\alpha} (\eta - P_K\eta) \ .$$

The functions ψ_α^* can be regarded as perturbations of ψ_p^* , for example, in the following sense : for all $\eta \in E$, $\psi_\alpha^*(\eta) \to \psi_p^*(\eta)$ when $\alpha \to 0$. In fact $P_K\eta = \eta$ and $\psi_\alpha^*(\eta) = \psi^*(\eta)$ if $\eta \in K$, while if $\eta \in K$, $\psi_\alpha^*(\eta) \to +\infty = \psi_p^*(\eta)$ for $\alpha \to 0$.

The function ψ_α^* is the conjugate of $\psi_\alpha = (\psi_\alpha^*)^*$ which is a convex and l.s.c. function from E into $\mathbb{R} \cup \{+\infty\}$ (cf. Sec. 2, Ch.I). It is not possible to give an explicit expression for ψ_α , but we can nevertheless note the following properties of the function ψ_α .

LEMMA 1.1

For any $\alpha>0$, ψ_α *is a continuous convex function mapping* E *into* $\mathbb{R}$ *; furthermore for any* $\xi \in E$:

$$(1.6) \qquad \psi_p(\xi) \leqq \psi_\alpha(\xi) \leqq \psi_e(\xi) \leqq \max\left(\mu, \frac{3\kappa}{2}\right).|\xi|^2 \ ,$$

$$(1.7) \qquad \psi_\alpha(\xi) \geqq \frac{\kappa}{2}(tr\ \xi)^2 + \frac{\alpha\mu}{\alpha+\mu} |\xi^D|^2 \ .$$

Proof

It is sufficient to prove the inequalities (1.6) and (1.7) ; (1.6) clearly implies that ψ_α is finite at every point of E and its convexity is enough to ensure its continuity; the inequalities (1.6) are easily proved since $\psi_e^* \leqq \psi^* \leqq \psi_p^*$. As for (1.7) we note that, since $K = K^D \oplus \mathbb{R} I$ (cf. I.(4.4)):

(1.8) $$P_K\eta = \frac{1}{3}\,\mathrm{tr}\,\eta\, I + P_{K^D}\,\eta^D , \quad \forall\, \eta \in E ,$$

where P_{K^D} is the projection operator from E^D onto K^D. Then since $0 \in K^D$, $|\eta - P_K\eta| = |\eta^D - P_{K^D}\eta^D| \leqq |\eta^D|$ and :

$$\begin{aligned}
\psi_\alpha(\xi) &= \sup_{\eta\in E} \{\xi.\eta - \psi_\alpha^*(\eta)\} \\
&= \sup_{\substack{\mathrm{tr}\,\in\mathbb{R} \\ \eta^D \in E^D}} \{\xi^D.\eta^D + \frac{1}{3}\,\mathrm{tr}\xi.\mathrm{tr}\eta - \frac{(\mathrm{tr}\eta)^2}{18\kappa} - \frac{1}{4\mu}|\eta^D|^2 - \frac{1}{4\alpha}|\eta^D - P_{K^D}\eta^D|^2 \\
&\geqq \frac{\kappa}{2}(\mathrm{tr}\xi)^2 + \sup_{\eta^D\in E^D} \{\xi^D.\eta^D - \left(\frac{1}{4\mu} + \frac{1}{4\alpha}\right)|\eta^D|^2\} \\
&\geqq \frac{\kappa}{2}(\mathrm{tr}\xi)^2 + \frac{\alpha\mu}{\alpha+\mu}\,|\eta^D|^2 .
\end{aligned}$$

□

The strain problem, denoted by $\mathscr{P}_\alpha$, is comparable to the problem I.(4.16), I.(4.17) (cf. also I.(3.18)) and can be written as

(1.9) $$\inf_{v \in \mathscr{C}_a} \{\Psi_\alpha(\varepsilon(v)) - L(v)\} \qquad (\mathscr{P}_\alpha) ,$$

where $\mathscr{C}_a$ and $L(v)$ are defined as in I.(4.13), I.(4.14), and

(1.10) $$\Psi_\alpha(\tau) = \int_\Omega \psi_\alpha(\tau(x))dx \quad ,$$

for all $\tau \in L^2(\Omega;E)$. It follows from (1.6) (1.7) that $\Psi_\alpha(\tau)$ is finite for all $\tau \in L^2(\Omega;E)$ and, by Proposition I.2.4 $\tau \mapsto \Psi_\alpha(\tau)$ is a finite and continuous function which maps $L^2(\Omega;E)$ into $\mathbb{R}$.

(1.11) $v \mapsto \Psi_\alpha(\varepsilon(v))$ is a convex continuous function which maps $H^1(\Omega)^3$ into $\mathbb{R}$.

To find the dual $\mathcal{P}_\alpha^*$ of $\mathcal{P}_\alpha$ we proceed exactly as in Sec.4.2, Ch.I; the choices of V, Y, Λ, F are the same, but in contrast we define G by

$$G(p) = \int_\Omega \psi_\alpha(p(x))dx \ , \quad \forall\, p \in Y = L^2(\Omega;E) \ . \tag{1.12}$$

To find $\mathcal{P}_{\alpha_*}^*$, we have to calculate $F^*(\Lambda^* p)$, which is done in I.(4.25), and also $G^*(p)$; $G^*(p)$ is found with the help of Proposition 2.5, Ch.I, which gives

$$\begin{aligned} G^*(p) = \Psi_\alpha^*(p) &= \int_\Omega \psi_\alpha^*(p(x))dx \\ &= \frac{1}{2}\mathscr{A}(p,p) + \frac{1}{4\alpha}\,|p - P_K p|^2_{L^2(\Omega;E)} \end{aligned} \tag{1.13}$$

$\forall\, p \in Y^* = L^2(\Omega;E)$. By introducing the tensor $\sigma = -p$, we can express $\mathcal{P}_\alpha^*$ in the explicit form given by I.(2.36):

$$\operatorname*{Sup}_\sigma \{-\frac{1}{2}\mathscr{A}(\sigma,\sigma) - \frac{1}{4\alpha}\,|\sigma - P_K\sigma|^2_{L^2(\Omega;E)} + \int_{\Gamma_0} (\sigma.\nu).u_0 d\Gamma\} \tag{1.14}$$

where the supremum is taken over all σ in $\mathcal{S}_a$ (cf. I.3.6)). This problem $\mathcal{P}_\alpha^*$ is a *penalised* version of the problem $\mathcal{P}^*$ in I.(4.29); the constraint $\sigma^D(x) \in K^D$ p.p. has been 'penalised' (assigned a penalty).

We suppose $\Gamma_0 \neq \emptyset$, to simplify matters a little, and we have the following

THEOREM 1.1

i) For all $\alpha>0$, *the problems* $\mathcal{P}_\alpha$ *and* $\mathcal{P}_\alpha^*$ *each have the unique solution,* u_α *and* σ_α *respectively, and*

$$\inf \mathcal{P}_\alpha = \sup \mathcal{P}_\alpha^* \in \mathbb{R} \tag{1.15}$$

(1.16) $$\varepsilon(u_\alpha) = A\sigma_\alpha + \frac{1}{2\alpha}(\sigma_\alpha - P_K\sigma_\alpha) \text{ , p.p. in } \Omega$$

ii) As $\alpha \to 0$,

(1.17) $$\inf \mathcal{P}_\alpha = \sup \mathcal{P}^*_\alpha \to \inf \mathcal{P} = \sup \mathcal{P}^* .$$

If $\sup \mathcal{P}^* > -\infty$, σ_α *converges in* $L^2(\Omega;E)$ *to the solution* σ *of* $\mathcal{P}^*$. *If, in addition, the safe load condition is satisfied, the sequence* u_α *is bounded in* $U(\Omega)$ *and every weak limit of a subsequence extracted from* u_α *is a generalised solution of* $\mathcal{P}$.

Proof

i) For a fixed $\alpha>0$, the set of σ admissible for (1.14) (= $\mathcal{P}_a$) is not void, and so $\sup \mathcal{P}^*_\alpha > -\infty$, and $\inf \mathcal{P}_\alpha \in \mathbb{R}$ by I.(2.28). We see, as we did for (1.11), that the function Ψ^*_α is continuous on $L^2(\Omega;E)$ and so the condition I.(2.29) is satisfied with, for example, $u^0 = u_0$; I.(2.30) then implies (1.15) and the existence of a solution for $\mathcal{P}^*_\alpha$. This solution σ_α is unique because of the strict convexit of $\sigma \mapsto \mathcal{A}(\sigma,\sigma)$ and consequently that of $\sigma \mapsto \Psi^*_\alpha(\sigma)$. The existence of a solution for $\mathcal{P}_\alpha$ follows from Proposition I.2.1; the conditions I.(2.38) and I.(2.42) are satisfied by reason of (1.11) and (1.7).

If u_α is a solution of $\mathcal{P}_\alpha$, the extremality relation I.(2.31) can be written here

(1.18) $$\Psi_\alpha(\varepsilon(u_\alpha)) + \Psi^*_\alpha(\sigma_\alpha) = (\sigma_\alpha,\varepsilon(u_\alpha)) ,$$

whence we have almost everywhere in Ω ,

$$\psi_\alpha(\varepsilon(u_\alpha)) + \psi^*_\alpha(\sigma_\alpha) = \sigma_\alpha.\varepsilon(u_\alpha) ,$$

which is equivalent (cf. I.(3.9), I.(3.10)) to one or other of the conditions

(1.19) $$\varepsilon(u_\alpha)(x) \in \partial\psi^*_\alpha(\sigma_\alpha(x)), \text{ p.p. } x \in \Omega ,$$

$$\sigma_\alpha(x) \in \partial\psi_\alpha(\varepsilon(u_\alpha)(x)), \text{ p.p. } x \in \Omega . \tag{1.20}$$

Because of (1.5), the condition (1.19) is in turn identical to (1.16).

Finally it now remains only to prove the uniqueness of the solution of $\mathscr{P}_\alpha$; σ_α is unique, and (1.16) shows that $\varepsilon(u_\alpha)$ is unique. It follows that u_α is unique to within a rigid displacement and the latter is determined by the boundary condition $u_\alpha = u_0$ on Γ_0 (Γ_0 is by hypothesis a nonvacuous open subset of Γ) .

ii) We now investigate the passage to the limit, $\alpha \to 0$. We begin by considering the case where $\sup \mathscr{P}^* > -\infty$, and there is a σ admissible for $\mathscr{P}^*$, in particular *the solution* (which exists and is unique to $\mathscr{P}^*$). Since σ is admissible for $\mathscr{P}^*_\alpha$ and since $\psi^*_\alpha \leq \psi^*$, we have

$$\begin{aligned} \Psi^*_\alpha(\sigma_\alpha) - \int_{\Gamma_0} (\sigma_\alpha.\nu).u_0 d\Gamma \\ \leqq \Psi^*_\alpha(\sigma) - \int_{\Gamma_0} (\sigma.\nu).u_0 d\Gamma \\ \leqq \Psi^*(\sigma) - \int_{\Gamma_0} (\sigma.\nu).u_0 d\Gamma . \end{aligned} \tag{1.21}$$

This implies that $\sup \mathscr{P}^*_\alpha$ is bounded below as $\alpha \to 0$ and

$$-\infty < \sup \mathscr{P}^* \leqq \liminf_{\alpha\to 0} (\sup \mathscr{P}^*_\alpha) . \tag{1.22}$$

The trace theorem I.(1.47) implies that there exists a constant c_1 , depending only on Ω , such that

$$\|\sigma.\nu\|_{H^{-1/2}(\Gamma)^3} \leqq c_1\{|\sigma|_{L^2(\Omega;E)} + |\text{div }\sigma|_{L^2(\Omega)^3}\} \tag{1.23}$$

for all σ in $L^2(\Omega;E)$ with div $\sigma \in L^2(\Omega)^3$; in particular if $\sigma \in \mathscr{S}_a$ then div $\sigma = f$ and

$$\|\sigma.\nu\|_{H^{-1/2}(\Gamma)^3} \leqq c_1\{|\sigma|_{L^2(\Omega;E)} + |f|_{L^2(\Omega)^3}\} . \tag{1.24}$$

This implies, for $\sigma \in \mathscr{S}_a$

(1.25)
$$\left|\int_\Gamma (\sigma.\nu).u_0 d\Gamma\right| = \left|\int_\Gamma (\sigma.\nu).u_0 d\Gamma\right| \leqq$$
$$\leqq \|\sigma.\nu\|_{H^{-1/2}(\Gamma)^3} \cdot \|\gamma_0(u_0)\|_{H^{1/2}(\Gamma)^3}$$
$$\leqq c_2\{|\sigma|_{L^2(\Omega;E)} + 1\} ,$$
$$\leqq \frac{1}{4}\mathscr{A}(\sigma,\sigma) + c_3 ,$$

where c_2 depends on Ω, f, u_0, while c_3 depends on these same quantities and also on κ and μ (but not on σ) . Substituting in (1.21) and taking account of (1.13) we obtain

(1.26)
$$\frac{1}{4}\mathscr{A}(\sigma_\alpha,\sigma_\alpha) + \frac{1}{4\alpha}|\sigma_\alpha - P_K\sigma_\alpha|^2_{L^2(\Omega;E)}$$
$$\leqq c_3 - \sup \mathscr{P}^*_\alpha \leqq c_3 - \sup \mathscr{P}^* .$$

The sequence σ_α is thus bounded in $L^2(\Omega;E)$; we can extract from it a subsequence, which we still denote by σ_α , which converges weakly in this space to a limit σ_0 . In fact $\sigma_0 \in \mathscr{S}_a$ since this set is closed and convex in $L^2(\Omega;E)$; furthermore, by (1.26) and by the lower semi-continuity of the convex function

$$\tau \mapsto |\tau - P_K\tau|^2_{L^2(\Omega;E)} ,$$

we have

(1.27)
$$|\sigma_0 - P_K\sigma_0|^2_{L^2(\Omega;E)} \leqq \liminf_{\sigma\to 0} |\sigma_\alpha - P_K\sigma_\alpha|^2_{L^2(\Omega;E)} = 0 .$$

This implies that $\sigma_0(x) = P_K\,\sigma_0(x)$ for almost all $x \in \Omega$ and thus $\sigma_0^D(x) \in K^D$ p.p., so that σ_0 is admissible for $\mathscr{P}^*$. We deduce from (1.21) and (1.13) that

$$(1.28) \qquad \frac{1}{2}\mathscr{A}(\sigma_\alpha,\sigma_\alpha) - \int_{\Gamma_0} (\sigma_\alpha.\nu).u_0 d\Gamma \leqq$$

$$\leqq \Psi_\alpha^*(\sigma_\alpha) - \int_{\Gamma_0} (\sigma_\alpha.\nu).u_0 d\Gamma = - \sup \mathscr{P}_\alpha^* \quad .$$

By the trace theorem I.(1.47),

$$(1.29) \qquad \sigma_\alpha.\nu \to \sigma_0.\nu \text{ weakly in } H^{-1/2}(\Gamma)^3$$

and

$$(1.30) \qquad \int_{\Gamma_0} (\sigma_\alpha.\nu).u_0 d\Gamma \to \int_{\Gamma_0} (\sigma_0.\nu).u_0 d\Gamma \ .$$

The inequality (1.28) then implies, by lower semicontinuity, that

$$(1.31) \qquad \frac{1}{2}\mathscr{A}(\sigma_0,\sigma_0) - \int_{\Gamma_0} (\sigma_0.\nu).u_0 d\Gamma \leqq \liminf_{\alpha\to 0} (- \sup \mathscr{P}_\alpha^*) \ .$$

Whence, with (1.22)

$$(1.32) \qquad \sup \mathscr{P}^* \leqq - \frac{1}{2}\mathscr{A}(\sigma_0,\sigma_0) + \int_{\Gamma_0} (\sigma_0.\nu).u_0 d\Gamma \ .$$

As σ_0 is admissible for $\mathscr{P}^*$, this means that σ_0 is a solution of $\mathscr{P}^*$, and therefore $\sigma_0 = \sigma$ since the solution of $\mathscr{P}^*$ is unique (Theorem 4.1, Ch.I); furthermore with (1.22) (1.31) (1.32) we obtain

$$(1.33) \quad \sup \mathscr{P}^* = - \frac{1}{2}\mathscr{A}(\sigma,\sigma) + \int_{\Gamma_0} (\sigma.\nu).u_0 d\Gamma \leqq$$

$$\leqq \liminf_{\alpha\to 0} \{- \frac{1}{2}\mathscr{A}(\sigma_\alpha,\sigma_\alpha) - \frac{1}{4\alpha}|\sigma_\alpha - P_K\sigma_\alpha|^2_{L^2(\Omega;E)}$$

$$- \int_{\Gamma_0} (\sigma_\alpha.\nu).u_0 d\Gamma\}$$

$$\leqq \limsup_{\alpha\to 0} (\mathrm{Sup}\, \mathscr{P}_\alpha^*) \leqq \mathrm{Sup}\, \mathscr{P}^* \ ,$$

and these inequalities are in fact equations. On the one hand this implies (1.17) in the present case $(\sup \mathcal{P}^* > -\infty)$, and on the other hand it shows that, as $\alpha \to 0$:

(1.34) $$\frac{1}{4\alpha}|\sigma_\alpha - P_K\sigma_\alpha|^2_{L^2(\Omega;E)} \to 0$$

(1.35) $$\mathcal{A}(\sigma_\alpha,\sigma_\alpha) \to \mathcal{A}(\sigma,\sigma) .$$

Thanks to (1.35) and the convergence of σ_α to σ in $L^2(\Omega;E)$ weakly we have $\mathcal{A}(\sigma_\alpha - \sigma, \sigma_\alpha - \sigma) \to 0$ as $\alpha \to 0$, which implies

(1.36) $$\sigma_\alpha \to \sigma \text{ in } L^2(\Omega;E) \text{ strongly}$$

Finally a classical *reductio ad absurdum* argument shows that the whole sequence σ_α, and not just a subsequence of it, converges to σ in the strong $L^2(\Omega;E)$.

Consider now the case where $\sup \mathcal{P}^* = -\infty$; $\sup \mathcal{P}^*_\alpha$ necessarily tends to $-\infty$. For, if this were not so, the above arguments (and (1.26)) would show that it would be possible to extract from σ_α a subsequence converging weakly in $L^2(\Omega;E)$ to an element σ_0 admissible for $\mathcal{P}^*$, which is impossible. Hence $\sup \mathcal{P}^*_\alpha \to -\infty$ and (1.17) has been established.

iii) The behaviour of the solution u_α of $\mathcal{P}_\alpha$ when the safe load condition is satisfied now remains to be investigated. From (1.6 and the foregoing we have:

(1.37) $$\Psi_p(\varepsilon(u_\alpha)) - L(u_\alpha) \leqq \Psi_\alpha(\varepsilon(u_\alpha)) - L(u_\alpha) = \inf \mathcal{P}_\alpha \to \inf \mathcal{P} .$$

As u_α belongs to $\mathcal{C}_a$, and therefore to $\tilde{\mathcal{C}}_a$, u_α is a minimising sequence for $\mathcal{P}$ and also for the generalised problem $\mathcal{Q}$; it is therefore sufficient to apply Theorem 8.1 Ch.II.

Remark 1.1. When $\inf \mathcal{P} = \sup \mathcal{P}^* = -\infty$, we have no information on the behaviour of u_α and σ_α .

Remark 1.2. We refer the reader to G. Duvaut - J.L. Lions [1] for other questions concerning the elasto-visco-plastic models and to J.L. Lions [2] for general results on penalty methods.

1.2. Perturbations of Norton-Hoff type

We recalled in I.(3.55) the definition of the gauge function of a convex set K ; namely the function

$$\theta(\xi) = \inf \{s, \xi \in sK, \ s>0\} \ , \quad \forall \xi \in E \ . \tag{1.38}$$

In the present case $K = K^D \oplus \mathbb{R} I$, $\xi \in sK$ if and only if $\xi^D \in sK^D$ and so

$$\theta(\xi) = \theta(\xi^D) \quad , \quad \forall \xi \in E \ , \tag{1.39}$$

and the restriction of θ to E^D is also the gauge function of K^D . It is a convex and finite function at every point of E since K^D contains a neighbourhood of 0 ; it is therefore continuous on E . Also since $0 \in K$

$$\theta(\xi) \leqq 1 \Leftrightarrow \xi \in K \ . \tag{1.40}$$

For a fixed α , with $0<\alpha\leqq 1$, we define, following A. Friaâ [1], the convex function ψ_α^* (compare I.(3.56) with $\alpha = 1/q$) :

$$\psi_\alpha^*(\eta) = \frac{1}{2} A\eta \, \eta + \alpha\kappa_1 \, \theta(\eta)^{1/\alpha} \tag{1.41}$$

where $\kappa_1 > 0$ is fixed. By the preceding considerations the function ψ_α^* is convex, finite, and continuous at every point of E ; it is also strictly convex since so is $\frac{1}{2} A \, \eta . \eta$. The function ψ_α^* is the conjugate of $\psi_\alpha = (\psi_\alpha^*)^*$ which is a convex l.s.c. function which maps E into $\mathbb{R} \cup \{+\infty\}$. We know of no explicit expression for ψ_α but we have the following inequalities which will be useful.

LEMMA 1.2

For all $\alpha>0$, ψ_α *is a convex continuous function mapping* E *into* $\mathbb{R}$ *and, for all* $\xi \in E$:

(1.42) $$\psi_p(\xi) - \alpha\kappa_1 \leqq \psi_\alpha(\xi) \leqq \frac{\kappa}{2}(tr\ \xi)^2 + c_1|\xi^D|^{1/(1-\alpha)}$$

(1.43) $$\psi_\alpha(\xi) \geqq \frac{\kappa}{2}(tr\ \xi)^2 - \frac{1}{4\mu} + (1-\alpha)\,c_2|\xi^D|^{1/(1-\alpha)}$$

where c_1, c_2 *are two constants not dependent on* ξ .

<u>Proof</u>

As in Lemma 1.1 it is sufficient to prove (1.42) and (1.43). We note that, since K^D contains a ball $B_{k_0'}(o)$ and is contained in a ball $B_{k_1'}(o)$ of E^D ,

(1.44) $$\frac{|\xi^D|}{k_1'} \leqq \theta(\xi^D) \leqq \frac{|\xi^D|}{k_0'}$$

Whence

$$\psi_\alpha(\xi) \leqq \sup_{\eta\in E} \{\xi.\eta - \frac{(tr\ \eta)^2}{18\kappa} - \kappa_1\alpha\left(\frac{|\eta^D|}{k_1'}\right)^{1/\alpha}\}$$

$$\psi_\alpha(\xi) \leqq \frac{\kappa}{2}(tr\ \xi)^2 + \sup_{\eta^D\in E^D} \{\xi^D.\eta^D - \kappa_1\alpha\left(\frac{|\eta^D|}{k_1'}\right)^{1/\alpha}\}$$

$$\psi_\alpha(\xi) \leqq \frac{\kappa}{2}(tr\ \xi)^2 + \sup_{t\geqq 0} \{|\xi^D|t - \kappa_1\alpha\left(\frac{t}{k_1'}\right)^{1/\alpha}\}$$

$$\leqq \frac{\kappa}{2}(tr\ \xi)^2 + (1-\alpha)\frac{(k_1'|\xi^D|)^{1/(1-\alpha)}}{\kappa_1^{\alpha/(1-\alpha)}} \quad ;$$

which proves the second inequality in (1.42), the first being a consequence of $\psi_\alpha^* \leqslant \psi_p^* + \alpha\kappa_1$. To prove (1.43) we write

$$\psi_\alpha(\xi) \geqq \sup_{\substack{(\mathrm{tr}\,\xi)\in\mathbb{R}\\ \eta^D\in E^D}} \{\xi^D.\eta^D + \frac{1}{3}\,\mathrm{tr}\,\xi.\mathrm{tr}\,\eta - \frac{(\mathrm{tr}\,\eta)^2}{18\kappa} - \frac{1}{4\mu}|\eta^D|^2 - \kappa_1\alpha\left(\frac{|\eta^D|}{k_o'}\right)^{1/\alpha}\}$$

$$\geqq \frac{\kappa}{2}(\mathrm{tr}\,\xi)^2 + \sup_{t\geqq 0}\{|\xi^D|t - \frac{t^2}{4\mu} - \kappa_1\alpha\left(\frac{t}{k_o'}\right)^{1/\alpha}\}$$

Since, by Young's inequality, $t^2 \leqslant 2\alpha t^{1/\alpha} + 1 - 2\alpha$, we have :

$$\psi_\alpha(\xi) \geqq \frac{\kappa}{2}(\mathrm{tr}\,\xi)^2 - \frac{1-2\alpha}{4\mu} + \sup_{t\geqq 0}\{|\xi^D|t - \left(\frac{\kappa_1}{(k_o')^{1/\alpha}} + 2\right)\alpha t^{1/\alpha}\}$$

$$\geqq \frac{\kappa}{2}(\mathrm{tr}\,\xi)^2 + \frac{2\alpha-1}{4\mu} + (1-\alpha)\left(\frac{\kappa_1}{(k_o')^{1/\alpha}} + 2\right)^{\alpha/(\alpha-1)}|\xi^D|^{1(1-\alpha)}$$

This proves the lemma.

□

The strain problem can be written

(1.45) $$\inf_{v\in\mathscr{C}_a}\{\Psi_\alpha(\varepsilon(v)) - L(v)\} \qquad (\mathscr{P}_\alpha),$$

where, as in (1.10)

(1.46) $$\Psi_\alpha(\tau) = \int_\Omega \psi_\alpha(\tau(x))dx , \qquad \forall\, \tau \in L^2(\Omega;E) .$$

By (1.42) (1.43) and Proposition I.2.4, the function Ψ_α is convex, finite, and continuous on

(1.47) $$\{\tau \in L^{1/(1-\alpha)}(\Omega;E) \ , \ \mathrm{tr}\,\tau \in L^2(\Omega)\} ,$$

and hence on $L^2(\Omega;E)$.

To obtain the dual $\mathcal{P}_\alpha^*$ of $\mathcal{P}_\alpha$ we proceed exactly as in Section 1.2 above and find in this case that

$$(1.48) \qquad G^*(p) = \Psi_\alpha^*(p) = \int_\Omega \psi_\alpha^*(p(x))dx$$

$$= \frac{1}{2}\mathcal{A}(p,p) + \alpha\kappa_1 \int_\Omega \theta(p(x))^{1/\alpha}dx \ ,$$

and, by putting $\sigma = -p$, the problem $\mathcal{P}_\alpha^*$ takes the form

$$(1.49) \qquad \sup_{\sigma\in\mathcal{S}_a} \{-\frac{1}{2}\mathcal{A}(\sigma.\sigma) - \alpha\kappa_1 \int_\Omega \theta(\sigma(x))^{1/\alpha}dx + \int_{\Gamma_0} (\sigma.\nu).u_0 d\Gamma\}$$

Before stating the main result, we formulate a slightly modified version of $\mathcal{S}_\alpha$, that in which $\mathcal{C}_a$ is replaced in (1.45) by

$$\mathcal{C}_a^{(\alpha)} = \{v \in L^1(\Omega)^3 \ , \ \varepsilon^D(v) \in L^{1/(1-\alpha)}(\Omega;E^D) \ ,$$

$$\operatorname{div} v \in L^2(\Omega) \ , \ v = u_0 \text{ on } \Gamma_0\} \ .$$

It can be shown, thanks to Proposition I.1.3 that the infima of $\Psi_\alpha(\varepsilon(v)) - L(v)$ in $\mathcal{C}_a^{(\alpha)}$ and $\mathcal{C}_a$ are the same.

We still assume that $\Gamma_0 \neq \emptyset$ and we then have a result analogous to Theorem 1.1:

THEOREM 1.2

i) For all $\alpha>0$ *, the problems* $\mathcal{P}_\alpha$ *and* $\mathcal{P}_\alpha^*$ *are dual to one another and*

$$(1.50) \qquad \inf \mathcal{P}_\alpha = \sup \mathcal{P}_\alpha^* \in \mathbb{R}$$

$\mathcal{P}_\alpha^*$ *has the unique solution* σ_α *and* $\mathcal{P}_\alpha$ *has a solution* u_α *provided that* $\mathcal{C}_a$ *be replaced by* $\mathcal{C}_a^{(\alpha)}$ *in (1.45).*

ii) As $\alpha \to 0$ *,*

$$(1.51) \qquad \inf \mathcal{P}_\alpha = \sup \mathcal{P}_\alpha^* \to \inf \mathcal{P} = \sup \mathcal{P}^* \ .$$

If $\sup \mathscr{P}^* > -\infty$, σ_α *converges in* $L^2(\Omega;E)$ *to the solution* σ *of* $\mathscr{P}^*$. *If, in addition, the safe load condition is satisfied then the family of* u_α *remains bounded in* $U(\Omega)$ *and every weak limit of a sequence extracted from the* u_α *is a generalised solution of* $\mathscr{P}$.

Proof

The proof is very much the same as that of Theorem 1.1 and we shall only bring out those points where the argument differs. As far as part i) is concerned the only difference comes from the fact that the extremality relation $\varepsilon(u_\alpha)(x) \in \partial\Psi^*_\alpha(\sigma_\alpha(x))$ p.p. cannot be put in explicit form as in (1.16) and we cannot therefore prove in this way the uniqueness of the solution of $\mathscr{P}_\alpha$.

For part ii) the argument is a little different when it comes to the matter of proving, after (1.26), that $\sigma^D_0(x) \in K^D$ p.p. and hence that σ_0 is admissible for $\mathscr{P}^*$. We have here

$$\text{(1.52)} \qquad \alpha \int_\Omega \theta(\sigma^D_\alpha)^{1/\alpha}\, dx \leqq \text{const.}$$

For any function $\phi \in \mathscr{C}(\bar{\Omega})$, $\phi \geqq 0$, we can write, by lower semi-continuity:

$$\int_\Omega \theta(\sigma^D_0(x))\phi(x)dx \leqq \liminf_{\alpha\to 0} \int_\Omega \theta(\sigma^D(x))\phi(x)dx$$

By Hölder's inequality, this is less than or equal to

$$\liminf_{\alpha\to 0} \left\{\left[\int_\Omega \phi(x)^{1/(1-\alpha)}dx\right]^{1-\alpha}\left[\int_\Omega \theta(\sigma^D)^{1/\alpha}dx\right]^\alpha\right\} .$$

By Lebesgue's theorem and (1.52), this is less than or equal to

$$\left[\int_\Omega \phi(x)dx\right] . \lim_{\alpha\to 0} \left(\frac{c}{\alpha}\right)^\alpha = \int_\Omega \phi(x)dx .$$

Thus

$$\int_\Omega \theta(\sigma^D_0)\phi dx \leqq \int_\Omega \phi dx ,$$

for all $\phi \geq 0$, $\phi \in \mathscr{C}(\bar{\Omega})$, and so

$$\theta(\sigma_0^D(x)) \leqq 1 \quad \text{p.p.} ;$$

which, with (1.40) proves that $\sigma_0^D(x) \in K^D$ p.p. and hence that σ_0 is admissible for $\mathscr{P}^*$. The rest of the proof then follows without any essential change.

Remark 1.3. We refer the reader to the original articles of H.J. Hoff [1] and F.H. Norton [1] for the mechanical aspects of the so-called Norton-Hoff constitutive law; and to A. Friâa [1] for other aspects of the generalised Norton-Hoff law which has been considered here. The results of Theorem 1.2 are in part contained in R. Temam [12] but the rest are new (cf. also A. Neveu [1]).

1.3. Models with hardening

We shall now consider perturbations of the perfectly plastic elastic problem which correspond to the taking into account of hardening phenomea (cf. Ch. I, Sec. 3.3, iv)). We introduce an auxiliary finite-dimensional Euclidean space E_1 and a function ϕ mapping Ω into E_1 , which represents the hardening parameter (cf. B. Halphen - N.Q. Son [1], N.Q. Son [1]). We assume that we are now given, for $\alpha \geq 0$, a family $\mathscr{K}_\alpha$ of convex sets in $E \times E_1$ such that

$$(1.53) \quad \mathscr{K}_0 = K \times E_1 \text{ and } \mathscr{K}_\alpha \cap (E \times \{o\}) = K \times \{o\} , \forall \alpha \geqq 0 .$$

For each $\alpha \geq 0$, we define a pair of conjugate l.s.c. proper convex functions from $E \times E_1$ into $\mathbb{R} \cup \{+\infty\}$; ψ_α^* is defined by

$$(1.54) \qquad \psi_\alpha^*(\xi,\eta) = \begin{cases} \frac{1}{2} A\xi.\xi + \frac{1}{2}|\eta|^2 & \text{if } (\xi,\eta) \in \mathscr{K}_\alpha \\ + \infty & \text{if not.} \end{cases}$$

This function is convex l.s.c., and not identically $+\infty (\mathscr{K}_\alpha \neq \emptyset)$; it is the conjugate of $\psi_\alpha = (\psi_\alpha^*)^*$ which has these same properties. The properties of ψ_α are specified in more detail in the following lemma

LEMMA 1.3

ψ_α is a convex continuous function from $E \times E_1$ into $\mathbb{R}$ for all $(\xi,\eta) \in E \times E_1$,

(1.55) $$\psi_\alpha(\xi,\eta) \leqq \frac{\kappa}{2}(tr\ \xi)^2 + \mu|\xi^D|^2 + \frac{1}{2}|\eta|^2$$

(1.55)' $$\psi_\alpha(\xi,o) \geqq \psi_p(\xi)\ .$$

Proof

By definition

$$\psi_\alpha(\xi,\eta) = \sup_{(\bar\xi,\bar\eta)\in\mathcal{K}_\alpha} \{\xi.\bar\xi + \eta.\bar\eta - \frac{1}{2}A\bar\xi.\bar\xi - \frac{1}{2}|\bar\eta|^2\}\ ,$$

$$\psi_\alpha(\xi,\eta) \leqq \sup_{\bar\xi\in E}\{\xi.\bar\xi - \frac{1}{2}A\bar\xi.\bar\xi\} + \sup_{\bar\eta\in E_1}\{\eta.\bar\eta - \frac{1}{2}|\bar\eta|^2\}\ ,$$

and (1.55) follows. Since $\psi_\alpha(\xi,\eta) \geqslant 0$ because $(0,0) \in \mathcal{K}_\alpha$, (1.55) shows as above that ψ_α is finite at every point of $E \times E_1$ and is therefore continuous there. To prove (1.55)' we note that $\mathcal{K}_\alpha \supset K \times \{0\}$ so that

$$\psi_\alpha(\xi,0) = \sup_{(\bar\xi,\bar\eta)\in\mathcal{K}_\alpha} \{\xi.\bar\xi - \frac{1}{2}A\bar\xi.\bar\xi - \frac{1}{2}|\eta|^2\}$$

$$\leqq \sup_{\bar\xi\in K}\{\xi.\bar\xi - \frac{1}{2}A\bar\xi.\bar\xi\} = \psi_p(\xi)\ .$$

□

We now make the further hypothesis in regard to ψ_α, which will be verified in the examples to be given later:

(1.56) $$\begin{cases} \text{There are two constants } c(\alpha)\ ,\ d(\alpha)\ , \text{ depending on } \alpha\ , \\ \text{with } c(\alpha) > 0\ , \text{ and such that} \\ \psi_\alpha(\xi,0) \geqq c(\alpha)|\xi|^2 - d(\alpha)\ , \quad \forall\ \xi \in E\ . \end{cases}$$

In the present case the problem for the displacement field u can be written

(1.57)
$$\operatorname*{Inf}_{u \in \mathscr{C}_a} \{\Psi_\alpha(\varepsilon(u),o) - L(u)\} \qquad (\mathscr{P}_\alpha) ,$$

where we have put

(1.58)
$$\Psi_\alpha(\tau,\tau_1) = \int_\Omega \psi_\alpha(\tau(x),\tau_1(x))dx ,$$

for all $\tau \in L^2(\Omega;E)$ and $\tau_1 \in L^2(\Omega;E_1)$.

To determine the dual $\mathscr{P}_\alpha^*$ of $\mathscr{P}_\alpha$ we again use the general framework of Chapter I, Section 2. We put :

$$V = H^1(\Omega)^3 \times L^2(\Omega;E_1) , \ V^* \text{ is the dual of } V ,$$

$$Y = Y^* = Y_1 \times Y_2 = L^2(\Omega;E) \times L^2(\Omega;E_1) ,$$

$$\Lambda = \Lambda_1 \times \Lambda_2 , \ \Lambda_1(u) = \varepsilon(u) , \ \Lambda_2 v = 0 .$$

For all $\{u,v\} \in V$, let

$$F(u,v) = \begin{cases} - L(u) & \text{if } u \in \mathscr{C}_a \\ + \infty & \text{if not} . \end{cases}$$

and for all $p = (p_1,p_2) \in Y$

$$G(p_1,p_2) = \Psi_\alpha(p_1,p_2) = \int_\Omega \psi_\alpha(p_1(x),p_2(x))dx .$$

It is then clear that the problem

$$\operatorname*{Inf}_{\{u,v\} \in V} \{F(u,v) + G(\Lambda(u,v))\}$$

which is of the form I.(2.26) is identical to $\mathscr{P}_\alpha$. To determine its dual I.(2.27), we have to calculate F^* and G^* ; we eventually find from Lemma I.2.2 that

(1.59) $F^*(\Lambda^* p) = \int_{\Gamma_0} (p_1 \cdot \nu) u_0 d\Gamma$ if $-p_1 \in \mathscr{S}_a$ and $= + \infty$ if not.

As for G^*, by applying Proposition I.2.5, we obtain, for $p = \{p_1,p_2\} \in Y$:

$$(1.60) \qquad G^*(p) = \int_\Omega \psi_\alpha^*(p_1(x),p_2(x))dx \quad ,$$
$$= \frac{1}{2}\mathscr{A}(p_1,p_2) + \frac{1}{2}\int_\Omega |p_2(x)|^2 dx \quad \text{(by (1.54))}.$$

The problem $\mathscr{P}_\alpha^*$ can then be written

$$(1.61) \qquad \sup_{\sigma,\phi} \{-\frac{1}{2}\mathscr{A}(\sigma,\sigma) - \frac{1}{2}\int_\Omega |\phi|^2 dx + \int_{\Gamma_0} (\sigma.\nu).u_0 d\Gamma\}$$

where we have put $\{p_1,p_2\} = \{-\sigma,\phi\}$, and the supremum in (1.61) is taken over

$$(1.62) \qquad \begin{cases} \sigma \in \mathscr{S}_a \ , \ \phi \in L^2(\Omega;E_1) \\ \{\sigma(x),\theta(x)\} \in \mathscr{K}_\alpha \quad \text{p.p. } x \in \Omega \end{cases}$$

□

We shall now study the problems $\mathscr{P}_\alpha$ and $\mathscr{P}_\alpha^*$ when $\alpha>0$ is fixed, and then investigate their behaviour as α tends to zero. The following heuristic remark gives us an insight into the link between $\mathscr{P}_\alpha^*$ and $\mathscr{P}^*$; if $\alpha = 0$ then, by reason of (1.53), the constraints (1.62) leave ϕ completely free, whereas $\sigma(x)$ must belong to K for almost all $x \in \Omega$. It is then easy to obtain in (1.61) a partial supremum with respect to ϕ : the supremum is attained at $\phi = 0$ and (1.61) becomes identical to the problem $\mathscr{P}^*$.

To put all this into a precise form we make the following hypotheses

$$(1.63) \qquad \begin{cases} \text{For all } \alpha>0 \text{ , the set } \mathscr{E}_\alpha \text{ of } (\sigma,\phi) \text{ admissible} \\ \text{for } \mathscr{P}_\alpha^* \text{ is non-vacuous :} \\ \mathscr{E}_\alpha = \{(\sigma,\phi) \in \mathscr{S}_a \times L^2(\Omega;E_1),(\sigma(x),\phi(x)) \in \mathscr{K}_\alpha \text{ p.p. } x \in \Omega\} \end{cases}$$

$$(1.64)\quad \begin{cases} \text{If } (\sigma_{\alpha_j}, \phi_{\alpha_j}) \in \mathscr{E}_{\alpha_j} \text{ and } \sigma_{\alpha_j} \to \sigma \text{ in } L^2(\Omega;E) \text{ weakly,} \\ \phi_{\alpha_j} \to \phi \text{ in } L^2(\Omega;E_1) \text{ weakly, when } \alpha_j \to 0 \text{, then} \\ (\sigma,\phi) \in \mathscr{E}_0 . \end{cases}$$

Our hypothesis are the general hypotheses of Section I.4, and the specific assumptions (1.53)(1.56)(1.63)(1.64) listed above, and we also suppose for simplicity that $\Gamma_0 \neq \emptyset$. We then have :

THEOREM 1.3

The hypotheses are those specified above.

i) For all $\alpha>0$ *, the problems* $\mathscr{P}_\alpha$ *and* $\mathscr{P}^*_\alpha$ *are dual to each other and*

$$(1.65)\qquad inf\ \mathscr{P}_\alpha = sup\ \mathscr{P}^*_\alpha \in \mathbb{R} .$$

$\mathscr{P}_\alpha$ *has at least one solution* u_α *and* $\mathscr{P}^*_\alpha$ *has a unique solution* $(\sigma_\alpha,\phi_\alpha)$*. These solutions are linked by the extremality relations (reflecting the constitutive law of the material):*

$$(1.66)\quad \begin{cases} (\varepsilon(u_\alpha)(x),0) \in \partial\psi^*_\alpha(\sigma_\alpha(x),\phi_\alpha(x)) \\ (\sigma_\alpha(x),\phi_\alpha(x)) \in \partial\psi_\alpha(\varepsilon(u_\alpha)(x),0) \text{ , p.p. } x \in \Omega \end{cases}$$

ii) We now suppose that, as α *tends to zero,*

$$(1.67)\qquad inf\ \mathscr{P}_\alpha = sup\ \mathscr{P}^*_\alpha \to inf\mathscr{P} = sup\ \mathscr{P}^* .$$

If $inf\,\mathscr{P} = sup\ \mathscr{P}^* > -\infty$ *, the family* σ_α *converges in* $L^2(\Omega;E)$ *to the solution* σ *of* $\mathscr{P}^*$*. Finally, if the safe load condition is satisfied,* u_α *is a minimising sequence for the problem* $\mathscr{P}$ *and its points of accumulation (for the weak topology of* $U(\Omega)$*) are generalised solutions of* $\mathscr{P}$*.*

Proof

i) For a fixed $\alpha>0$ the set of (σ,ϕ) admissible for $\mathscr{P}^*_\alpha$ is the set $\mathscr{E}_\alpha$ which is, by hypothesis, not empty; hence $\sup\mathscr{P}^*_\alpha > -\infty$ and

$\inf \mathscr{P}_\alpha \in \mathbb{R}$. By reason of (1.54) and (1.56) the function $G(p) = \psi_\alpha(p_1,p_2)$ is finite at every point of Y, and is therefore convex and continuous on Y. Condition I.(2.29) is satisfied with $u^0 = (u_0,o)$ and I.(2.30) then implies (1.65) and the existence of a solution $(\sigma_\alpha,\phi_\alpha)$ for $\mathscr{P}_\alpha^*$; the uniqueness of the solution of $\mathscr{P}^*$ is of course due to the strict convexity of

$$(\sigma,\phi) \to \mathscr{A}(\sigma,\sigma) + |\phi|^2_{L^2(\Omega;E_1)} .$$

The existence of a solution u_α to $\mathscr{P}_\alpha$ is proved by applying Proposition I.2.1. The continuity of the function Ψ_α on Y implies I.(2.38) and I.(2.42) follows easily from (1.56). Lastly (1.66) is simply another way of writing the optimality condition I.(2.31) which here gives

$$\Psi_\alpha(\varepsilon(u_\alpha),o) + \Psi_\alpha^*(\sigma_\alpha,\phi_\alpha) = (\sigma_\alpha,\varepsilon(u_\alpha))$$

which is equivalent to

$$\psi_\alpha(\varepsilon(u_\alpha)(x),o) + \psi_\alpha^*(\sigma_\alpha(x),\phi_\alpha(x)) = \sigma_\alpha(x).\varepsilon(u_\alpha)(x) \qquad \text{p.p. } x \in \Omega ,$$

and this is none other than (1.66).

ii) We now investigate the passage to the limit, $\alpha \to 0$, under the hypothesis (1.67).

If $\sup \mathscr{P}^* > -\infty$, then the sequence $\sup \mathscr{P}_\alpha^*$ is bounded and

$$\sup \mathscr{P}_\alpha^* = \frac{1}{2}\mathscr{A}(\sigma_\alpha,\sigma_\alpha) + \frac{1}{2}|\phi_\alpha|^2_{L^2(\Omega;E_1)} - \int_{\Gamma_0} (\sigma_\alpha.\nu).u_0 d\Gamma \leqq \text{const.}$$

The inequality (1.25) enables us to conclude that $(\sigma_\alpha,\phi_\alpha)$ is bounded in $L^2(\Omega;E) \times L^2(\Omega;E_1)$. There exist therefore (σ,ϕ) and a subsequence $\alpha_j \to 0$, such that $\sigma_{\alpha_j} \to \sigma_0$ weakly in $L^2(\Omega;E)$ and $\phi_{\alpha_j} \to \phi_0$ weakly in $L^2(\Omega;E_1)$. Since $(\sigma_{\alpha_j},\phi_{\alpha_j}) \in \mathscr{E}_{\alpha_j}$, $(\sigma_0,\phi_0) \in \mathscr{C}_0$ and so, thanks to (1.53), σ_0 is admissible for $\mathscr{P}^*$. By (1.30) and semi-continuity,

$$- \frac{1}{2} \mathscr{A}(\sigma_o,\sigma_o) - \frac{1}{2} |\phi_o|^2_{L^2(\Omega;E_1)} + \int_{\Gamma_o} (\sigma_o.\nu).u_o d\Gamma \geq$$

$$\geq \limsup_{\alpha_j \to 0} \{\sup \mathscr{P}^*_{\alpha_j}\} = \sup \mathscr{P}^* .$$

By a remark made before Theorem 1.3, $\mathscr{P}^*_\alpha$ for $\alpha = 0$ is identical to $\mathscr{P}^*$ and its unique solution is (σ,o) where σ is the solution of $\mathscr{P}^*$; thus $\sigma_o = \sigma$ and $\phi_o = o$. It can be shown, as in Theorem 1.1 that $\sigma_{\alpha_j} \to \sigma$ in $L^2(\Omega;E)$ strongly and $\phi_{\alpha_j} \to 0$ in $L^2(\Omega;E_1)$ strongly and lastly that it is the whole family $(\sigma_\alpha , \phi_\alpha)$ which converges to (σ, o) in $L^2(\Omega;E) \times L^2(\Omega;E_1)$ strongly.

By (1.55) and (1.67) and since $u_\alpha \in \mathscr{C}_a$:

$$\inf \mathscr{P} \leq \Psi_p(\varepsilon(u_\alpha)) - L(u_\alpha) \leq \Psi_\alpha(\varepsilon(u_\alpha),o) - L(u_\alpha)$$

$$= \inf \mathscr{P}_\alpha \to \inf \mathscr{P} .$$

We deduce from this that u_α is a minimising sequence for $\mathscr{P}$ and if the safe load condition is satisfied we can draw the same conclusions as in Theorem 1.1.

Theorem 1.3 has thus been proved.

□

We now turn to the study of the two particular cases mentioned in I.(3.63) and I.(3.64), for which we shall give a direct verification of the hypotheses (1.53) (1.56) (1.63) (1.64) (1.67).

Example 1

This is I.(3.63); K is the von Mises convex set

$$(1.68) \qquad K = \{\xi \in E , \quad |\xi^D| \leq \sqrt{2}\, k\} ,$$

$E_1 = \mathbb{R}$ and, for $\alpha \geq 0$,

$$\mathscr{K}_\alpha = \{(\xi,\eta) \in E \times \mathbb{R} , \quad |\xi^D| < \sqrt{2}\, k + \alpha\eta\}$$

These are closed convex sets, $\mathscr{K}_o = K \times E_1$ and $\mathscr{K}_\alpha \cap (E \times \{o\}) = K \times \{o\}$. To prove (1.56) we note that

$$\psi_\alpha(\xi,o) = \sup_{(\bar\xi,\bar\eta)\in\mathscr{K}_\alpha} \{\xi.\bar\xi - \frac{1}{2} A\bar\xi.\bar\xi - \frac{1}{2}|\bar\eta|^2\}$$

$$= \sup_{\bar\xi \in E} \{\xi.\bar\xi - \frac{1}{2} A\bar\xi.\bar\xi - \frac{1}{2\alpha}(|\bar\xi^D| - \sqrt{2}k)_+^2\}$$

$$\geqq \sup_{\bar\xi \in E} \{\xi.\bar\xi - \frac{1}{2} A\bar\xi.\bar\xi - \frac{1}{2\alpha}(|\bar\xi^D| - \sqrt{2}k)_+^2\}$$

$$= \frac{\kappa}{2}(\mathrm{tr}\,\xi)^2 + \sup_{r\geqq 0}\{|\xi^D|r - \frac{r^2}{4\mu} - \frac{1}{2\alpha}(r-\sqrt{2}k)_+^2\}$$

$$= \frac{\kappa}{2}(\mathrm{tr}\,\xi)^2 + \frac{\mu(\alpha|\xi^D| + \sqrt{2}k)^2}{\alpha(\alpha+2\mu)} - \frac{k^2}{\alpha}$$

which proves (1.56). The condition (1.63) is satisfied; for if σ is any element of $\mathscr{S}_a$ and if ϕ is defined by

$$\phi(x) = \frac{(|\sigma^D(x)| - \sqrt{2}k)_+}{\alpha} ,$$

then $(\sigma,\phi) \in \mathscr{E}_\alpha$ and this set is therefore not empty. Lastly, as to (1.64), the inequality $|\sigma^D_{\alpha_j}| \leqq \sqrt{2}k + \alpha_j\,\phi_{\alpha_j}$ obviously implies, in the limit, $|\sigma^D| \leqq \sqrt{2}k$.

It now remains only to verify (1.67). We can take in (1.61) the supremum with respect to ϕ which is constrained to belong to $L^2(\Omega;E_1)$ and to satisfy (for fixed σ) .

$$\phi(x) \geqq \frac{1}{\alpha}(|\sigma^D(x)| - \sqrt{2}k) \quad \text{p.p.} \quad x\in\Omega .$$

We thus obtain

(1.69) $$\sup_\sigma \{-\frac{1}{2}\mathscr{A}(\sigma,\sigma) - \frac{1}{2\alpha^2}\int_\Omega (|\sigma^D(x)| - \sqrt{2}k)_+^2 dx + \\ + \int_{\Gamma_0} (\sigma.\nu).u_0 d\Gamma\} ,$$

where the supremum has to be taken over all $\sigma \in \mathcal{S}_a$. This problem (like (1.14)) is a *penalised* form of $\mathcal{P}^*$, the penalty for non-compliance with the constraint $|\sigma^D| \leqq \sqrt{2}k$ being represented by the term

$$\frac{1}{2\alpha^2}\int_\Omega (|\sigma^D(x)| - \sqrt{2}k)_+^2 dx \ .$$

The argument which allows one to deduce that $\sup \mathcal{P}_\alpha^* \to \sup \mathcal{P}^*$ is very similar to the one used in the proof of Theorem 1.1 and need not be repeated.

Example 2

This is I.(3.64); K is again the von Mises convex set (1.68); but now $E_1 = E^D$ and for $\alpha \geqq 0$, $\mathcal{K}_\alpha$ is the set defined by

(1.70) $$\mathcal{K}_\alpha = \{(\xi,\eta) \in E \times E^D \ , \ |\xi^D - \alpha\eta| \leqq \sqrt{2}k\} \ .$$

The $\mathcal{K}_\alpha$ are closed convex sets, $\mathcal{K}_0 = K \times E_1$ and $\mathcal{K}_\alpha \cap (E\times\{o\}) = K\times\{o\}$. The condition (1.63) is satisfied because if $\sigma \in \mathcal{S}_a$ and $\phi(x) = (1/\alpha)\sigma^D(x)$, $(\sigma,\phi) \in \mathcal{E}_\alpha$; (1.64) is obtained by a semicontinuity argument.

The verification of (1.56) and (1.67) requires the following lemma.

LEMMA 1.4

For all $\xi \in E$

(1.71) $$0 \leqq \psi_\alpha(\xi,o) - \psi_p(\xi) \leqq \frac{\alpha^2}{2(\alpha^2+1)}(|\xi^D| + 1)^2$$

(1.72) $$\psi_\alpha(\xi,o) - \psi_p(\xi) \geqq \frac{\alpha^2}{4(\alpha^2+1)}(|\xi^D|^2 - 4)$$

Proof

By definition

$$\psi_\alpha(\xi,o) = \sup_{\substack{\bar\xi \in E,\eta\in E^D \\ |\bar\xi^D-\alpha\eta| \leqq \sqrt{2}k}} \{\xi.\bar\xi - \frac{1}{2}A\bar\xi.\bar\xi - \frac{1}{2}|\eta|^2\} \ ,$$

and taking the supremum with respect to $\operatorname{tr}\bar{\xi}$, and on putting $\zeta = \bar{\xi}^D - \alpha\eta$, we have

$$\psi_\alpha(\xi,o) = \frac{\kappa}{2}(\operatorname{tr}\xi)^2 + \sup_{\substack{\zeta,\eta\in E^D\\ |\zeta|<\sqrt{2}k}} \{\xi^D.\zeta + \alpha\xi^D.\eta - \frac{1}{2}|\zeta+\alpha\eta|^2 - \frac{1}{2}|\eta|^2\} .$$

We can now work out the supremum with respect to ζ and we find

$$\psi_\alpha(\xi,o) = \frac{\kappa}{2}(\operatorname{tr}\xi)^2 + \sup_{\eta\in E^D} \{\alpha\xi^D.\eta - \frac{1}{2}(\alpha^2+1)|\eta|^2 + \psi_p(\xi^D-\alpha\eta)\}$$

The calculation of ψ_p for the von Mises convex set was done in I.(4.11), I.(4.12); the explicit form of ψ_p shows that this function is Lipschitzian with a ratio 2k ; thus

$$(1.73)\qquad |\psi_p(\xi^D - \alpha\eta) - \psi_p(\xi^D)| \leqq \alpha|\eta| .$$

Whence

$$\begin{aligned}\psi_\alpha(\xi,o) - \psi_p(\xi) &= \psi_\alpha(\xi,o) - \frac{\kappa}{2}(\operatorname{tr}\xi)^2 - \psi_p(\xi^D)\\ &\leqq \sup_{\eta\in E^D}\{\alpha\xi^D.\eta - \frac{1}{2}(\alpha^2+1)|\eta|^2 + \alpha|\eta|\}\\ &\leqq \frac{\alpha^2}{2(\alpha^2+1)}\,|\xi^D| + 1)^2 .\end{aligned}$$

Similarly

$$\begin{aligned}\psi_\alpha(\xi,o) - \psi_p(\xi) &\geqq \sup_{\eta\in E^D}\{\alpha\xi^D.\eta - \frac{1}{2}(\alpha^2+1)|\eta|^2 - \alpha|\eta|\}\\ &= \frac{\alpha^2}{2(\alpha^2+1)}\,(|\xi^D| - 1)_+^2\\ &\geqq \frac{\alpha^2}{4(\alpha^2+1)}\,(|\xi^D|^2 - 4) .\end{aligned}$$

The validity of the hypothesis (1.56) is a consequence of (1.73). To prove (1.67) we note, using (1.71), that

$$\inf \mathscr{P} \leqq \Psi_p(\varepsilon(u_\alpha),o) - L(u_\alpha) \leqq$$

$$\leqq \Psi_\alpha(\varepsilon(u_\alpha),o) - L(u_\alpha) = \inf \mathscr{P}_\alpha .$$

Also, for all $v \in \mathscr{C}_a$,

$$\inf \mathscr{P}_\alpha \leqq \Psi_\alpha(\varepsilon(v),o) - L(v)$$

$$\leqq \Psi_p(\varepsilon(v)) - L(v) + \frac{\alpha^2}{2(\alpha^2+1)} \int_\Omega (|\varepsilon^D(v)| + 1)^2 dx ,$$

by (1.71).

When $\alpha \to 0$, this gives us

$$\limsup_{\alpha \to 0} \{\inf \mathscr{P}_\alpha\} \leqq \Psi_p(\varepsilon(v)) - L(v) .$$

Since $v \in \mathscr{C}_a$ is arbitrary, this upper limit is necessarily less than or equal to $\inf \mathscr{P}$ and this proves (1.67).

2. Some problems in the theory of plates

We now propose to look at a variational problem in plate theory derived from a perfectly-plastic elastic model. The study, which follows F. Demengel [1-2-3], parallels that made in Chapters I and II for the three-dimensional problem. In Section 2.1, after a few remarks on some useful functional spaces, we recall the formulation of the mechanical model of the plate in question, and we consider briefly the case where the plate is elastic: the variational problems for the stress and strain fields, existence and uniqueness of solutions, duality. In Section 2.2 we describe the plastic model, properly so called, after giving some properties of the surface potential ψ . We formulate the variational problems for stresses and strains, study their duality, and the existence and uniqueness of the solutions for the stress problem. We then go on to consider the associated limit analysis problem; at the end of Section 2.2 we begin to tackle the generalised strain problem: at that particular point we shall be considering only a relaxed problem analogous to the one in Section I.6 where the boundary conditions were partially relaxed but the strain field was still required to be regular in the interior of the domain. Section 2.3 is devoted to a study of the finite-energy space for the problem, that is to say the space $HB(\Omega)$ consisting of functions which are summable within the open set $\Omega \subset \mathbb{R}^2$, as are their first derivatives, and whose second derivatives (and therefore Hessian) are bounded measures. Finally in Section 2.4 we formulate the generalised variational problem in the space $HB(\Omega)$; we study its duality in relation to the variational problem relating to stresses and obtain the result on the existence of a solution.

To avoid too lengthy developments we shall content ourselves in this section with describing the main steps and refer the reader to F. Demengel [1] for the full details.

2.1. Some reminders. The elastic model

2.1.1. Functional spaces

Let Ω be an open set in $\mathbb{R}^2$ whose boundary Γ we shall suppose to be of class $\mathscr{C}^2$. We shall be concerned with the spaces $H^1(\Omega)$ and $H^2(\Omega)$ (cf. I.(1.26)). If $u \in H^2(\Omega)$, we write ∇u for the gradient of u and $\nabla\nabla u$ for the Hessian matrix of u; the norm of u in $H^2(\Omega)$ can therefore be written as

$$|u|_2 = \{|u|^2 + |\nabla u|^2 + |\nabla\nabla u|^2\}^{1/2} \tag{2.1}$$

where, as before, $|.|$ denotes the norm in a space $L^2(\Omega)^m$. We recall that $H^1(\Omega)$ is included in $L^p(\Omega)$ for any finite $p \geq 1$ (cf. I.(1.28)), and that there exist two continuous linear mappings γ_0, γ_1, defined on $H^2(\Omega)$ and with values in $L^2(\Gamma)$, such that

$$\gamma_0(u) = u_{|\Gamma}\ ,\ \gamma_1(u) = \frac{\partial u}{\partial \nu}\Big|_\Gamma\ , \tag{2.2}$$

for all u in $\mathscr{C}^2(\bar{\Omega})$ (cf. I.(1.31)). The image of $H^2(\Omega)$ under the mapping γ_1 is $H^{1/2}(\Gamma)$ and we call $H^{3/2}(\Gamma) \subset H^{1/2}(\Gamma)$ the image of $H^2(\Omega)$ under γ_0; it is a Hilbert space, for example for the structure carried over by γ_0 and it can be shown that $\gamma_0 \times \gamma_1$, is a surjective mapping of $H^2(\Omega)$ on to $H^{3/2}(\Gamma) \times H^{1/2}(\Gamma)$; we refer the reader to R.S. Adams [1], J.L. Lions - E. Magenes [1] for further details. By I.(1.27) and the Proposition I.1.3, $\mathscr{C}^\infty(\bar{\Omega})$ is dense in $H^2(\Omega)$. If a distribution u on Ω possesses second derivatives in $L^2(\Omega)$ then u belongs to $H^2(\Omega)$: this follows easily by applying

Proposition I.1.2 to D_1u , D_2u and u . Furthermore, if Π_1 is the space of polynomials of degree not exceeding 1 (i.e. the kernel of $\nabla\nabla$) , $|\nabla\nabla u|$ is, on $H^2(\Omega)/\Pi_1$, a norm equivalent to the natural norm; it can also be verified that if p is a seminorm on $H^2(\Omega)$ which is a norm on Π_1 , then

$$(2.3) \qquad p(u) + |\nabla\nabla u|$$

is, on $H^2(\Omega)$, a norm equivalent to the natural norm. This applies for example to $|u| + |\nabla\nabla u|$ which corresponds to $p(u) = |u|$.

Let E be, as before, the space of symmetric tensors of order 2. If $M \in L^2(\Omega;E)$, we write $\nabla.M$ for the divergence of M , that is to say the vector whose components are

$$(2.4) \qquad M_{ij,j} = D_1M_{i1} + D_2M_{i2} \ , \ i = 1,2,$$

and we write $\nabla.\nabla.M$ for the divergence of $\nabla.M$, i.e. the distribution

$$(2.5) \qquad \nabla.\nabla.M = M_{ij,ij} = \sum_{i,j=1}^{2} D_{ij}\, M_{ij} \ .$$

The space

$$\mathscr{H} = \{M \in L^2(\Omega;E) \ , \nabla.\nabla.M \in L^2(\Omega)\} \ ,$$

is a Hilbert space for the natural norm

$$(2.6) \qquad ||M||_{\mathscr{H}} = \{|M|^2 + |\nabla.\nabla.M|^2\}^{1/2}$$

It follows from Proposition I.1.3 that $\mathscr{C}^\infty(\bar{\Omega};E)$ is dense in $\mathscr{H}$ (for the norm (2.6)).

If $M \in \mathscr{C}^2(\bar{\Omega};E)$ and $v \in \mathscr{C}^2(\bar{\Omega})$, we have a Green's formula (cf. G. Duvaut - J.L. Lions [1]):

$$(2.7) \qquad \int_\Omega v.(\nabla.\nabla.M)dx - \int_\Omega M.(\nabla\nabla v)dx = \int_\Gamma b_0(M)v ds - \int_\Gamma b_1(M)\, \frac{\partial v}{\partial \nu}\, ds$$

where, for $M \in \mathscr{C}^2(\bar{\Omega};E)$,

$$b_o(M) = (\nabla.M).\nu + \frac{\partial}{\partial s}((M.\nu).\tau) \tag{2.8}$$

$$b_1(M) = (M.\nu).\nu \quad ; \tag{2.9}$$

in these formulae ν is the outward normal unit vector on Γ , τ is a tangent unit vector and s the curvilinear abscissa on Γ measured positively in the direction of τ . These trace operators and the Green's formula (2.7) are extended in F. Demengel [1], as follows:

PROPOSITION 2.1.

i) There is continuous linear surjective mapping, $b = b_o \times b_1$ *of* $\mathscr{H}$ *on to* $H^{-3/2}(\Omega) \times H^{-1/2}(\Omega)$ *such that* $b_o(M)$ *and* $b_1(M)$ *coincide with the expressions (2.8) (2.9) for all* M *in* $\mathscr{C}(\bar{\Omega};E)$.

ii) The mapping b *has a continuous 'lifting', that is to say there exists a continuous linear mapping* r *of* $H^{-3/2}(\Gamma) \times H^{-1/2}(\Gamma)$ *into* $\mathscr{H}$, *such that* $b_o r$ *is the identity mapping.*

iii) For all $M \in \mathscr{H}$ *and all* $v \in H^2(\Omega)$, *we have the generalised Green's formula*

$$\int_\Omega v.(\nabla.\nabla.M)dx - \int_\Omega M.(\nabla\nabla v)dx = \tag{2.10}$$

$$= \langle b_o(M),\gamma_o(v)\rangle - \langle b_1(M),\gamma_1(v)\rangle$$

The spaces $H^{-3/2}(\Gamma)$ and $H^{-1/2}(\Gamma)$ are the spaces of distributions on Γ , which are respectively the duals of $H^{3/2}(\Gamma)$ and $H^{1/2}(\Gamma)$, $(L^2(\Gamma) \subset H^{-1/2}(\Gamma) \subset H^{-3/2}(\Gamma))$; the bracketed symbols $\langle .,. \rangle$ on the right-hand side of (2.10) are duality pairings between $H^{-3/2}(\Gamma)$ and $H^{3/2}(\Gamma)$, and $H^{-1/2}(\Gamma)$ and $H^{1/2}(\Gamma)$ respectively. To simplify the notation we shall in future write (2.10) conventionally in the form (2.7).

The proof of Proposition 2.1 will be found in the reference quoted above.

2.1.2. Mechanical formulation of the plate problem

We consider a thin plate occupying the given surface Ω. This plate is subjected to forces, of surface density f, acting in a direction at right angles to the surface. The boundary Γ of Ω, consists of three open disjunct portions $\Gamma_0, \Gamma_1, \Gamma_2$, whose complement Γ_* contains by hypothesis only a finite number of points; we shall suppose the plate to be subjected on Γ_2 to forces of linear density g^0 and on $\Gamma_1 \cup \Gamma_2$ to couples whose moments have a linear density g^1. The plate's state of stress is entirely determined by a field of tensors M of order 2, defined over Ω (which represent certain averages of the stress tensors in the plate). Since displacements of the plate parallel to its plane are negligible and will be ignored, the deformation of the plate can be entirely determined by a scalar function $w : \Omega \to \mathbb{R}$, where $w(x)$ represents the displacement of the point $x \in \Omega$ in the direction orthogonal to the plate.

We shall suppose the vertical displacement w of the plate to be imposed on $\Gamma_0 \cup \Gamma_1$, and the normal derivative of w to be imposed on Γ_0, so that the boundary conditions are

$$(2.11) \qquad w = u_0 \text{ on } \Gamma_0 \cup \Gamma_1 , \quad \frac{\partial w}{\partial \nu} = u_1 \text{ on } \Gamma_0 , \quad \Gamma_0 \neq \emptyset .$$

We shall suppose, for simplicity, that Γ_0 is not void and contains three non-collinear points ([1]). If $u_0 = u_1 = 0$, the plate is firmly embedded along Γ_0, but only simply supported along Γ_1.

$$(2.12) \qquad f \in L^2(\Omega) , \quad g^0 \in L^2(\Gamma_2) , \quad g^1 \in L^2(\Gamma_1 \cup \Gamma_2) ;$$

and the functions u_0 and u_1 will be assumed to have been given as the traces of U and $\partial U/\partial \nu$ on Γ,

---

([1]) This ensures that polynomials of degree $\leqslant 1$ which vanish on Γ_0 are identically zero. This hypothesis, which is of course not essential, makes for a certain amount of simplification.

(2.13) $u_0 = \gamma_0(U)$ on $\Gamma_0 \cup \Gamma_1$, $u_1 = \gamma_1(U)$ on Γ_0 , $U \in H^2(\Omega)$

We shall suppose, for technical reasons, that

(2.14) $$\gamma_0(U) = 0 \text{ on } \Gamma_2 \text{ , } \gamma_1(U) = 0 \text{ on } \Gamma_1 \cup \Gamma_2 \text{ .}$$

Kinematically admissible and statically admissible sets

We now define the kinematically admissible strain fields, and the statically admissible stress fields. The former comprise the set

(2.15) $$\mathscr{C}_a = \mathscr{C}_a(u_0,u_1) = \{v \in H^2(\Omega) \text{ , } v_{|\Gamma_0 \cup \Gamma_1} = u_0 \text{ , } \frac{\partial v}{\partial \nu}\Big|_{\Gamma_0} = u_1\} \text{ .}$$

As regards the stresses we note that by the fundamental laws of mechanics (cf. G. Duvaut - J.L. Lions [1]), the tensor field M must satisfy the following boundary conditions

(2.16) $$\begin{cases} b_0(M) + g^0 = 0 \text{ on } \Gamma_2 \\ b_1(M) + g^1 = 0 \text{ on } \Gamma_1 \cup \Gamma_2 \end{cases}$$

and, within Ω , the equation

(2.17) $$\nabla.\nabla.M + f = 0 \text{ .}$$

This leads us to define

(2.18) $$\mathscr{S}_a = \mathscr{S}_a(f,g^0,g^1) =$$

$$= \{M \in L^2(\Omega;E)\text{, } M \text{ satisfies (2.16) and (2.17)}\} \text{ .}$$

Of course (2.17) must be interpreted in the sense of distributions in Ω , and (2.16) is then to be understood in the sense of the trace theorem given above (Proposition 2.1).

The set $\mathscr{C}_a$ is not void; we shall give later on some conditions which ensure that $\mathscr{S}_a$ is not void. It is also useful to introduce the sets $\mathscr{C}_o = \mathscr{C}_a(o,o)$ and $\mathscr{S}_o = \mathscr{S}_a(o,o,o)$,

$$(2.19) \qquad \mathscr{C}_o = \{v \in H^2(\Omega)\ ,\ v_{|\Gamma_o \cup \Gamma_1} = 0\ ,\ \frac{\partial v}{\partial \nu}\Big|_{\Gamma_o} = 0\}$$

$$(2.20) \qquad \mathscr{S}_o = \{M \in L^2(\Omega;E)\ ,\ \nabla.\nabla.M = 0\ ,\ b_o(M) = 0 \text{ on } \Gamma_2\ ,$$
$$b_1(M) = 0 \text{ on } \Gamma_1 \cup \Gamma_2\}\ .$$

Constitutive law and variational problems

The unknowns of the problem are the functions w and M which determine the geometrical and mechanical state of the loaded plate; $w \in \mathscr{C}_a$ and $M \in \mathscr{S}_a$. To determine these functions completelv a further relation between them has to be given, namely the constitutive law of the material which specifies, for each point x of Ω , the relationship between $M(x)$ and $\nabla\nabla w(x)$. As in Section I.3, this constitutive law is given by a surpotential ψ , a proper l.s.c. convex function from E into $[0, +\infty]$, and its conjugate ψ^* having the same properties. At each point $x \in \Omega$,

$$(2.21) \qquad M(x) \in \partial\psi(\nabla\nabla w(x))$$

which is equivalent (cf. I.(2.21) and I.(2.23)) to :

$$(2.22) \qquad \nabla\nabla w(x) \in \partial\psi^*(M(x))\ .$$

From this as starting point we now obtain, exactly as in Section I.3, the variational principles which yield problems whose solutions are w and M . We put for $v \in H^2(\Omega)$,

$$(2.23) \qquad L(v) = \int_\Omega f\ v\ dx + \int_{\Gamma_1 \cup \Gamma_2} g^1 . \frac{\partial v}{\partial \nu}\, ds - \int_{\Gamma_2} g^o . v\ ds$$

and for all $\tau \in \mathscr{H}$,

(2.24) $$\bar{L}(\tau) = \int_{\Gamma_0 \cup \Gamma_1} b_0(M)u_0 \, ds - \int_{\Gamma_0} b_1(M)u_1 \, ds$$ [1]

Also, for $\tau \in L^2(\Omega)$,

(2.25) $$\Psi(\tau) = \int_\Omega \psi(\tau(x))dx$$

(2.26) $$\Psi^*(\tau) = \int_\Omega \psi^*(\tau(x))dx \; .$$

Thus, exactly as in Section I.3, it can be verified that

(2.27) w attains the minimum of $\Psi(\nabla\nabla v) - L(v)$ in $\mathscr{C}_a$

(2.28) M attains the minimum of $\Psi^*(\tau) - \bar{L}(\tau)$ in $\mathscr{S}_a$

The case of linear elasticity.

Assuming that the plate has an elastic behaviour, we take

(2.29) $$\psi(\xi) = \frac{1}{2} B\xi.\xi \quad , \quad \forall \xi \in E \; ,$$

where B is the continuous linear operator from E into E given by

(2.30) $$B \cdot \begin{pmatrix} \xi_{11} & \xi_{12} \\ \xi_{21} & \xi_{22} \end{pmatrix} = D(1-\alpha) \begin{pmatrix} \xi_{11} & \xi_{12} \\ \xi_{21} & \xi_{22} \end{pmatrix} + D\alpha \begin{pmatrix} \xi_{22} & 0 \\ 0 & \xi_{11} \end{pmatrix}$$

---

(1) $\int_{\Gamma_0 \cup \Gamma_1} b_0(M)u_0$ and $\int_{\Gamma_0} b_1(M)u_1 \, ds$ are 'abusive' notations for the duality pairings $\langle b_0(M),\gamma_0(U)\rangle$ and $\langle b_1(M),\gamma_1(U)\rangle$, on $H^{-3/2}(\Gamma) \times H^{3/2}(\Gamma)$ and $H^{-1/2}(\Gamma) \times H^{1/2}(\Gamma)$ respectively We make use here of (2.14).

where $D, \alpha,$ are two positive constants, with $0<\alpha<1/2$, ([1]), so that

$$\psi(\xi) = D\{(1-\alpha)|\xi|^2 + \alpha(\xi_{11}^2 + \xi_{22}^2)\} . \tag{2.31}$$

The conjugate function ψ^* is easily found; since B is an invertible operator, we have, for all $\eta \in E$,

$$\begin{aligned}\psi^*(\eta) &= \operatorname*{Sup}_{\xi \in E} \{\eta.\xi - \frac{1}{2} B\xi.\xi\} = \frac{1}{2} A\eta.\eta \\ &= \frac{1}{2D(1-\alpha^2)} \{(1-\alpha)|\eta|^2 + \alpha(\operatorname{tr} \eta)^2\} ,\end{aligned} \tag{2.32}$$

where $A = B^{-1}$.

2.1.3. Investigation of the elastic problem

By (2.25), w is the solution of the problem

$$\operatorname*{Inf}_{v \in \mathscr{C}_a} \{\Psi(\nabla\nabla v) - L(v)\} , \qquad (\mathscr{P}_{e\ell}) \tag{2.33}$$

where $\mathscr{C}_a$ is defined in (2.15) and Ψ in (2.25) (2.31). We write this problem in the general form envisaged in I.(2.26) and to this end we put

$$V = H^2(\Omega) , \quad V^* = \text{its dual}, \quad Y = Y^* = L^2(\Omega;E) ;$$

Λ is the linear operator $v \to \nabla\nabla v$ and the functions F, G are given by

$$F(v) = - L(v) \text{ if } v \in \mathscr{C}_a , = +\infty \text{ otherwise},$$
$$G(p) = \Psi(p) , \quad \forall\, p \in Y .$$

To find the dual I.(2.27), we have to calculate F^* and G^* ; by Proposition I.2.5,

---

([1]) As shown in G. Duvaut - J.L. Lions [1], $D = Eh^3/12(1-\alpha^2)$, where α and E are respectively, Poisson's ratio and Young's modulus $(E>0)$, and h is the thickness of the plate.

$$G^*(p) = \Psi^*(p) \ , \ \forall \ p \in L^2(\Omega;E) \ .$$

The following lemma is concerned with the calculation of F^*.

LEMMA 2.1

For all $p \in Y = L^2(\Omega;E)$

$$F^*(\Lambda^* p) = -\bar{L}(p) \ \text{ if } \ p \in \mathscr{S}_a \ , \ = +\infty \ \text{ if not} \ . \tag{2.34}$$

Proof

We apply Lemma I.2.1 with $V = H^2(\Omega)$, $v_0^* = -L$, $V_0 = U$, and $\mathscr{B} = \mathscr{C}_0$ (cf. (2.19)). We deduce from this that $F^*(\Lambda^*p)$ is equal to

$$\langle \Lambda^*p + L, U\rangle = \int_\Omega (p.\nabla\nabla U + fU)dx + \int_{\Gamma_1\cup\Gamma_2} g^1 \frac{\partial U}{\partial \nu} ds - \int_{\Gamma_2} g^0 \, U \, ds \ , \tag{2.35}$$

if $\Lambda^*p + L$ is in $\mathscr{B}^0$, and is $+\infty$ otherwise.

To say that $\Lambda^*p + L \in \mathscr{B}^0$ is equivalent to saying that

$$\int_\Omega (p.\nabla\nabla v + f.v)dx + \int_{\Gamma_1\cup\Gamma_2} g^1 \frac{\partial v}{\partial \nu} ds - \int_{\Gamma_2} g^0 \, v \, ds = 0 \ , \tag{2.36}$$

for all $v \in \mathscr{B}$. By writing (2.36) in the first place for $v \in \mathscr{C}_0^\infty(\Omega)$, we find that $\nabla.\nabla.p + f = 0$ in the sense of distributions on Ω^0 . This implies that $p \in \mathscr{H}$ and we can then apply the generalised Green's formula (2.10); (2.36) becomes

$$-\int_\Gamma b_0(p)v \, ds + \int_\Gamma b_1(p) \frac{\partial v}{\partial \nu} ds + \int_{\Gamma_1\cup\Gamma_2} g^1 \frac{\partial v}{\partial \nu} ds - \int_{\Gamma_2} g^0 v \, ds = 0 \ ,$$

$\forall \ v \in \mathscr{B}$, or in other words

$$\int_{\Gamma_2} (b_0(p) + g^0)v \, ds - \int_{\Gamma_1\cup\Gamma_2} (b_1(p) + g^1)ds = 0 \ , \ \forall \ v \in \mathscr{B} \ .$$

Since $b_0 \times b_1$ is surjective (cf. Proposition 2.1), this means that

$b_0(p) + g^0 = 0$ on Γ_2 , $b_1(p) + g^1 = 0$ on $\Gamma_1 \cup \Gamma_2$ and $p \in \mathscr{S}_a$. Conversely if $p \in \mathscr{S}_a$, then $\Lambda^* p + L \in \mathscr{B}^0$ (1).

When $p \in \mathscr{S}_a$, we apply Green's formula (2.10) to the expression (2.35) for $F^*(\Lambda^* p)$ and we obtain, as stated, $-\bar{L}(p)$.

□

We can now express the dual of (2.33) in an explicit form; by I.(2.27) and what precedes it, it can be written (we replace p by M):

$$\text{(2.37)} \qquad \operatorname*{Sup}_{M \in \mathscr{S}_a} \{-\tfrac{1}{2}\mathscr{A}(M,M) + \bar{L}(M)\} , \qquad (\mathscr{P}^*_{e\ell}) ,$$

where we have put

$$\text{(2.38)} \qquad \mathscr{A}(M,M) = \frac{1}{2}\int_\Omega A\, M(x).M(x)\, dx .$$

PROPOSITION 2.2

Problems (2.31) and (2.36) are dual to one another

$$\text{(2.39)} \qquad \inf\{\mathscr{P}_{e\ell}\} = \sup\{\mathscr{P}^*_{e\ell}\} \in \mathbb{R} \cup \{-\infty\}$$

The number defined by (2.39) is finite ($>-\infty$), if and only if $\mathscr{S}_a$ is non-void. In this case $\mathscr{P}^*_{e\ell}$ has the unique solution M, $\mathscr{P}_{e\ell}$ has a unique solution modulo $\Pi_1 \cap \mathscr{C}_0$, and w and M are bound by the relation

$$\text{(2.40)} \qquad M(x) = A(\nabla\nabla w(x)) \quad p.p. \quad x \in \Omega$$

Proof

When $\inf \mathscr{P}^*_{e\ell} = -\infty$, the equation (2.39) follows from I.(2.28); since $\sup \mathscr{P}^*_{e\ell} = -\infty$, this means that $\mathscr{S}_a$ is void. When $\inf \mathscr{P}^*_{e\ell} > -\infty$ we note that the criterion I.(2.29) applies with, for example, $u^0 = U$, because the function G is continuous at every point of Y . Hence, by I.(2.30), the equation (2.39) holds and $\mathscr{P}^*_{e\ell}$

(1) We use here a result similar to I.(2.106): $\mathscr{C}^2(\Omega) \cap \mathscr{C}_0$ is dense in $\mathscr{C}_0$.

has a solution (so that $\mathscr{S}_a \neq \emptyset$) . This solution is unique because the function $M \to \mathscr{A}(M,M)$ is strictly convex.

Since $v \in \mathscr{C}_0 \cap \Pi_1$, we see that $w = U + v$ is in $\mathscr{C}_a$ and thus

$$\Psi(\nabla\nabla w) - L(w) = \Psi(\nabla\nabla U) - L(U) - L(v) \geq \inf \mathscr{P}_{e\ell} > -\infty ,$$

which shows that the restriction of the linear form L to $\mathscr{C}_0 \cap \Pi_1$ is bounded below; hence we must have

$$L(v) = 0 \quad , \quad \forall\, v \in \mathscr{C}_0 \cap \Pi_1 \tag{2.41}$$

Putting $J(w) = \Psi(\nabla\nabla w) - L(w)$, we can define on $H^2(\Omega)/\mathscr{C}_0 \cap \Pi_1$

$$\dot{J}(\dot{w}) = J(w) \quad , \quad \forall\, w \in \dot{w} \ ,$$

(thanks to (2.41)). If $\dot{\mathscr{C}}_a$ denotes the quotient set of $\mathscr{C}_a$ in $H^2(\Omega)/\mathscr{C}_0 \cap \Pi_1$, we can now consider the problem

$$\operatorname*{Inf}_{\dot{w} \in \dot{\mathscr{C}}_a} \dot{J}(\dot{w}) \ .$$

It is clear from Proposition I.2.1 and I.2.2 that this problem has an unique solution $\dot{w}$ (we use Proposition 2.1 to verify condition I.(2.43)). Each element w of the equivalence class, $w \in \dot{w}$, is a solution of $\mathscr{P}_{e\ell}$ and there are no other solutions.

We conclude by observing that (2.40) is a translation of the optimality condition I.(2.31).

Remark 2.1. It emerges from the proof of Proposition 2.2 that the two following conditions are equivalent:

$$\mathscr{S}_a \neq \emptyset \Leftrightarrow L(v) = 0 \ , \quad \forall\, v \in \mathscr{C}_0 \cap \Pi_1 \ ; \tag{2.42}$$

they are also necessary and sufficient for $\inf \mathscr{P}_{e\ell} = \sup \mathscr{P}^*_{e\ell} > -\infty$.

□

2.2. The plastic model

2.2.1 Variational problems for stresses and strains

We take a set $K \subset E$, where

(2.43) K is a bounded, closed, convex set containing a neighbourhood of 0 .

The most usual examples (cf. W.H. Yang [3]) are the convex sets

$$K = \{M \in E \ , \ |M|^2 \leqq k\}$$

or

$$K = \{M \in E \ , \ \max|\lambda_i(M)| \leqq k' \ , \ i = 1, 2, 3,\}$$

where the $\lambda_i(M)$ are the eigenvalues of M .

In the case of perfectly elastic-plastic behaviour (with the convex plasticity set K) , the function ψ is defined by

$$(2.44) \quad \begin{cases} \psi^*(\xi) = \frac{1}{2} A\xi.\xi \ , \ \text{if} \ \xi \in K \ , \ \text{and} \ = +\infty \ \text{if not}, \forall \ \xi \in E, \\ \psi = (\psi^*)^* \ . \end{cases}$$

Clearly ψ and ψ^* are both proper convex l.s.c. functions mapping E into $\mathbb{R} \cup \{+\infty\}$; as regards ψ , by proceeding exactly as in Lemma I.4.1, the following lemma can be proved:

LEMMA 2.2.

i) The function ψ *is a convex continuous function which maps* E *into* $\mathbb{R}$ *and there exist two constants* $k_0 \ , \ k_1, \ 0 < k_0 < k_1 < +\infty$ *such that*

$$(2.45) \qquad k_0(|\xi| - 1) \leqq \psi(\xi) \leqq k_1|\xi| \qquad \forall \ \xi \in E \ .$$

ii) The function Ψ *defined by (2.25) is a convex continuous function which maps* $L^2(\Omega;E)$ *into* $\mathbb{R}$.

By (2.27) the field w of the strain problem is the solution of

$$(2.46) \qquad \operatorname*{Inf}_{v \in \mathscr{C}_a} \{\Psi(\nabla\nabla v) - L(v)\} \,, \qquad (\mathscr{P}).$$

Our object now is to determine the dual of (2.46) in order to arrive at the problem whose solution gives the stress field (i.e. (2.28)) and to *specify precisely the relationship between these two problems.* We write (2.46) in the form I.(2.26) and for this purpose we proceed exactly as in the elastic case; V, Y, Λ, F, G, are defined exactly as above; the only difference is that the function ψ occurring in the definition of G is the one given by (2.44); and by Proposition I.2.5 its conjugate is

$$(2.47) \quad G^*(p) = \Psi^*(p) = \int_\Omega \psi^*(p(x))dx =$$

$$= \frac{1}{2}\mathscr{A}(p,p) \text{ if } p(x) \in K \text{ for almost all } x \in \Omega, \text{ and}$$

$$= +\infty \quad \text{otherwise.}$$

We replace p by M and the dual problem becomes

$$(2.48) \qquad \operatorname*{Sup}_{M} \{-\frac{1}{2}\mathscr{A}(M,M) + \tilde{L}(M)\} \,, \qquad (\mathscr{P}^*)$$

where the supremum is taken over $M \in \mathscr{S}_a \cap \mathscr{K}_a$, and

$$(2.49) \qquad \mathscr{K}_a = \{p \in L^2(\Omega;E) \,,\ p(x) \in K \,,\ \text{p.p. } x \in \Omega\}$$

PROPOSITION 2.3.

The problems $\mathscr{P}$ *and* $\mathscr{P}^*$ *are dual to one another and*

$$(2.50) \qquad \inf \mathscr{P} = \sup \mathscr{P}^* \in \mathbb{R} \cup \{-\infty\}$$

The number defined by (2.50) is finite if and only if there are M *which are admissible for* $\mathscr{P}^*$ $(\mathscr{S}_a \cap \mathscr{K}_a \neq \emptyset)$. *In this case the problem* $\mathscr{P}^*$ *has a unique solution.*

The proof is similar to that of Proposition 2.2 and will not be spelled out in detail.

2.2.2. Limit Analysis

Our object here is to introduce the limit analysis problem $\mathscr{P}_{AL}$ and its dual $\mathscr{P}^*_{AL}$ associated with the problem $\mathscr{P}$; we shall deduce from this a necessary and sufficient condition for $\inf\mathscr{P}>-\infty$ to hold.

We consider, as in Section I.5, the function ψ_∞ , the 'principal part' of ψ , which is the support function of the convex set K ,

(2.51)
$$\psi_\infty(\xi) = \sup_{\eta\in K} \xi.\eta \quad .$$

It is easy to see that ψ_∞ is positively homogeneous, i.e.

$$\psi_\infty(r\xi) = r\,\psi_\infty(\xi) \quad , \quad \forall\, r > 0 \ ,$$

and also that there are two constants k'_0, k'_1, $0<k'_0<k'_1<+\infty$, such that

(2.52)
$$k'_0|\xi| \leq \psi_\infty(\xi) \leq k'_1|\xi| \ ,$$

for all $\xi \in E$. It follows that ψ_∞ is a continuous convex function mapping E into $\mathbb{R}$. For all τ of $L^1(\Omega;E)$ we put

(2.53)
$$\Psi_\infty(\tau) = \int_\Omega \psi_\infty(\tau(x))dx \ .$$

By (2.52) Ψ_∞ is a continuous convex function from $L^1(\Omega;E)$ into $\mathbb{R}$.

The limit analysis problem can be written as

(2.54)
$$\inf_v \Psi_\infty(\nabla\nabla v) \quad , \qquad (\mathscr{P}_{AL}) \ ,$$

where the infimum is taken over the following v:

(2.55)
$$\{v \in \mathscr{C}_0 \ , \ L(v) = 1\} \ .$$

To establish the link between this problem, and the problems $\mathscr{P}, \mathscr{P}^*$, we shall determine the dual of $\mathscr{P}_{AL}$. To do this we write the problem $\mathscr{P}_{AL}$ in the form I.(2.26), and for this we put:

$$V = \mathscr{C}_0 \quad (= \text{a Hilbert subspace of } H^2(\Omega)) \ ,$$

$$V^* = \text{the dual of } V, \ Y = Y^* = L^2(\Omega;E) \ ,$$

Λ is the mapping $v \to \nabla\nabla v$ and the functions F and G are chosen as follows:

$$F(v) = 0 \ \text{ if } \ v \in \mathscr{C}_0 \ \text{ and } \ L(v) = 1 \ , \ = +\infty \ \text{ otherwise}, \ \forall\, v \in V \ ,$$

$$G(p) = \Psi_\infty(p) \ , \ \forall\, p \in Y \ .$$

By Proposition I.2.5, we have for all $p \in Y$

$$G^*(p) = \Psi^*_\infty(p) = 0 \ \text{ if } \ p(x) \in K \ \text{ p.p.}, \ = +\infty \ \text{ if not.}$$

To calculate $F^*(\Lambda^*p)$, $p \in Y$, we apply Lemma I.2.1: let v_0 be an arbitrary element of $\mathscr{C}_0$ such that $L(v_0) = 1$; if we denote by $\mathscr{B}$, the kernel of L in V , F is simply the indicator function of $v_0 + \mathscr{B}$ and Lemma I.2.1 implies that

$$F^*(\Lambda^*p) = \langle \Lambda^*p, v_0 \rangle \ ,$$

if $\Lambda^*p \in \mathscr{B}^0$ and $= +\infty$ otherwise. To say that $\Lambda^*p \in \mathscr{B}^0$ is equivalent to saying that $\langle \Lambda^*p, v \rangle = 0$ for all $v \in V$ such that $L(v) = 0$. By an elementary theorem of linear algebra this means that there exists a $\lambda \in \mathbb{R}$ such that $\Lambda^*p + \lambda L = 0$,

$$\langle \Lambda^*p + \lambda L \ , \ v \rangle = 0 \ , \ \forall\, v \in \mathscr{C}_0 \ .$$

This condition was characterised, in connection with Lemma 2.1, for $\lambda = 1$. Replacing L by λL we see that it is equivalent to $\lambda^{-1}p \in \mathscr{S}_a$.

Lastly $F^*(\Lambda^*p) = \langle \Lambda^*p, v_0 \rangle$ when there is a $\lambda \in \mathbb{R}$ such that $\Lambda^*p + \lambda L = 0$ and $= +\infty$ otherwise; but when $\Lambda^*p + \lambda L = 0$, $\langle \Lambda^*p, v_0 \rangle$ has the value $-\lambda L(v_0) = -\lambda$. Hence

$$(2.56) \quad F^*(\Lambda^*p) = \begin{cases} -\lambda \ \text{ if there is a } \lambda \in \mathbb{R} \text{ such that } \lambda^{-1}p \in \mathscr{S}_a \\ +\infty \ \text{ otherwise.} \end{cases}$$

The dual problem I.(2.27) can now be stated explicitly; it can be written

$$(2.57)\qquad \operatorname{Sup}_{p} \{\lambda\} \ , \qquad (\mathscr{P}^*_{AL}) \ ,$$

where the supremum is taken over those p such that

$$(2.58)\qquad \begin{cases} \nabla.\nabla.p + \lambda f = 0 \ \text{ in } \Omega \ , \ b_o(p) + \lambda g^o = 0 \ \text{ on } \Gamma_2 \ , \\ b_1(M) + \lambda g_1 = 0 \ \text{ on } \Gamma_1 \cup \Gamma_2 \ \text{ and } \ p(x) \in K \ \text{ p.p. in } \Omega \ . \end{cases}$$

We have

PROPOSITION 2.4

The problems (2.54) and (2.57) are dual to one another and

$$(2.59)\qquad \bar{\lambda} = \inf \mathscr{P}_{AL} = \sup \mathscr{P}^*_{AL} \geqq 0 \ .$$

Furthermore $\mathscr{P}^*_{AL}$ *has at least one solution.*

This is an immediate consequence of I.(2.29), I.(2.30). The fact that $\mathscr{P}^*_{AL}$ has a solution means that there is a $p \in L^2(\Omega;E)$ satisfying (2.58) with λ replaced by $\bar{\lambda}$. We deduce from this

PROPOSITION 2.5

The following three conditions are equivalent

(i) $\inf \mathscr{P} = \sup \mathscr{P}^* > -\infty$

(ii) $\mathscr{S}_a \cap \mathscr{K}_a \neq \emptyset$

(iii) $\bar{\lambda} = \inf \mathscr{P}_{AL} \geqq 1$.

If $\bar{\lambda} > 1$, *then every minimising sequence of* $\mathscr{P}$ *is of bounded Hessian in* $L^1(\Omega;E)$.

Proof

Let $\bar{p}$ be a solution of $\mathscr{P}^*_{AL}$: $\bar{p}$ satisfies (2.58) with $\lambda = \bar{\lambda}$, and therefore for all $0<\lambda<\bar{\lambda}$, $p = (\lambda/\bar{\lambda})\bar{p}$ is in $\mathcal{S}_a$ and $p(x)\in(\lambda/\bar{\lambda})K$ for almost all x in Ω. Now $(\lambda/\bar{\lambda})K \subset K$ since $0 \in K$, which implies that p satisfies (2.58). For all $\lambda \in [0,\bar{\lambda}]$, there is a p satisfying (2.58), and by the definition of λ , there is no such p when $\lambda > \bar{\lambda}$. Taking Proposition 2.3 into account, it follows that (i) (ii) (iii) are equivalent to one another.

Let us denote by $\mathcal{P}_\lambda$ the problem $\mathcal{P}$ in which L is replaced by λL . The preceding argument shows that $\inf \mathcal{P}_\lambda > -\infty$ if and only if $0\leqq\lambda\leqq\bar{\lambda}$. In the case where $\bar{\lambda}>1$, there is a λ such that $1<\lambda<\bar{\lambda}$. Let us consider therefore a minimising sequence v_m for $\mathcal{P}$:

$$\Psi(\nabla\nabla v_m) - L(v_m) = \frac{1}{\lambda}\{\Psi(\nabla\nabla v_m) - \lambda L(v_m)\} + (1 - \frac{1}{\lambda})\,\Psi(\nabla\nabla v_m)$$

$$\geqq \inf \mathcal{P}_\lambda + (1 - \frac{1}{\lambda})\,\Psi(\nabla\nabla v_m) \; .$$

The sequence $\Psi(\nabla\nabla v_m)$ is thus bounded, and because of (2.45), the sequence $\nabla\nabla v_m$ is bounded in $L^1(\Omega;E)$.

Remark 2.2. The results of Propositions 2.3 and 2.4 can easily be extended to the problems $\mathcal{P}_\lambda$, be replacing L by λL .

□

Orientation

We shall see in the following section that a minimising sequence of $\mathcal{P}$ is in fact bounded in $W^{2,1}(\Omega)$ (when $\bar{\lambda}>1$) . The behaviour of such a sequence $\{v_m\}$ will be completely clarified in Sections 3 and 4. For the moment let us merely note that as the injection of $W^{2,1}(\Omega)$ into $W^{1,1}(\Omega)$ (like the injection of the latter into $L^1(\Omega)$, cf. I.(1.30)), is compact, the sequence v_m contains a subsequence v_{m_j} which converges in norm to a limit v in $W^{1,1}(\Omega)$. Thanks to E. Gagliardo's trace theorem (cf. I.(1.32)), $\gamma_o(v_m)$ converges to $\gamma_o(v)$ in $L^1(\Omega)$ and hence $\gamma_o(v_m) = u_o$ on $\Gamma_o \cup \Gamma_1$. It is by no means obvious however that $v \in W^{2,1}(\Omega)$, and still less so that $\gamma_1(v_m) = \frac{\partial v_m}{\partial \nu}$ converges to $\gamma_1(v) = \frac{\partial v}{\partial \nu}$: we therefore do not know whether v satisfies the boundary condition $\frac{\partial v}{\partial \nu} = u_1$ on Γ_o .

To overcome this difficulty we are led, as in Section I.6, to introduce a relaxed problem $\mathscr{PR}$, in which the boundary conditions of the problem $\mathscr{P}$ are partially abandoned. This is the object of Section 2.2.3.

2.2.3. The relaxed problem in $H^2(\Omega)$

As the existence of a solution to the problem $\mathscr{P}$ cannot in general be assumed (even when $\inf \mathscr{P}_{AL} > 1$), we shall be led later on to introduce a problem generalising $\mathscr{P}$ which will be *formulated in a finite-energy space.* As a first step in this direction we shall introduce here a problem $\mathscr{PR}$ in which the boundary conditions are partially relaxed.

We consider the problem

$$(2.60) \qquad \operatorname*{Inf}_{v} \{\Psi(\nabla\nabla v) + \int_{\Gamma_0} \psi_\infty(\mathscr{F}(u_1 - \gamma_1(v)))d\Gamma - L(v)\} , \qquad (\mathscr{PR}) ,$$

where ψ_∞ is the function introduced in (2.51), where the infimum is taken over the following v

$$(2.61) \qquad \{v \in H^2(\Omega) , \quad v = u_0 \text{ on } \Gamma_0 \cup \Gamma_1\} ,$$

and where, for $p \in L^1(\Gamma)$, $\mathscr{F}(p)$ is the tensor whose components are

$$(2.62) \qquad \mathscr{F}_{ij}(p) = p\, \nu_i\, \nu_j , \quad i,j = 1,2,$$

$\nu = (\nu_1,\nu_2)$ being the outward unit normal on Γ.

The problem $\mathscr{PR}$ is similar to the problem $\mathscr{P}$: the boundary condition $\gamma_1(v) = u_1$ on Γ_0 has been abandoned, but on the other hand the term

$$\int_{\Gamma_0} \psi_\infty(\mathscr{F}(u_1 - \gamma_1(v))d\Gamma$$

has been added to the functional to be minimised. Since $\psi_\infty(o) = o$, $\mathscr{PR}$ is an extension of $\mathscr{P}$, and involves the minimising of the same functional (the one in (2.60)) over a larger set (namely (2.61) instead of $\mathscr{C}_a$) ; hence

$$\mathscr{PR} \supset \mathscr{P} \ , \quad \inf \mathscr{PR} \leqq \inf \mathscr{P} \ . \tag{2.63}$$

We put this into a more precise form in the following proposition.

PROPOSITION 2.6

i) Problems $\mathscr{P}$ *and* $\mathscr{PR}$ *admit of the same dual and,*

$$\inf \mathscr{P} = \inf \mathscr{PR} = \sup \mathscr{P}^* = \sup \mathscr{PR}^* \quad (\in \mathbb{R} \cup \{-\infty\}) \ . \tag{2.64}$$

ii) Every minimising sequence for $\mathscr{P}$ *is also a minimising sequence for* $\mathscr{PR}$

iii) If the number defined by (2.64) is $> -\infty$, *then* $\mathscr{P}^*$ *and* $\mathscr{PR}^*$ *have an unique solution* M *which defines the state of stress in the plate.*

We shall determine the dual problem $\mathscr{PR}^*$ and we shall then be in a position to prove the proposition.

To determine $\mathscr{PR}^*$ we put

$$V = H^2(\Omega) \ , \ V^* = \text{its dual}$$
$$Y = Y^* = Y_1 \times Y_2 \ , \ Y_1 = L^2(\Omega;E) \ , \ Y_2 = L^2(\Gamma_0)$$
$$\Lambda = \Lambda_1 \times \Lambda_2 \ , \ \Lambda_1 v = \nabla\nabla v \ \text{ and } \ \Lambda_2 v = \gamma_1(v)_{|\Gamma_0} \ .$$

We put

$$F(v) = -L(v) \ \text{ if } v \text{ is in the set (2.61)}$$
$$\text{and } = +\infty \ , \text{ otherwise, for all } v \in V \ ,$$
$$G(p) = G_1(p_1) + G_2(p_2) \ , \quad \text{for } p = (p_1,p_2) \in Y \ ,$$
$$G_1(p_1) = \Psi(p_1) \ , \ G_2(p_2) = \int_{\Gamma_0} \psi_\infty(\mathscr{F}(u_1-p_2))d\Gamma \ .$$

The problem I.(2.26) is now identical to $\mathscr{PR}$. We have for $p = (p_1,p_2) \in Y$,

$$G_1^*(p_1) = \Psi^*(p_1) =$$
$$= \frac{1}{2}\mathscr{A}(p_1,p_1) \quad \text{if} \quad p_1 \in \mathscr{K}_a \ , \ = +\infty \quad \text{otherwise.}$$

Also, as we shall prove later on,

$$(2.65) \qquad F^*(\Lambda^* p) = \begin{cases} -\langle b_o(p_1),U\rangle & \text{if} \ \ p_1 \in \mathscr{S}_a \ \text{ and} \\ & b_1(p_1) + p_2 = 0 \ \text{ on } \ \Gamma_o \\ +\infty & \text{otherwise.} \end{cases}$$

The problem I.(2.27) now gives us $\mathscr{PR}^*$:

$$(2.66) \qquad \operatorname*{Sup}_p\{-\frac{1}{2}\mathscr{A}(M,M) - G_2^*(b_1(M)) + \int_{\Gamma_o \cup \Gamma_1} b_o(M)u_o\, ds\} \ ,$$

where the supremum is taken over the following p

$$(2.67) \qquad \begin{cases} p = (p_1,p_2) = (M,-b_1(M)) \in L^2(\Omega;E) \times L^1(\Gamma_o) \\ M \in \mathscr{S}_a \cap \mathscr{K}_a \ . \end{cases}$$

If $M \in L^2(\Omega;E) \cap \mathscr{S}_a$, then $b_o(M)$ and $b_1(M)$ have a meaning by Proposition 2.1 and all the conditions (2.67) are meaningful.

<u>Proof of Proposition 2.6</u>

We shall prove later on that

$$(2.68) \qquad G_2^*(b_1(M)) = \int_{\Gamma_o} b_1(M)u_1\, ds \ , \quad \text{when} \ \ M \ \text{ satisfies (2.67)}$$

It is then clear that $\mathscr{P}^*$ and $\mathscr{PR}^*$ are identical; the equation

$$\inf \mathscr{PR} = \sup \mathscr{PR}^*$$

and (2.64) are consequences of I.(2.28)-I.(2.30), and the other assertions follow easily.

□

To conclude we prove (2.65) and (2.68).

Proof of (2.65)

We apply Lemma I.2.1 with $V = H^2(\Omega)$, $v_0 = U$, $v_0^* = -L$, $v^* = \Lambda^* p$,

(2.69) $$\mathscr{B} = \{v \in H^2(\Omega) \ , \ v = 0 \ \text{ on } \ \Gamma_0 \cup \Gamma_1\} \ .$$

It follows that $F^*(\Lambda^* p) = \langle \Lambda^* p+L, U\rangle$ if

(2.70) $$\langle \Lambda^* p+L, v\rangle = 0 \ , \quad \text{for all} \quad v \in \mathscr{B} \ ,$$

and $+\infty$ otherwise. The characterisation of the p satisfying (2.70) is done as in Lemma 2.1 and we find

$$p_1 \in \mathscr{S}_a \quad \text{and} \quad b_1(p_1) + p_2 = 0 \quad \text{on} \quad \Gamma_0 \ .$$

Lastly, if p satisfies these conditions, $b_0(p_1)$ and $b_1(p_1)$ are defined, thanks to Proposition 2.1, and we can use Green's formula (2.10) which gives

$$\langle \Lambda^* p+L, U\rangle = - \langle b_0(p_1), U\rangle \ ;$$

which proves (2.65).

Proof of (2.68)

We have to prove that

$$\int_{\Gamma_0} b_1(M) u_1 ds = \sup_{p_2 \in L^2(\Gamma_0)} \{\int_{\Gamma_0} b_1(M).p_2 ds - \int_{\Gamma_0} \psi_\infty(\mathscr{F}(u_1-p_2))ds\} \ ,$$

or, putting $q = p_2 - u_1$,

$$\sup_{q \in L^2(\Gamma_0)} \{\int_{\Gamma_0} b_1(M) q ds - \int_{\Gamma_0} \psi_\infty(\mathscr{F}(q))ds\} = 0 \ .$$

As this supremum is obviously non-negative, it is sufficient to show that

(2.71) $$b_1(M).q - \psi_\infty(\mathscr{F}(q)) \leqq 0 \ ,$$

almost everywhere on Γ_0, for all $q \in L^2(\Gamma_0)$. This inequality is easily proved if, in addition, $M \in \mathscr{C}(\bar{\Omega};E)$ for we then have

$$b_1(M).q = M_{ij}\, \nu_i\, \nu_j\, q = M.\mathscr{F}(q) \ ,$$

and (2.71) then expresses the fact that $\psi_\infty^*(M) = 0$ on Γ_0 (cf. the definition of ψ_∞ and the fact that by continuity, $M(x) \in K$ for all $x \in \Gamma_0$). If M satisfies only the weaker condition (2.67) then (2.71) can be proved by an approximation argument which can be found in F. Demengel [2] (cf. also Lemma I.6.3).

It is not known whether $\mathscr{PR}$ always has a solution and the same applies to $\mathscr{P}$, even when the safe load condition is satisfied, i.e. $\inf \mathscr{P}_{AL} > 1$. The existence of a solution will however be proved for the extension of $\mathscr{PR}$ to a finite-energy space. *Our immediate object will now be to introduce and study the finite-energy space* $HB(\Omega)$ *corresponding to our plate problem.*

2.3. Finite-energy space: the space $HB(\Omega)$

We shall introduce and study a functional space $HB(\Omega)$ which will play a role analogous to that of $BD(\Omega)$ and which, as we shall see, is the finite-energy space for the plate problem.

2.3.1. The space $HB(\Omega)$

The space $HB(\Omega)$ is defined as the space of the v which, together with their first derivatives, belong to $L^1(\Omega)$ and whose second derivatives are bounded measures. We can imagine considering initially a space $HL(\Omega)$ analogous to the space $LD(\Omega)$: this would be the space of functions v which together with their first and second derivatives belong to $L^1(\Omega)$ and we should then simply have $HL(\Omega) = W^{2,1}(\Omega)$.

We shall not dwell on the space $W^{2,1}(\Omega)$ but merely content ourselves with recalling its main properties:

(2.72) $$\mathscr{C}^2(\bar{\Omega}) \text{ is dense in } W^{2,1}(\Omega)$$

(2.73) The injection of $W^{2,1}(\Omega)$ into $W^{1,1}(\Omega)$ is compact.

(2.74) $W^{2,1}(\Omega)$ can be embedded by a continuous injection in $\mathscr{C}(\bar{\Omega})$.

(2.75) There exist two continuous linear mappings γ_0, γ_1 of $W^{2,1}(\Omega)$ into $L^1(\Gamma)$ such that

$$\gamma_0(v) = v_{|\Gamma} \quad , \quad \gamma_1(v) = \frac{\partial v}{\partial \nu}\bigg|_\Gamma$$

for all $v \in \mathscr{C}^2(\bar{\Omega})$, and γ_1 is surjective.

The open set $\Omega \subset \mathbb{R}^2$ has been assumed to be bounded and of class $\mathscr{C}^2$; (2.72) and (2.73) merely reproduce I.(1.27) and I.(1.30); (2.74) is proved in R.S. Adams [1]. As for (2.75) the existence of γ_0 , γ_1 follows by applying I.(1.32) to the function v and its first derivatives. We refer to F. Demengel [1] for the proof that γ_1 is surjective and other additional results (in particular the fact that $\gamma_0(W^{2,1}(\Omega)) \subsetneq W^{1,1}(\Gamma))$.

□

Now let $HB(\Omega)$ be defined by

(2.76) $$HB(\Omega) = \{v \in L^1(\Omega) , D_i v \in L^1(\Omega) , D_{ij} v \in M_1(\Omega) , i,j = 1,2\}$$
$$= \{v \in W^{1,1}(\Omega) , D_{ij} v \in M_1(\Omega) , i,j = 1,2\} .$$

This is a Banach space for the natural norm

$$||v||_{HB(\Omega)} = ||v||_{W^{1,1}(\Omega)} + ||\nabla\nabla v||_{M_1(\Omega;E)} .$$

We shall see that if $v \in HB(\Omega)$, then v is continuous on $\bar{\Omega}$ but on the other hand its first derivatives may be discontinuous in crossing certain lines, which cannot happen with functions belonging to $W^{2,1}(\Omega)$. This will enable us to take account of the flexure of plastic plates which fold without breaking. We shall see also that $HB(\Omega)$ is the largest space of functions for which the energy function $\Psi(\nabla\nabla v)$ may be defined (with a value $< +\infty$) .

If $v \in HB(\Omega)$ then its first derivatives belong to $BV(\Omega)$ the space of functions with bounded variation in Ω (see the end of Section II.2.1). The vector grad $v = \nabla v$ is in $BD(\Omega)$; several properties of $HB(\Omega)$ will therefore result from known properties of $BV(\Omega)$ and $BD(\Omega)$. For example, since $BV(\Omega)$ can be embedded by a continuous injection in $L^2(\Omega)$ and by a compact injection in $L^p(\Omega)$, $1 \leqq p < 2$, we have

(2.77) $HB(\Omega) \subset H^1(\Omega)$ with continuous injection

(2.78) The injection of $HB(\Omega)$ in $W^{1,p}(\Omega)$, $1 \leqq p < 2$ is compact.

Similarly if u is a distribution on Ω such that $\nabla\nabla u \in M_1(\Omega;E)$ then $\nabla u \in \mathscr{D}'(\Omega)^2$ with $\varepsilon(\nabla u) \in M_1(\Omega;E)$; this implies, by Theorem II.2.3 that $\nabla u \in L^1(\Omega)^2$. Thus the first derivatives of u are in $BV(\Omega)$ and since $BV(\Omega) \subset L^2(\Omega)$ $(n=2)$, $u \in L^2(\Omega)$ and hence $u \in HB(\Omega)$, so that we have proved:

(2.79) If $u \in \mathscr{D}'(\Omega)$ and $\nabla\nabla u \in M_1(\Omega;E)$, then $u \in HB(\Omega)$.

From the foregoing and using arguments similar to those used in proving Propositions II.2.3, II.2.4, II.2.5, ([1]) , we can also show that

---

([1]) We use here the analogue, for Hessians, of Proposition II.1.1: Let g_{ij}, $i,j = 1,2$ be a family of distributions on Ω . Then the two following conditions are equivalent:

- there is a distribution u on Ω such that $g_{ij} = D_{ij}u$, $i, j=1,2$;
- for every family of functions $\phi_{i,j} \in \mathscr{C}_0^\infty(\Omega)$, $i, j=1,2$, such that

  $\sum_{i,j=1}^{2} D_{ij}\phi_{ij} = 0$ $(\nabla.\nabla.\phi = 0$ for $\phi = \{\phi_{ij}\} \in \mathscr{C}_0^\infty(\Omega;E))$ we have

  $\sum_{i,j=1}^{2} \langle g_{ij},\phi_{ij}\rangle = 0$.

(2.80) $||\nabla\nabla u||_{M_1(\Omega;E)}$ induces on $HB(\Omega)/\Pi_1$ a norm equivalent to the one induced by $HB(\Omega)$.

(2.81) For all $u \in HB(\Omega)$, there is an $r = r(u) \in \Pi_1$ such that $||u-r||_{HB(\Omega)} \leqq c(\Omega)||\nabla\nabla u||_{M_1(\Omega;E)}$, where the constant c depends only on Ω .

(2.82) If p is a continuous semi-norm on $HB(\Omega)$ and also a norm on Π_1 , then

$$p(u) + ||\nabla\nabla u||_{M_1(\Omega;E)}$$

is a norm on $HB(\Omega)$, equivalent to the initial norm.

Trace of functions on $HB(\Omega)$

As, by (2.77), $HB(\Omega) \subset H^1(\Omega)$, the trace $\gamma_0(u)$ of a function $u \in HB(\Omega)$ is defined and belongs to $H^{1/2}(\Gamma)$. Since $D_i u \in BV(\Omega)$, $i = 1,2$, it follows from the trace theorem of M. Miranda [1] (recalled in Section II.2.1) that $\gamma_0(\nabla u)$ is defined and belongs to $L^1(\Gamma)^2$. We put $\gamma_1(u) = \nu.\gamma_0(\nabla u) \in L^1(\Gamma)$ and note that it can be shown (using an approximation argument) that $\tau.\gamma_0(\nabla u) \in L^1(\Gamma)$ is the same as the tangential derivative $\dfrac{\partial\gamma_0(u)}{\partial s}$ of $\gamma_0(u)$. The mappings γ_0 and γ_1 are moreover continuous :

(2.83) There exist two continuous linear transformations γ_0, γ_1 which map $HB(\Omega)$ into $W^{1,1}(\Gamma)$ and $L^1(\Gamma)$ respectively, such that

$$\gamma_0(u) = u_{|\Gamma} \quad , \quad \gamma_1(u) = \frac{\partial u}{\partial\nu}\Big|_{\Gamma} \quad ,$$

for all $u \in \mathscr{C}^2(\bar{\Omega})$.

By using the Green's formula II.(2.36) and the one for $BV(\Omega)$, we can obtain a Green's formula which generalises (2.7) :

(2.84) For all $v \in HB(\Omega)$ and all $\phi \in \mathscr{C}^2(\Omega;E)$,

$$\int_\Omega v.(\nabla.\nabla.\phi)dx - \int_\Omega \phi.(\nabla\nabla v) =$$

$$= \langle b_0(\phi) , \gamma_0(v)\rangle - \langle b_1(\phi) , \gamma_1(v)\rangle .$$

Extension and continuity of functions of $HB(\Omega)$

It is natural to ask what are the conditions which a piece-wise regular function must satisfy if it is to belong to $HB(\Omega)$. We shall answer this question in more general terms.

PROPOSITION 2.7

Let Ω_1 *be a relatively compact subset of* Ω *whose boundary* Σ *is of class* $\mathscr{C}^2$ *and let* $\Omega_2 = \Omega\backslash\bar{\Omega}_1$. *Let* u *be a function of* $L^1(\Omega)$ *whose restriction* u_i *to* Ω_i *is in* $HB(\Omega)$, $i = 1,2$. *Then* $u \in HB(\Omega)$ *if and only if the traces of* u_1 *and* u_2 *on* Σ *are equal*

(2.85) $$\gamma_0(u_1) = \gamma_0(u_2) \quad on \quad \Sigma .$$

In this case

(2.86) $$\nabla\nabla u = \sum_{i=1}^{2} \theta_{\Omega_i} \nabla\nabla u_i + \mathscr{F}\left(\frac{\partial u_2}{\partial \nu} - \frac{\partial u_1}{\partial \nu}\right)\delta_\Sigma$$

where θ_{Ω_i} *is the characteristic function of* Ω_i *and* $\frac{\partial u_i}{\partial \nu}$ *is the value on* Σ *of* $\gamma_1(u_i)$, *and* ν *is the normal on* Σ *in the direction from* Ω_1 *to* Ω_2 .

Proof

If $u \in HB(\Omega)$ then $u \in W^{1,1}(\Omega)$ and (2.85) is therefore a necessary condition. Assuming this condition to be satisfied, we calculate the distributional derivative $\nabla\nabla u$. Let $\phi \in \mathscr{C}_0^\infty(\Omega;E)$, we write

$$\langle \nabla\nabla u, \phi\rangle = \langle u, \nabla.\nabla.\phi\rangle$$

$$= \sum_{i=1}^{2} \int_{\Omega_i} u_i \; \nabla.\nabla.\phi \; dx =$$

$$= \sum_{i=1}^{2} \int_{\Omega_i} \phi.(\nabla\nabla u_i) + \int_{\Sigma} \left(\frac{\partial u_2}{\partial \nu} - \frac{\partial u_1}{\partial \nu}\right) (\phi.\nu.\nu) d\Gamma \; . \quad \text{(by Green's formula (2.84))}.$$

This proves the formula (2.86); it shows that $u \in HB(\Omega)$ because

$$\int_{\Omega} |\nabla\nabla u| \leqq \sum_{i=1}^{2} \int_{\Omega_i} |\nabla\nabla u_i| + c \int_{\Sigma} \left|\frac{\partial u_2}{\partial \nu} - \frac{\partial u_1}{\partial \nu}\right| ds < +\infty \; .$$

□

The condition (2.85) shows that a function u of $HB(\Omega)$ has no discontinuity across lines in Ω. In fact a much more precise result has been proved by J. Rauch and B.A. Taylor [1] (cf. also F. Demengel [2]) :

(2.87) $\qquad HB(\Omega) \subset \mathscr{C}(\bar{\Omega})$ with continuous injection.

2.3.2 Approximation of functions of $HB(\Omega)$

We shall use the expression, strong topology on $HB(\Omega)$ for the topology defined by the norm; the weak topology will be the topology defined by the following family of norms and seminorms (cf. N. Bourbak [2]) :

- the norm $||u||_{W^{1,1}(\Omega)}$,

- the seminorms $\left|\int_{\Omega} D_{ij}u\phi\right|$, $\phi \in \mathscr{C}_0^{\infty}(\Omega)$, $i,j = 1,2$.

Thus for example a sequence u_m converges to u under the weak topology if

$$(2.88) \qquad \begin{cases} u_m \to u & \text{in the strong } W^{1,1}(\Omega) \\ \nabla\nabla u_m \to \nabla\nabla u & \text{weakly in } M_1(\Omega;E) \; . \end{cases}$$

If u_m is a bounded sequence in $HB(\Omega)$ then by (2.78) it is relatively compact in $W^{1,1}(\Omega)$; it therefore contains a subsequence u_{m_j} which converges to a limit u in $W^{1,1}(\Omega)$. It is easy to see that u_{m_j} converges to u under the weak topology (in the sense of (2.88)) ; we have thus proved :

(2.89) Every bounded sequence in $HB(\Omega)$ contains a weakly convergent subsequence.

As with $BD(\Omega)$, we can define on $HB(\Omega)$ an intermediate topology determined by the distance

$$d(u,v) = \|u,v\|_{W^{1,1}(\Omega)} + \left| \int_\Omega |\nabla\nabla u| - \int_\Omega |\nabla\nabla v| \right| . \tag{2.90}$$

This distance defines a topology intermediate between the weak topology and the strong topology.

Furthermore the function ψ given in (2.29) satisfies the hypotheses of Section II.4 and this enables us to define a bounded measure on Ω , $\psi(\nabla\nabla u) \geq 0$, where $u \in HB(\Omega)$. We can therefore define another intermediate topology corresponding to the distance

$$d_\psi(u,v) = d(u,v) + \left| \int_\Omega \psi(\nabla\nabla u) - \int_\Omega \psi(\nabla\nabla v) \right| . \tag{2.91}$$

It is possible to consider other distances analogous to (2.91) associated with other functions f_i satisfying the hypotheses of Section II.4, but we shall confine ourselves to (2.91).

Accordingly a sequence u_m converges to u for this intermediate topology associated with (2.91), if, as $m \to \infty$,

$$\begin{cases} u_m \to u \text{ in } W^{1,1}(\Omega) \text{ strongly} \\ \nabla\nabla u_m \to \nabla\nabla u \text{ weakly as a measure} \\ \int_\Omega |\nabla\nabla u_m| \to \int_\Omega |\nabla\nabla u| \\ \psi(\nabla\nabla u_m) \to \psi(\nabla\nabla u) \text{ weakly as a measure} \\ \int_\Omega \psi(\nabla\nabla u_m) \to \int_\Omega \psi(\nabla\nabla u) \end{cases} \tag{2.92}$$

The weak convergence of $\psi(\nabla\nabla u_m)$ to $\psi(\nabla\nabla u)$ follows from the other properties of convergence and Lemma II.5.1.

Finally we state a result on the approximation of functions of $HB(\Omega)$ which is analogous to Theorems II.5.2 and II.5.3, and whose proof is to be found in F. Demengel [1].

PROPOSITION 2.8

For all $u \in HB(\Omega)$, *there exists a sequence* $u_m \in \mathscr{C}^\infty(\Omega) \cap W^{2,1}(\Omega)$ *with the following properties*

(2.93) $$\gamma_o(u_m) = \gamma_o(u) \text{ and } \gamma_1(u_m) = \gamma_1(u) \text{ , } \forall m .$$

(2.94) *As* $m \to \infty$, u_m *converges to* u *for the distance (2.91) (that is to say when convergence is defined as in (2.92))*

Remark 2.3. The preceding result enables us to verify that $HB(\Omega)$ *is the finite-energy space* for the problem (2.46) in the following sense : *if* u_j *is a sequence of functions of* $\mathscr{C}^2(\bar{\Omega})$ *which converge uniformly (or in* $L^1(\Omega)$*) to* u *, and which is such that*

$$\int_\Omega \psi(\nabla\nabla u_j)dx \leq const. ,$$

then u *is in* $HB(\Omega)$. To see that this is so, we need only use (2.89) noting that, because of (2.45), the sequence u_j is bounded in $HB(\Omega)$.

Remark 2.4. The sequence u_m given by Proposition 2.8 can be constructed so that it converges uniformly to u in $\bar{\Omega}$; cf. F. Demengel [3].

2.4. Generalised solution of the strain problem

We now propose to introduce a new form for the variational problem for the displacements, and to establish the existence of solutions and to study the duality between this generalised variational problem for the displacements and the variational problem for the stresses.

2.4.1. The strain problem in $HB(\Omega)$

The generalised version of the strain problem is the extension to $HB(\Omega)$ of the problem $\mathscr{PR}$. The function ψ corresponding to the plastic model (i.e. (2.29)) satisfies the hypotheses (4.1) (4.6) (4.8) of Section II.4 and we can define $\psi(\mu)$ as a bounded measure on Ω when μ is a bounded measure with values in E, $\mu \in M_1(\Omega;E)$. In particular $\psi(\nabla\nabla v)$ is a well-defined bounded measure when $v \in HB(\Omega)$, and we can extend the definition (2.25) of the function Ψ by putting

$$(2.95)\qquad \Psi(\mu) = \int_\Omega \psi(\mu) \quad \forall\, \mu \in M_1(\Omega;E),$$
$$\Psi(\nabla\nabla v) = \int_\Omega \psi(\nabla\nabla v), \quad \forall\, v \in HB(\Omega).$$

By the trace theorem (2.83), the expression $L(v)$ in (2.23) is well-defined when $v \in HB(\Omega)$:

$$(2.96)\qquad L(v) = \int_\Omega fv\,dx + \int_{\Gamma_1\cup\Gamma_2} g^1.\gamma_1(v)ds - \int_{\Gamma_2} g^0.\gamma_0(v)ds,$$

provided that (2.12) be replaced by

$$(2.97)\qquad f \in L^2(\Omega),\ g^0 \in L^\infty(\Gamma_2),\ g^1 \in L^\infty(\Gamma_1\cup\Gamma_2)\ (^1).$$

Similarly, for $v \in HB(\Omega)$, the function $\psi_\infty(\mathscr{F}(u_1 - \gamma_1(v))$ is well-defined on Γ and is summable by reason of (2.52); the term

$$\int_{\Gamma_0} \psi_\infty(\mathscr{F}(u_1 - \gamma_1(v))d\Gamma$$

is thus meaningful.

---

(1) We could equally well have supposed only that $f \in L^1(\Omega)$ or even, by making use of (2.86), that $f \in \mathscr{C}(\bar\Omega)'$: *this would allow us to take account of forces localised at a point on the plate (cf. F. Demengel [2] [3]).*

We thus define the generalised strain problem as

$$(2.98) \qquad \operatorname*{Inf}_{v \in \mathscr{C}_a} \{\psi(\nabla\nabla v) + \int_{\Gamma_0} \psi_\infty(\mathscr{F}(u_1 - \gamma_1(v)))\, ds - L(v)\} \qquad (\mathscr{Q}),$$

where

$$(2.99) \qquad \mathscr{C}_a = \{v \in HB(\Omega) \ , \ v = u_0 \ \text{ on } \ \Gamma_0 \cup \Gamma_1\} .$$

The problem $\mathscr{Q}$ is an extension of the problem $\mathscr{PR}$ which is itself an extension of the problem $\mathscr{P}$: the function to be minimised is the same, but the field of admissible displacements is greater :

$$\mathscr{P} = \{(2.46)\} \subset \mathscr{PR} = \{(2.60)\} \subset \mathscr{Q} = \{(2.98)\} .$$

PROPOSITION 2.9

The problems $\mathscr{P}$, $\mathscr{PR}$ *and* $\mathscr{Q}$ *have the same infimum, which is finite* ($> -\infty$) *if and only if* $\inf \mathscr{P}_{AL} \geqq 1$. *Furthermore if* $\inf \mathscr{P}_{AL} > 1$ *then every minimising sequence for any one of the three problems is bounded in* $HB(\Omega)$.

Proof

Since $\inf \mathscr{P} = \inf \mathscr{PR}$ and the inequality $\inf \mathscr{PR} \geqq \inf \mathscr{Q}$ is obvious, we have to prove that $\inf \mathscr{PR} \leqq \inf \mathscr{Q}$.

Let $v \in \mathscr{C}_a$; by Proposition 2.8 there is a sequence $v_m \in \mathscr{C}^\infty(\Omega) \cap W^{2,1}(\Omega) \cap \mathscr{C}_a$ which converges to v in the sense of (2.91) (2.92); this implies

$$\Psi(\nabla\nabla v_m) + \int_{\Gamma_0} \psi_\infty(\mathscr{F}(u_1 - \gamma_1(v_m)))ds - L(v_m) =$$

$$= \Psi(\nabla\nabla v_m) + \int_{\Gamma_0} \psi_\infty(\mathscr{F}(u_1 - \gamma_1(v)))ds - L(v)$$

$$\to \Psi(\nabla\nabla v) + \int_{\Gamma_0} \psi_\infty(\mathscr{F}(u_1 - \gamma_1(v)))ds - L(v) .$$

This proves that the infimum in (2.98) is the same as that on

$\mathscr{C}^\infty(\Omega) \cap W^{2,1}(\Omega) \cap \tilde{\mathscr{C}}_a$. If now $v \in W^{2,1}(\Omega) \cap \tilde{\mathscr{C}}_a$, then $v - u_0 \in W^{2,1}(\Omega)$ and vanishes on Γ_0 (in the sense of traces). We can now prove, with the aid of an approximation argument which has already been used on several occasions (1) , that

(2.100)
$$\{w \in \mathscr{C}^\infty(\Omega)\ ,\ w = 0 \text{ on } \Gamma_0\} \text{ is dense in } \{w \in W^{2,1}(\Omega)\ ,\ w = 0 \text{ on } \Gamma_0\}$$

Thus $w = v - u_0$ is the limit in $W^{2,1}(\Omega)$ of a sequence of functions w_m in $\mathscr{C}^\infty(\Omega)$, which vanish on Γ_0 . The functions $v_m = u_0 + w_m$ belong to the set defined by (2.61) and converge to v in $W^{2,1}(\Omega)$. As the function to be minimised in (2.60) or (2.98) is continuous on $W^{2,1}(\Omega)$ under the norm topology, the infimum of (2.98) on $W^{2,1}(\Omega) \cap \tilde{\mathscr{C}}_a$ is the same as on $H^2(\Omega) \cap \tilde{\mathscr{C}}_a$, which latter is in fact the set (2.61).

Proposition 2.5 shows that $\inf \mathscr{Q} = \inf \mathscr{P} > -\infty$ if and only if $\inf \mathscr{P}_{AL} \geqq 1$. Furthermore every minimising sequence for $\mathscr{P}$, $\mathscr{PR}$ or $\mathscr{Q}$ is a minimising sequence for $\mathscr{Q}$. If $\inf \mathscr{P}_{AL} > 1$ and if v_m is a minimising sequence for $\mathscr{Q}$, then it can be shown, just as in Proposition 2.5, that

$$\Psi(\nabla\nabla v_m) \leqq \text{const.}\ ,$$

which, with (2.45), implies that

$$\int_\Omega |\nabla\nabla v_m| \leqq \text{const.}$$

Since Γ_0 is not void and contains three non-collinear points, (2.82) implies that

$$\int_{\Gamma_0} |\gamma_0(v)|ds + \int_\Omega |\nabla\nabla v|$$

---

(1) The argument is similar to the one used in obtaining I.(2.105) and I.(2.106) but is simpler since the boundary of Γ is of dimension 1.

defines, on $HB(\Omega)$, a norm equivalent to the natural norm. The sequence of functions v_m is in $\tilde{\mathscr{C}}_a$, and $v_m = u_0$ on Γ_0 ; it follows that v_m is, as stated, bounded in $HB(\Omega)$.

2.4.2. Duality between generalised stresses and strains

In considering the generalised strains v which are in $HB(\Omega)$, we come up against a difficulty similar to the one encountered in Section II.7 : when $v \in HB(\Omega)$ and $M \in L^2(\Omega;E)$, the product $M.\nabla\nabla v$ has no meaning and we have *a priori* no Green's formula such as (2.7) (2.10) or (2.89) at our disposition. Our immediate object now is to overcome this difficulty : we shall see that, by making use of all the available information, it is possible to define the product $M.\nabla\nabla v$ as a bounded measure, when $M \in \mathscr{S}_a \cap \mathscr{K}_a$ and $v \in \tilde{\mathscr{C}}_a$. We shall also obtain a generalisation of the Green's formula (2.7) and, in the final analysis, we shall be able to exhibit a duality between (2.48) and (2.98).

□

We introduce the space

$$S(\Omega) = \{M \in L^\infty(\Omega;E) \ , \ \nabla.\nabla.M \in L^2(\Omega)\} \ (^1) \ . \tag{2.101}$$

This is a Banach space for the natural norm

$$\|M\|_{L^\infty(\Omega;E)} + \|\nabla.\nabla.M\|_{L^2(\Omega)} \ . \tag{2.102}$$

The set of regular functions (for example $\mathscr{C}^\infty(\bar{\Omega};E)$) is not dense in M ; on the other hand, by proceeding exactly as in Lemma 7.1 (cf. also Remark 7.1), it can be shown that for all

---

(1) At the expense of a few additional complications, and using (2.87) (which has not been proved here) we could take

$$S(\Omega) = \{M \in L^\infty(\Omega;E) \ , \ \nabla.\nabla.M \in L^1(\Omega) \ \text{or even} \in M_1(\Omega)\}$$

$M \in S(\Omega)$, there is a sequence M_j in $\mathscr{C}_0^\infty(\bar{\Omega};E)$ such that

(2.103) $$\|M_j\|_{L^\infty(\Omega;E)} \leqq c(\Omega)\|M\|_{L^\infty(\Omega;E)} ,$$

where $c(\Omega)$ depends only on Ω and where, as $j \to \infty$,

(2.104) $$\begin{cases} M_j \to M \text{ in } L^p(\Omega;E) \text{ strongly, } \forall\, p , 1 \leqq p < \infty \\ \text{and in } L^\infty(\Omega;E) \text{ weak-star} \\ \nabla.\nabla.M_j \to \nabla.\nabla.M \text{ in } L^2(\Omega) . \end{cases}$$

Incidentally (2.103) can be sharpened as follows : it is possible to choose the sequence M_j in such a way that, for all j ,

(2.105) $$\|M_j\|_{L^\infty(\Omega \cap \mathscr{O};E)} \leqq \|M\|_{L^\infty(\Omega;E)}$$

where $\mathscr{O}$ is any fixed relatively compact open subset of Ω or where $\mathscr{O}$ is any open ball of sufficiently small radius centred on an arbitrary arbitrary point x on Γ .

□

When $v \in HB(\Omega)$ and $M \in S(\Omega)$, we can define a distribution $M.\nabla\nabla v$ by putting

(2.106) $$\langle M.\nabla\nabla v,\phi\rangle =$$
$$- \int_\Omega vM.(\nabla\nabla\phi)dx - 2\int_\Omega M.(\nabla v \otimes \nabla\phi)dx + \int_\Omega v\phi(\nabla.\nabla.M)$$

for all $\phi \in \mathscr{C}_0^\infty(\Omega)$. The following proposition gives some properties of $M.\nabla\nabla v$ and shows that it coincides with the usual product when the latter has a meaning $(M \in \mathscr{C}(\bar{\Omega};E)$ or $v \in H^2(\Omega))$:

Proposition 2.10

i) For $v \in HB(\Omega)$ and $M \in S(\Omega)$, the distribution $M.\nabla\nabla v$ defined by (2.106) is a bounded measure on Ω , is absolutely continuous with respect to $|\nabla\nabla v|$ and ,

(2.107) $$\left|\int_\Omega \phi M.\nabla\nabla v\right| \leq \|M\|_{L^\infty(\Omega;E)} \int_\Omega |\phi|\ |\nabla\nabla v| ,$$

for all $\phi \in \mathscr{C}_o(\Omega)$.

ii) If $M_j \in S(\Omega)$ *is a sequence of functions converging to* M *in the sense of (2.103) - (2.105), then* $M_j.\nabla\nabla v$ *converges to* $M.\nabla\nabla v$ *in the following sense :*

$$\int_\Omega \phi M_j.\nabla\nabla v \to \int_\Omega \phi M.\nabla\nabla v , \ \forall\ \phi \in \mathscr{C}(\bar{\Omega}) .$$

iii) If $v_j \in HB(\Omega)$ *is a sequence of functions converging to* v *under the metric (2.90) then* $M.\nabla\nabla v_j$ *converges to* $M.\nabla\nabla v$ *in the following sense :*

$$\int_\Omega \phi M.\nabla\nabla v_j \to \int_\Omega \phi M.\nabla\nabla v , \ \forall\ \phi \in \mathscr{C}(\bar{\Omega})$$

Proof

Using Proposition 2.8 we approximate v by a sequence of functions v_j in $\mathscr{C}^\infty(\Omega) \cap W^{2,1}(\Omega)$, which converge to v under the metric (2.90). By integrating by parts we see that $M.\nabla\nabla v_j$ coincides with the usual product for all j and that

(2.108) $$\left|\int_\Omega \phi M.\nabla\nabla v_j\right| \leq \|M\|_{L^\infty(\Omega;E)} \int_\Omega |\phi|\ |\nabla\nabla v_j| .$$

By II.(2.15), when $j \to \infty$, we have, for all $\theta \in \mathscr{C}(\bar{\Omega})$:

$$\int_\Omega \theta\ |\nabla\nabla v_j| \to \int_\Omega \theta\ |\nabla\nabla v| .$$

The inequality (2.108) thus implies (2.107) in the limit ($\phi \in \mathscr{C}_o^\infty(\Omega)$ being fixed), and $M.\nabla\nabla v$ is indeed a bounded measure.

The proof of the statements ii) and iii) is now identical to that of Lemmata II.7.4 and II.7.2.

Proposition 2.8, and in particular (2.93), shows that $\gamma_0(HB(\Omega)) = \gamma_0(W^{2,1}(\Omega))$ (for which no characterisation is known) and $\gamma_1(HB(\Omega)) = \gamma_1(W^{2,1}(\Omega)) = L^1(\Gamma)$. The following proposition allows us to define $b_0(M) \times b_1(M) \in \gamma_0(W^{2,1}(\Omega))' \times L^\infty(\Gamma)$ when $M \in S(\Omega)$, and thus to express (2.89) in a more precise form.

Proposition 2.11

There exist two continuous linear surjective mappings β_0, β_1, *from* $S(\Omega)$ *on to* $\gamma_0(W^{2,1}(\Omega))' \times L^\infty(\Omega)$, *such that*

$$\beta_0(M) = b_0(M) \ , \ \beta_1(M) = b_1(M) \ , \tag{2.109}$$

for all $M \in \mathscr{C}^2(\bar{\Omega})$; *furthermore for all* $v \in HB(\Omega)$, *the following Green's formula holds :*

$$\int_\Omega M.\nabla\nabla v - \int_\Omega v(\nabla.\nabla.M)dx = \tag{2.110}$$

$$= \langle\beta_1(M), \gamma_1(v)\rangle - \langle\beta_0(M), \gamma_0(v)\rangle .$$

Proof

We first define β_0, β_1, satisfying (2.110) when $v \in W^{2,1}(\Omega)$ and we shall then extend the validity of (2.110) to v in $HB(\Omega)$.

i) Let M be given in $S(\Omega)$. For any given (w_0, w_1) in $\gamma_0(W^{2,1}(\Omega)) \times L^1(\Gamma)$, there are, by (2.75), $w \in W^{2,1}(\Omega)$ such that $\gamma_0(w) = w_0$, $\gamma_1(w) = w_1$. We put

$$X_M(w_0, w_1) = \int_\Omega w(\nabla.\nabla.M)dx - \int_\Omega M.(\nabla\nabla w)dx . \tag{2.111}$$

As $\mathscr{C}_0^\infty(\Omega)$ is dense in the set defined by

$$W_0^{2,1}(\Omega) = \{v \in W^{2,1}(\Omega) \ , \ \gamma_0(v) = \gamma_1(v) = 0\} ,$$

it is easily verified that $X_M(w_0, w_1)$ does not depend on the choice of w satisfying $\gamma_0(w) = w_0$, $\gamma_1(w) = w_1$. It is also easily proved that the mapping

$$\{w_0,w_1\} \to X_M(w_0,w_1)$$

is linear. Now it is proved in F. Demengel [2] that for all $\{w_0,w_1\} \in \gamma_0(W^{2,1}(\Omega)) \times L^1(\Gamma)$, there is a $w \in W^{2,1}(\Omega)$ such that

$$(2.112) \qquad \|w\|_{W^{2,1}(\Omega)} \leqq c(\|w_0\|_{\gamma_0(W^{2,1}(\Omega))} + \|w_1\|_{L^1(\Gamma)})$$

where the constant c depends only on Ω . With this choice of w we find

$$(2.113) \quad |X_M(w_0,w_1)| \leqq c(\|w_0\|_{\gamma_0(W^{2,1}(\Omega))} + \|w_1\|_{L^1(\Gamma)}).\|M\|_{S(\Omega)} .$$

It follows that X_M is a continuous linear form on $\gamma_0(W^{2,1}(\Omega)) \times L^1(\Gamma$ and that there are two elements $\beta_0(M)$, $\beta_1(M)$ in $\gamma_0(W^{2,1}(\Omega))' \times L^\infty(\Gamma$ such that

$$X_M(w_0,w_1) = \langle\beta_0(M),w_0\rangle - \langle\beta_1(M), w_1\rangle .$$

This gives us the Green's formula (2.110) when $v \in W^{2,1}(\Omega)$, and (2.109) then follows with (2.7).

ii) We now consider the mapping $M \mapsto \{\beta_0(M) , \beta_1(M)\}$ of $S(\Omega)$ into $\gamma_0(W^{2,1}(\Omega)' \times L^\infty(\Gamma)\}$. We see from (2.111) that it is linear, and (2.113) shows that it is continuous. Let therefore $L_0,L_1 \in \gamma_0(W^{2,1}(\Omega))' \times L^\infty(\Gamma)$; the mapping ℓ from $W^{2,1}(\Omega)$ into $\mathbb{R}$ defined by

$$v \mapsto \ell(v) = - \langle L_0,\gamma_0(v)\rangle + \langle L_1,\gamma_1(v)\rangle$$

is a continuous linear form on $W^{2,1}(\Omega)$. By (2.82) (and for $p(u) = |u|_{L^1(\Omega)}$) , the mapping $v \mapsto \{v, \nabla\nabla v\}$ is an isomorphism of $W^{2,1}(\Omega)$ into $L^1(\Omega) \times L^1(\Omega;E)$; by virtue of the Hahn-Banach theorem, the form ℓ can be extended to a continuous linear form with the same norm on $L^1(\Omega) \times L^1(\Omega;E)$; there exist therefore $m,M \in L^\infty(\Omega) \times L^\infty(\Omega;E)$ such that

$$\ell(v) = \int_\Omega M.\nabla\nabla v \, dx - \int_\Omega v\, m\, dx \tag{2.114}$$

for all $v \in W^{2,1}(\Omega)$. The equation (2.114) with $v \in \mathscr{C}_0^\infty(\Omega)$ shows that

$$\nabla.\nabla.M = m \in L^\infty(\Omega) .$$

Thus $M \in S(\Omega)$ and (2.114) together with the preceding statements imply that

$$L_0 = \beta_0(M) \ , \ L_1 = \beta_1(M) .$$

Finally since ℓ is extended into a linear mapping with the same norm

$$\|\ell\| = \{\|L_0\|_{W^{2,1}(\Omega)'} + \|L_1\|_{L^1(\Gamma)}\}$$

$$= \|M\|_{L^\infty(\Omega;E)} + |m|_{L^\infty(\Omega)}$$

$$\geqq c'\|M\|_{S(\Omega)} ;$$

and this completes the proof of surjectivity.

iii) It now remains only to extend (2.110) to $v \in HB(\Omega)$. If $v \in HB(\Omega)$, we use an approximating sequence v_j given by Proposition 2.8. The equation (2.110) holds with v replaced by v_j , and by (2.93),

$$\int_\Omega M.\nabla\nabla v_j dx - \int_\Omega v_j(\nabla.\nabla.M)dx =$$

$$= \langle\beta_1(M), \gamma_1(v_j)\rangle - \langle\beta_0(M), \gamma_0(v_j)\rangle$$

$$= \langle\beta_1(M), \gamma_1(v)\rangle - \langle\beta_0(M), \gamma_0(v)\rangle .$$

We obtain (2.110) in the limit by using part iii) of Proposition 2.10

(with $\phi = 1$).

Remark 2.5. The Green's formula (2.110) shows that the operators (β_0, β_1) which extend (b_0, b_1) are not necessarily unique. For simplicity we shall continue to denote them by b_0 and b_1.

Generalised duality

We define, for $v \in HB(\Omega)$

(2.115) $$J(v) = \Psi(\nabla\nabla v) + \int_{\Gamma_0} \psi_\infty(\mathscr{F}(u_1 - \gamma_1(v)))ds .$$

By proceeding exactly as in Proposition II.7.2 (and with an easy extension of Lemma II.7.8), we prove that

(2.116) $$J(v) - L(v) \geqq \bar{L}(M) - \frac{1}{2}\mathscr{A}(M,M) ,$$

for all $v \in \tilde{\mathscr{C}}_a$ and all $M \in \mathscr{S}_a \cap \mathscr{K}_a$. We can then introduce the Lagrangian of the problem, defined on $\tilde{\mathscr{C}}_a \times (\mathscr{S}_a \cap \mathscr{K}_a)$:

$$\mathscr{L}(v,M) = - L(v) - \frac{1}{2}\mathscr{A}(M,M) +$$
$$+ \int_\Omega M.\nabla\nabla v + \langle b_1(M), u_1 - \gamma_1(v)\rangle .$$

The classical primal and dual problems associated with this Lagrangian are (cf. I. Ekeland and R. Temam [1], chapter 6) :

$$\operatorname*{Inf}_{v \in \tilde{\mathscr{C}}_a} \{ \operatorname*{Sup}_{M \in \mathscr{S}_a \cap \mathscr{K}_a} \mathscr{L}(v,M)$$

and

$$\operatorname*{Sup}_{M \in \mathscr{S}_a \cap \mathscr{K}_a} \{ \operatorname*{Inf}_{v \in \tilde{\mathscr{C}}_a} \mathscr{L}(v,M)\} .$$

By reason of (2.116) and the results stated above, these two problems are now seen to be identical to the problems (2.98) and (2.48) respectively (i.e. the generalised strain problem in $HB(\Omega)$, and the 'natural' stress problem.

2.4.3. Existence of a strain in $HB(\Omega)$

The hypotheses are those assumed above, in particular (2.13) (2.14) (2.43) and (2.97); the open set Ω is assumed to be of class $\mathscr{C}^2$ and $\Gamma_0 \neq \emptyset$, Γ_0 containing three non-collinear points. We shall call the hypothesis

$$\text{(2.117)} \qquad \inf \mathscr{P}_{AL} > 1 \iff \exists\, M \in \mathscr{S}_a \cap \lambda \mathscr{K}_a \quad \text{with} \quad 0 \leqq \lambda < 1$$

the safe load condition.

This hypothesis implies that every minimising sequence for $\mathscr{Q}$ (or $\mathscr{P}$ or $\mathscr{PR}$) is bounded in $HB(\Omega)$ (cf. Proposition 2.9). Proceeding exactly as in Lemmata II.8.1 and II.8.2 we can show (for the complete proof, cf. F. Demengel [1] [3]) that the following holds :

Lemma 2.3

Let $v_m \in \tilde{\mathscr{C}}_a$ *be a sequence which converges to* $v \in \tilde{\mathscr{C}}_a$ *in the sense of (2.88). If* $\mathscr{S}_a \cap \mathscr{K}_a$ *is not void then*

$$\text{(2.118)} \qquad J(v) - L(v) \leqq \liminf_{m\to\infty} \{J(v_m) - L(v_m)\} .$$

This enables us to prove

THEOREM 2.1

If (2.13) (2.14) (2.43) (2.97) hold, and if $\Gamma_0 \neq \emptyset$ *and the safe load condition is satisfied, then the strain problem (2.98) has a solution in* $HB(\Omega)$.

Under these same hypotheses, if $v \in \tilde{\mathscr{C}}_a$ *and* $M \in \mathscr{S}_a \cap \mathscr{K}_a$, *then* v *is a solution of (2.98) and* M *a solution of (2.48) if and only if*

$$\text{(2.119)} \qquad \psi(\nabla\nabla v) + \frac{1}{2} AM.M = M.\nabla\nabla v \quad \text{in} \quad \Omega \text{, and}$$

$$\text{(2.120)} \qquad \psi_\infty(\mathscr{F}(u_1-\gamma_1(v))) = (u_1-\gamma_1(v)).b_1(M) \quad \text{on} \quad \Gamma_0 .$$

The proof is similar to that of Theorem II.8.1; cf. F. Demengel [1] [3].

Principal Notations

The page indicated is that on which the notation appears for the first time.

$\mathcal{H}(\mathcal{A}_1, \ldots, \mathcal{A}_k ; \alpha_0, \ldots, \alpha_k ; \Omega)$		21
$\chi_{\mathcal{A}}$,	*the indicator function of* $\mathcal{A}$	27
∂ ,	*the ordinary symbol for partial differentiation or the symbol for the sub-differential*	3 34
$\mathcal{P}, \mathcal{P}^*$,	*variational problem and its dual or strain problem and its dual*	37 77,88,307
λ, μ,	*Lamé constants*	44
μ	*is also used to denote measures, from page 138 onwards*	
κ	*the bulk modulus*	44
ψ, ψ^*,	*the surpotential function and its conjugate*	44,60,75
$L(v)$		45
$\mathcal{A}(\sigma,\tau)$		48
$\mathcal{C}_a, \mathcal{S}_a, \mathcal{C}_0, \mathcal{S}_0$		59,298
ψ_e, ψ_e^*		64
$A\xi$		64 & 75
K, K^D ,	*convex plasticity domains (cf. also* $K = \text{dom } f^*$ *, p.176 and p.305 for plates).*	65 & 74
ψ_∞, f_∞		88,176
$\mathcal{P}_{AL}, \mathcal{P}^*_{AL}$		88,307
$\mathcal{PR}, \mathcal{PR}^*$,	*the problem* $\mathcal{P}$ *with partially relaxed boundary conditions and its dual*	100,103,311
$\mathcal{T}, \mathcal{T}^D$,		100
n^*		117
$BV(\Omega)$		143
$BD(\Omega)$		144
$U(\Omega)$		167
$\mathcal{Q}$,	*the relaxed version of* $\mathcal{P}$ *and* $\mathcal{PR}$	217

Index

Bibliography

R.S. Adams

[1] Sobolev Spaces, Academic Press, New-York, 1975.

G. Anzelotti

[1] On the existence of the rates of stress and displacement for Prandtl-Reuss plasticity, (to appear).

G. Anzelotti and M. Giaquinta

[1] Funzioni BV e trace. *(Functions of bounded variation and traces)* Rend. Sem. Mat. Univ. Padova, 60, 1978, p.1-21.

[2] On the existence of the fields of stresses and displacements for an elasto-perfectly plastic body in static equilibrium. J. Math. Pures Appl., (to appear).

M.E. Bogovski

[1] Solution of the first boundary value problem for the equation of continuity of an incompressible medium. Soviet Mat. Doklady, 20, 1979, p.1094-1098.

N. Bourbaki

[1] Eléments de Mathématiques, Livre III, Topologie Générale, *(Elements of Mathematics, Book III, General Topology)*, Hermann, Paris, 1964.

[2] Eléments de Mathématiques, Livre V, Espaces Vectoriels Topologiques, *(Elements of Mathematics, Book V, Topological Vector Spaces)*, Hermann, Paris, 1974.

[3] Eléments de Mathématiques, Livre VI, Intégration, *(Elements of Mathematics, Book VI, Integration)*, Hermann, Paris, 1965.

H. Brézis

[1] **Inéquations Variationnelles**, *(Variational Inequalities)*, J. Math Pures Appl., 1971.

[2] **Multiplicateur de Lagrange en torsion élasto-plastique**, *(Lagrange multiplier in elasto-plastic torsion)*, Arch, Rational Mech. Anal., 49, 1972-73, p.32-40.

[3] **Intégrales Convexes dans les Espaces de Sobolev**, *(Convex integrals in Sobolev spaces)*, Isr. J. Math., 13, 1972, p.9-23.

H. Brézis and G. Stampacchia

[1] **Sur la régularité de la solution d'inéquations élliptiques**, *(On the regularity of the solution to a set of elliptic inequalities)*, Bull, Soc. Math. France, 96, 1968, p.153-180.

H.D. Bui

[1] **La Mécanique de la Rupture Fragile**, *(The mechanics of fracture)*, Masson, Paris, 1977.

R. Caccioppoli

[1] **Misura a integrazione sugli insiemi dimensionalmente orientati.** *(Integration measure on dimensionally-oriented sets)*, Rend. Accad. Naz. Lincei., Cl. Sc. fis. mat. nat., 8, vol. 12, 1952, p.3-11 and 137-146.

L. Caffarelli and N. Rivière

[1] **The Lipschitz Character of the stress tensor, when twisting an elastic-plastic bar.** Arch. Rational Mech. Anal., 69, 1979, p.31-36.

L. Cattabriga

[1] **Su un problema al contorno relativo al sistema di equazioni di Stokes.** *(On a boundary value problem relating to the system of Stokes equations)*, Rend. Sem. Mat. Univ. Padova, 31, 1961, 1961, p.308-340.

L. Cesari

[1] **Surface area**, Princeton University Press, 1956.

[2] **Asymptotic behaviour and stability problems in ordinary differential equations**, Academic Press, New-York, 1963.

E. Christiansen

[1] **Limit analysis in Plasticity as a mathematical programming problem**, Calcolo, 17, 1980, p.41-65.

[2] **Limit analysis for plastic plates**, SIAM J. Math. Anal., 11, 1980, p.514-522.

[3] Computation of Limit Loads, Internat. J. Numer. Methods Engrg., 17, 1981, p.1547-1570.

[4] Examples of collapse solutions in limit analysis, (to appear) in Utilitas Math.

See also H. Matthies, G. Strang and E. Christiansen.

D. Clément

[1] **Résolution numérique d'un problème modèle en plasticité.** *(Numerical solution of a plasticity model problem).* Thèse de 3e Cycle, Université de Paris-Sud, 1982.

R. Courant

[1] **Variational methods for the solution of problems of equilibrium and vibrations.** Bull. Amer. Math. Soc., 49, 1943, p.1-23.

F. Crépel

[1] **Résolution de problèmes de plasticité par des méthodes de dualité et d'éléments finis: calcul à la rupture et déformation.** *(Solution of plasticity problems by duality methods and finite element methods : breakdown and deformation analysis).* Thèse de 3e Cycle, Université de Paris-Sud, 1982.

O. Debordes

[1] **Dualité des théorèmes statique et cinématique dans la théorie de l'adaptation des milieux continus élasto-plastiques,** *(Duality of static and kinematic theorems in the theory of the adaptation of continuous elasto-plastic media)*, C.R. Acad. Sc. Paris, 282, série A, 1976, p.535-537.

[2] **Contribution à la théorie et au calcul de l'élastoplasticité asymptotique,** *(Contribution to the theory and computation of asymptotic elastoplasticity)*, Thèse, Université de Marseille, 1977.

[3] **Compléments au cadre fonctionnel classique en élastoplasticité parfaite et existence de contraintes dans ce cadre,** *(Complements to the classical functional framework in perfect elasto-plasticity and the existence of constraints in this framework)*, C.R. Acad. Sc. Paris, série I, 294, 1982, p.349-352.

O. Debordes and B. Nayroles

[1] **Sur la théorie et le calcul à l'adaptation des structures élasto-plastiques,** *(On the theory and computation of the adaptation of elasto-plastic structures)*, J. Mécanique, 15, 1976, p.1-53.

F. Demengel

[1] **Problèmes variationnels en plasticité parfaite des plaques,** *(Variational problems in the perfect plasticity of plates)*, Thèse de 3e Cycle, Université de Paris-Sud, Orsay, 1982.

[2] **Fonctions à hessien borné**, *(Functions with bounded Hessian)*, Ann. Inst. Fourier, 1984.

[3] **Problèmes variationnels en plasticité parfaite des plaques**, *(Variational problems in the perfect plasticity of plates)*, Num. Funct. Anal. Optim., 6, 1983, p.73-119.

F. Demengel and R. Temam

[1] **Convex Functions of a Measure and Applications**, Indiana Univ. Math. Journ. 33, 1984, p.673-709.

J. Deny and J.L. Lions

[1] **Les espaces du type de Beppo Levi**, *(Spaces of Beppo-Levi type)*, Ann. Inst. Fourier, 5, 1954, p.305-370.

N. Dinculeanu

[1] **Integration on locally compact spaces.** Noordhoff Intern. Publ., Leyden, 1974.

[2] **Vector Measures**, Pergamon Press, Oxford, New-York, 1966.

N. Dunford and J.T. Schwartz

[1] **Linear Operators**, Interscience Publishers, New-York, 1958.

G. Duvaut and J.L. Lions

[1] **Les inéquations en mécanique et en physique**, Dunod, Paris, 1972 *(Inequalities in mechanics and physics)*, Springer Verlag.

I. Ekeland and R. Temam

[1] **Analyse convexe et problèmes variationnels**, *(Convex analysis and variational problems)*, Dunod, Paris, 1974. (See [2]).

[2] **Convex analysis and variational problems**, *(translation from the French, revised edition of [1])*, North Holland, Amsterdam, New-York, 1976.

W. Fenchel

[1] **On Conjugate Convex Functions**, Canad. J. Math., 1, 1949, p.73-77.

[2] **Convex Cones, Sets and Functions**, Notes de cours polycopiées, *(mimeographed lecture notes)*, Princeton University, 1951.

[3] **A remark on convex sets and polarity**, Modd. Lunds. Univ. Mat. Sem., *(Supplementary volume, 1952)*, p.82-89.

W.H. Fleming

[1] **Functions with generalized gradient and generalized surfaces**, Ann. Mat. Pura Appl., séries 4, 44, 1957, p.93-103.

M. Frémond and A. Friaâ

[1] **Analyse limite. Comparaison des méthodes statique et cinématique,** *(Limit analysis, comparison between the static and kinematic methods)*, C.R. Acad. Sci. Paris, 286, série A, 1978, p.170-110.

[2] **Limit Analysis of physical Models,** in Nonlinear Problems, Present and Future, A.R. Bishop, D.K. Campbell, B. Nicolaenko, Ed., North-Holland, 1982.

A. Friaâ

[1] **La loi de Norton-Hoff généralisée en plasticité et visco-plasticité,** *(The generalised Norton-Hoff law in plasticity and visco-plasticity)*, Thèse, Université de Paris VI, 1979.

See also M. Frémond and A. Friaâ.

E. Gagliardo

[1] **Caratterizzazioni delle trace sulla frontiera relative alcune classi di funzioni in n variabili,** *(Characterisation of traces on the boundary with respect to certain classes of functions of n variables)*, Rend. Sem. Mat. Univ. Padova, 27, 1957, p.284-305.

P. Germain

[1] **Cours de Mécanique des Milieux Continus,** *(A course on the mechanics of continuous media, Volume I)*, Tome I, Masson, Paris, 1973.

[2] **Cours de Mécanique,** *(A course on mechanics)*, Ecole Polytechnique, 1979.

[3] **Commentaires sur l'Article "Duality and relaxation in the variational problems of plasticity de R. Temam and G. Strang",** *(Remarks on the article "Duality and relaxation in the variational problems of plasticity by R. Temam and G. Strang")*, J. Mécanique, 19, 1980, p.529-538.

M. Giaquinta

See G. Anzelotti and M. Giaquinta, M. Giaquinta and G. Modica.

M. Giaquinta and G. Modica

[1] Nonlinear systems of the type of the stationary Navier-Stokes system. (To appear).

E. de Giorgi

[1] **Su una teoria generale della misura (r-1) dimensionale in uno spazio ad r dimensioni,** *(On a general theory of (r-1)-dimensional measure in an r-dimensional space)*, Annali di Mat., série 4, vol. 36, 1954, p.191-213.

[2] **Nuovi teoremi relativi alla misure (r-1)-dimensionali in uno spazio ad r dimensioni,** *(New theorems relating to (r-1)-dimensional measures in a space of dimensions r)*, Ricerche di Mat., 4, 1955, p.95-113.

E. Giusti

[1] Minimal surfaces and functions of bounded variation. Notes de cours rédigées par *(Lecture notes written by)* G.H. Williams, Depart. of Math., Australian National University, Canberra, 10, 1977.

J. Gobert

[1] Une inéquation fondamentale de la théorie de l'élasticité, *(A fundamental inequality in elasticity theory)*, Bull. Soc. Roy. Sci. Liège, no 3 and 4, 1962.

[2] Sur une inégalité de coercivité, *(On an inequality relating to coercivity)*, J. Math. Anal. Appl., 36, 1971, p.518-528.

C. Goffman and J. Serrin

[1] Sublinear functions of Measures and Variational Integrals, Duke Math. J., 31, 1964, p.159-178.

H.J. Greenberg

[1] Complementary minimum principle for an elastic-plastic material. Quart. Appl. Math., 7, 1948, p.85-95.

M. Gurtin

[1] An Introduction to Continuum Mechanics, Academic Press, New York, 1981.

M. Gurtin and R. Temam

[1] On the antiplane shear problem in finite elasticity, J. Elasticity, 11, 1981, p.197-206.

P.R. Halmos

[1] Measure Theory, Van Nostrand, New-York, 1951.

B. Halphen and N.Q. Son

[1] Sur les matériaux standards généralisés, *(On generalised standard materials)*, J. Mécanique, 14, 1975, p.39-63.

H. Hencky

[1] Z. Angew. Math. Phys., 4, 323, 1924.

R. Hill

[1] Mathematical Theory of Plasticity, University Press, Oxford, 1950.

P. Hodge

[1] Plastic Analysis of Structures. New edition, R.E. Krieger, Pub. Comp., Malabar, Florida, 1981.

See also W. Prager and P. Hodge.

H.J. Hoff

[1] Approximate analysis of structures in the presence of moderately large creep deformations, Quart. Appl. Math., 12, 1954, p.49.

C. Johnson

[1] Existence theorems for plasticity problems. J. Math. Pures Appl., 55, 1976, p.431-444.

[2] On plasticity with hardening. J. Math. Anal. Appl., 62, 1978, p.325-336.

[3] A mixed finite element method for plasticity problems with hardening, SIAM J. Numer. Anal., 14, 1977, p.575-583.

R. Kohn

[1] Ph.D. Thesis, Princeton University, 1979.

[2] New integral estimates for deformations in terms of their non-linear strains, Arch. Rational Mech. Anal., (to appear).

R. Kohn and G. Strang

[1] Optimal design for torsional rigidity, Proc. Int. Symp. on Mixed and Hybrid Finite Element Methods, Atlanta, 1981.

[2] Optimal design of cylinders in shear, MAFELAP Conference, Brunel, 1981.

[3] Structural design optimization, homogenization, and relaxation of variational problems, Proc. Conf. on Disordered Media, Lecture Notes in Physics 154, Springer-Verlag, New-York, 1982.

[4] Hencky-Prandtl nets and constrained Michell trusses, Conf. on Optimum Structural Design, Tucson, 1981, (to appear) in Comput. Methods Appl Mech. Engrg.

R. Kohn and R. Temam

[1] Principes variationnels duaux et théorèmes de l'énergie dans le modèle de plasticité de Hencky, *(Dual variational principles and finite-energy theorems relating to the Hencky plasticity model)*, C.R. Acad. Sci. Paris, Série I, 294, 1982, p.205-208.

[2] Dual spaces of stresses and strains with applications to Hencky plasticity, Appl. Math. Optim., 10: 1, 1983, p.1-35.

W.T. Koiter

[1] General theorems for elastic plastic solids in Progress in Solid Mechanics, North-Holland, Amsterdam, 1960.

M.A. Krasnosel'skii

[1] Topological Methods in the Theory of Nonlinear Integral Equations, Pergamon Press, London 1964.

O.A. Ladyzhenskaya

[1] The mathematical theory of viscous imcompressible flow, Gordon and Breach, New-York, *English version*, Second Edition, 1969.

H. Lanchon

[1] **Solution du problème de torsion élasto-plastique d'une barre de section quelconque**, *(Solution of the elasto-plastic torsion problem for a bar of arbitrary cross-section)*, C.R. Acad. Sci. Paris, 269, 1969, p.791-794.

[2] **Sur la solution du problème de torsion élastoplastique d'une barre cylindrique de section multiconnexe**, *(On the solution of the elasto-plastic torsion problem for a cylindrical bar with multiply connected cross-section)*, C.R. Acad. Sci. Paris, 271, 1970. p.1137-1140.

[3] **Problème d'élastoplasticité statique pour un matériau régi par la loi de Hencky**, *(A problem in elasto-plastic statics for a material obeying Hencky's law)*, C.R. Acad. Sci. Paris, 271, 1970, p.888-891.

[4] **Torsion élastoplastique d'une barre de section simplement ou multiplement connexe**, *(Elasto-plastic torsion of a bar with simply or multiply-connected cross-section)*, J. Mécanique, 12, 1973, p.151-171.

A. Lichnewsky

[1] **Un problème modèle en théorie de la plasticité : analogie avec le problème des surfaces minimales**, *(A model problem in the theory of plasticity : analogy with the minimal surface problem.)* Seminar Goulaouic-Schwartz, 1978-1979, Ecole Polytechnique.

J.L. Lions

[1] **Problèmes aux limites dans les équations aux dérivées partielles**, *(Boundary value problems in partial differential equations)*, Séminaire de Mathématiques Supérieures, *(Seminar on advanced mathematics)*, University of Montreal, 1962.

[2] **Quelques méthodes de résolution des problèmes aux limites non linéaires**, *(Some methods of solving non-linear boundary value problems)*, Dunod and Gauthier-Villars, Paris, 1969.

[3] **Some methods in the mathematical analysis of systems and their control**, Academia Sinica, (to appear).

See also, J.L. Lions and E. Magenes, G. Duvaut and J.L. Lions, J. Deny and J.L. Lions.

J.L. Lions and E. Magenes

[1] **Problèmes aux limites non homogènes et applications**, *(Non-homogeneous boundary value problems and applications)*, Dunod, Paris, Vol. 1 and 2, 1968. (Translation from the French), Springer Verlag, Heidelberg, New-York, 1972.

E. Magenes

See J.L. Lions and E. Magenes, E. Magenes and G. Stampacchia.

E. Magenes and G. Stampacchia

[1] I problemi al contorno per le equazioni differenziali di tipo ellittico, *(Boundary value problems for differential equations of elliptic type)*, Ann. Scuola Norm. Sup. Pisa, 12, 1958, p.247-357, note (27), p.320.

H. Matthies, G. Strang and E. Christiansen

[1] **The saddle point of a differential program**, in **Energy Methods in Finite Element Analysis**, R. Glowinski, E. Rodin, O.C. Zienkiewicz Ed., John Wiley, New-York, 1979.

B. Mercier

[1] **Sur la Théorie et l'Analyse Numérique de Problèmes de Plasticité.** *(On the theory and numerical analysis of plasticity problems)*, Thesis, University of Paris, 1977.

M. Miranda

[1] **Comportamento delli successioni convergenti di frontiere minimali.** *(Behaviour of converging sequences of minimal boundaries)*, Rend Sem. Math. Univ. Padova, 1967, p.238-257.

G. Modica

See M. Giaquinta and G. Modica.

J.J. Moreau

[1] **Fonctionnelles convexes, Séminaire Equations aux Dérivées Partielles,** *(Convex functionals, Seminar on partial differential equations)*, Collège de France, 1966.

[2] **La notion de sur-potentiel et les liaisons unilatérales en élastostatique,** *(The concept of surpotential and unilateral constraints in elastostatics)*, C.R. Acad. Sci. Paris, Série A, 267, 1968, p.954-957.

[3] **Application of Convex Analysis to the treatment of elastoplastic systems,** in **Applications of Methods of Functional Analysis to Problems of Mechanics**, P. Germain, B. Nayroles Ed., Lecture Notes in Math., Vol. 503, Springer Verlag, 1976, p.56-89.

[4] **On unilateral constraints, friction and plasticity,** in New **Variational Techniques in Mathematical Physics**, G. Capriz, G. Stampacchia, Ed., Edizioni Cremonese, Rome, 1974, p.175-322.

[5] **Champs et distributions de tenseurs déformation sur un ouvert de connexité quelconque.** *(Fields and distributions of strain tensors on an open domain of arbitrary connexity)*, Séminaire d'Analyse Convexe, University of Montpellier, 6, 1976.

[6] **Isotropie et convexité dans l'espace des tenseurs symétriques,** *(Isotropy and convexity in the space of symmetric tensors)*, Séminaire d'Analyse Convexe, 12, no. 6, 1982.

J. Naumann

[1] On variational principles in perfect plasticity, Lecture at the Conference on "Problemi matematici nella meccanica dei continui", *(Mathematical problems in the mechanics of continuous media)*, Trente, January 12-17, 1981.

[2] **An existence theory for displacement field in perfect plasticity under general yield conditions**, Rend. Mat. Roma, (to appear).

B. Nayroles

[1] **Essai de théorie fonctionnelle des structures rigides plastiques parfaites**, *(Attempt at a functional theory of perfectly plastic rigid structures)*, J. Mécanique, 9, 1970, p.491-506.

[2] **Quelques applications variationnelles de la théorie des fonctions duales à la mécanique des solides**, *(Some variational applications of the theory of dual functions to solid mechanics)*, J. Mécanique 10, 1971, p.263-289.

See also, O. Debordes and B. Nayroles.

J. Něcas

[1] **Equations aux Dérivées Partielles**, *(Partial differential equations*, Presses de l'Université de Montréal, 1965.

A. Neveu

[1] **Mémoire non publié**, *(Unpublished memoir)*.

F.H. Norton

[1] **The creep of steel at high temperature**, McGraw-Hill, New-York, 1929.

D. Ornstein

[1] **A non-inequality for differential operators in the L^1 norm.** Arch. Rational Mech. Anal., 11, 1962, p.40-49.

L. Paris

[1] **Etude de la régularité d'un champ de vecteurs à partir de son tenseur déformation.** *(Study of the regularity of a vector field defined by it's strain tensor)*, Séminaire d'Analyse Convexe, University of Montpellier, 12, 1976.

W. Prager and P. Hodge

[1] **Theory of Perfectly Plastic Solids**, J. Wiley, New-York, 1951.

G. de Rham

[1] **Variétés Différentiables**, *(Differentiable varieties)*, Hermann, Paris, 1960.

N. Rivière

See L. Caffarelli and N. Rivière.

R.T. Rockafellar

[1] **Duality and stability in extremum problems involving convex functions**, Pacific J. Math., 21, 1967, p.167-187.

[2] **Convex Analysis**, Princeton University Press, 1970.

[3] **Integrals which are convex functionals I and II**, Pacific J. Math., 24, 1966, p.523-539 and 39, 1971, p.439-469.

[4] **Integrals functionals, normal integrals and measurable selections**, **in Nonlinear Operators and the Calculus of Variations**, Lecture Notes in Mathematics, Vol. 543, Springer Verlag, 1976.

J. Salençon

[1] **Applications of the Theory of Plasticity in Soil Mechanics**, J. Wiley, New-York, 1974.

J.T. Schwartz

See N. Dunford and J.T. Schwartz.

L. Schwartz

[1] **Théorie des distributions**, *(Theory of distributions)*, I, II, Hermann, Paris, 1950-51 (2nd edition, 1957).

J. Serrin

See C. Goffman and J. Serrin.

S.L. Sobolev

[1] **Sur un théorème de l'analyse fonctionnelle**, *(On a theorem in functional analysis)*, Dokl. Akad. Nauk SSSR, 20, 1938, p.5-9.

N.Q. Son

[1] **Contribution à la théorie macroscopique de l'élastoplasticité avec écrouissage**, *(Contributions to the macroscopic theory of elasto-plasticity with hardening)*, Thesis, Paris, 1973.

[2] **Méthodes énergétiques en mécanique de la rupture**, *(Energy methods in the mechanics of break-down)*, J. Mécanique, 19, 1980, p.363-386.

[3] **Problèmes de Plasticité et de Rupture**, *(Problems of plasticity and break-down, a graduate course)*, Cours de 3e Cycle, University Paris-Sud, Orsay, 1980-81.

See also B. Halphen and N.Q. Son.

G. Stampacchia

See H. Brézis and G. Stampacchia, E. Magenes and G. Stampacchia.

G. Strang

[1] **A family of model problems in plasticity.** Proc. Symp. Comp. Meth in Appl. Sc., R. Glowinski and J.L. Lions Ed., Lecture Notes in Computer Sciences, 704, 1979, p.1979, Springer Verlag.

See also R. Kohn and G. Strang, H. Matthies, G. Strang and E. Christiansen, R. Temam and G. Strang.

M. Strauss

[1] **Variations of Korn's and Sobolev's inequalities**, Proc. Symp. Pure Math., Vol. 23, D. Spencer Ed., AMS, Providence R.I., 1973.

P. Suquet

[1] **Sur un nouveau cadre fonctionnel pour les équations de la plasticité**, *(On a new functional framework for the plasticity equations*, C.R. Acad. Sci. Paris, 286, série A, 1978, p.1129-1132.

[2] **Existence et régularité des solutions des équations de la plasticité**, *(Existence and regularity of solutions to plasticity equations)*, C.R. Acad. Sci. Paris, 286, série A, 1978, p.1201-1204.

[3] **Existence et régularité des solutions des équations de la plasticité parfaite**, *(Existence and regularity of solutions to perfect plasticity equations)*, Thèse de 3e Cycle, Université de Paris, 1978.

[4] **Un espace fonctionnel pour les équations de la plasticité**, *(A functional space for the plasticity equations)*, Ann. Fac. Sci. Toulouse, 1, 1979, p.77-87.

[5] **Existence and regularity of solutions for plasticity problems** in **Variational Methods in Solid Mechanics**, S. Nemat Nasser Ed., Pergamon Press, 1980.

[6] **Sur les équations de la Plasticité : existence et régularité des solutions**, *(On the plasticity equations : existence and regularity of solutions)*, J. Mécanique, 20, 1981, p.3-39.

[7] **Plasticité et homogénéisation**, *(Plasticity and homogenisation)*, Thesis, University of Paris, VI, 1982.

R. Temam

[1] **Solutions généralisées de certains problèmes de calcul des variations**, *(Generalised solutions of certain problems in the calculus of variations)*, C.R. Acad. Sci. Paris, 271, 1970, p.1116-1119.

[2] **Solutions géneralisées de certaines équations du type hypersurfaces minima**, *(Generalised solutions to certain equations of the "minimal hypersurface" type)*, Arch. Rational Mech. Anal., 44, 1971, p.121-156.

[3] **Existence et unicité de solutions pour des problèmes du type de Neumann coercifs dans des espaces non réflexifs.** *(Existence and uniqueness of solutions to problems of the Neumann coercive type in non-reflexive spaces)*, C.R. Acad. Sci. Paris, 273, série A, 1971, p.609-611.

[4] Sur un problème non linéaire lié aux équations de Maxwell, *(On a non-linear problem relating to Maxwell's equations)*, in the Actes du Congrès d'Analyse Numérique, E. Magenes, Ed., Zanichelli and Academic Press, Rome, 1972.

[5] Applications de l'analyse convexe au calcul des variations, *(Applications of convex analysis to the calculus of variations)*, in Nonlinear operators and the calculus of variations, Lecture Notes in Mathematics, vol.543, Springer Verlag, 1976.

[6] Navier-Stokes Equations, Theory and Numerical Analysis, 3rd edition, North-Holland, Amsterdam, New-York, 1984.

[7] Mathematical Problems in Plasticity Theory, in Complementary Problems and Variational Inequalities, R.W. Cottle, F. Gianessi and J.L. Lions, Ed., J. Wiley, New-York, 1980.

[8] Mathematical Problems in Plasticity, in Computational Methods in Nonlinear Mechanics, J.T. Oden Ed., North-Holland, Amsterdam, 1980.

[9] On the continuity of the trace of vector functions with bounded deformations. Applicable Anal., 11, 1981, p.291-302.

[10] Existence theorems for the variational problems of plasticity, in Nonlinear Problems of Analysis in Geometry and Mechanics, M. Atteia, D. Bancel and I. Gumowski, Ed., Pitman, London, 1981.

[11] Approximation de fonctions convexes sur un espace de mesures et applications, *(Approximation of convex functions on a space of measures and applications)*, Canad. Math. Bull., 25, 1982, p.392-413.

[12] Some asymptotic problems in mechanics, in Analytic and Numerical Approaches to asymptotic problems in Analysis, O. Axelson, L.S. Frank, A. Vand der Sluis, Ed., North-Holland, Amsterdam, 1981.

[13] Calculus of variations, in Encyclopedia of Systems and Control, Madan Singh Editor, Pergamon Press, (to appear).

See also F. Demengel and R. Temam, I. Ekeland and R. Temam, M. Gurtin and R. Temam, R. Kohn and R. Temam, R. Temam and G. Strang.

R. Temam and G. Strang

[1] Existence de solutions relaxées pour les équations de la plasticité : étude d'un espace fonctionnel. *(Existence of relaxed solutions for the equations of plasticity : study of a functional space)*. C.R. Acad. Sci. Paris, 287, série A, 1978, p.515-518.

[2] Duality and relaxation in the variational problems of plasticity. J. Mécanique, 19, 1980, p.493-527.

[3] Functions of bounded deformation. Arch. Rational Mech. Anal., 75, 1980, p.7-21.

F. Treves

[1] Topological Vector Spaces, Distributions and Kernels, Academic Press, New-York, 1967.

A.I. Vol'pert

[1] The spaces BV and quasilinear equations, Math. USSR-Sb, 2, 1967, p.225-267.

W.H. Yang

[1] A generalized von Mises Criterion for Yield and Fracture, J. Appl. Mech. Tech. Phys., 47, 1980, p.297-300.

[2] A useful theorem for constructing convex yield functions, J. Appl. Mech. Tech. Phys., 47, 1980, p.301-303.

[3] Simplified limit analysis of plates, preprint.

[4] How to optimally support a plate? J. Appl. Mech. Tech. Phys., (1981).

L.C. Young

[1] Surfaces paramétriques généralisées. *(Generalised parametric surfaces)*, Bull. Soc. Math. France, 79, 1951, p.59-85.

[2] Champs vectoriels attachés à une mesure plane, *(Vector fields associated with a plane measure)*, Journ. de Math., 35, 1956.

[3] Generalized curves and the existence of an attained absolute minimum in the calculus of variations. C.R. Soc. Sci et Lettres Varsovie, C1.III, vol. 20, 1937, p.212-235.

E.H. Zarantonello

[1] Projections on convex sets in Hilbert spaces and spectral theory, in Contributions to Non-linear Functional Analysis, E.H. Zarantonello Ed., Academic Press, New York, 1971.

Appendix

BY JEAN-FRANÇOIS BABADJIAN AND ROGER TEMAM

This appendix is intended to briefly describe more recent advances on plasticity problems, which appeared after the initial publication of *Mathematical Problems in Plasticity* in 1983. It is not aimed to be exhaustive but to give a bibliographical guide to further developments of the theory.

Plasticity as a variational evolution. The introduction of an abstract variational theory to study quasi-static evolution of rate independent processes in [56] has revitalized the interest and the study of plasticity models at the beginning of the twenty-first century. The first fundamental contribution is contained in the seminal paper [17], where standard linearly elastic-perfectly plastic problems are studied in the variational framework of rate independent processes. In this formulation, the model turns out to be equivalent to global stability and energy balance principles. The existence of a variational evolution and the extent to which that evolution is "equivalent" to the original one is the main focus of [17].

The derivation of perfect plasticity models as the limit of regularized models such as linear isotropic hardening models has been performed in [8].

Softening phenomena. Many contributions then arise starting from much less standard models such as plasticity in the presence of softening. In contrast with hardening where the model is regularized, leading to much simpler mathematical arguments, softening induces a nonconvex dissipation potential with respect to the additional internal variables, which leads to the introduction of generalized Young measure valued solutions [18, 19] or to a relaxed formulation [20].

Nonassociative plasticity. Classical models of plasticity usually apply to most metals and alloys. In that case, the flow follows a normality rule, which means that the plastic strain rate is oriented in a direction normal to the elasticity set. These models are referred to as “associative” models. If one is interested in the plasticity of geomaterials, one has to account for the possibility of considering nonassociative models where the flow rule does not follow a normality rule (see [50, 51]). The investigation of Cam-Clay elastoplastic evolutions has been performed in [21, 22]. In these papers, a model of elastoplastic clay, which exhibits hardening and/or softening, is studied. The most striking feature of this model is that it exhibits a certain degree of nonassociativity, which means that the flow rule does not follow anymore a standard normality rule. This model is usually referred to as one with a nonassociative hardening rule. One of the key features is that it generates time discontinuities as shown in the study of the space homogeneous case [24]. It induces a well-posedness result in a suitable rescaled time. Such is also the case for other types of nonassociative models such as Mohr-Coulomb or Drucker-Prager models, arising in the study of soils, sand, or concrete, which have been studied in [64, 4, 40]. These materials have the particularity to be sensitive to hydrostatic pressure leading to permanent volumetric changes. The Armstrong-Frederick model studied in [41], which represents a class of nonlinear kinematic hardening models, also shares this nonassociativity, generating time discontinuities.

Multiscale plasticity problems. Periodic homogenization in perfect plasticity has been addressed in [36]. It rests on a series of papers accounting for discontinuities (see [67, 35, 68, 37]) and, more specifically, how to define the flow rule at the interface of two different phases. The resulting homogenization result is obtained by means of two-scale convergence techniques. It leads to a thermodynamically consistent elastoplastic model with an infinite number of internal variables since, in the resulting limit evolution, the dependence with respect to the fast variable cannot be averaged. Homogenization results in the case of strain gradients plasticity are obtained in [43, 39, 45]. An extension to the setting of Hencky plasticity with nonconvex potentials and vanishing hardening is the subject of [46]. The framework of layered materials in single-slip finite crystal plasticity is analyzed in [16].

The study of plate models in its original formulation has been revisited in [32], using modern tools of the variational theory of rate independent processes introduced in [17]. The derivation of reduced models starting from three-dimensional elastoplasticity has been studied in [54], in the presence of isotropic hardening by means of abstract Γ-convergence of rate independent processes, and in [28], in the case of perfect plasticity. As the thickness of the plate tends to zero, Γ-convergence techniques show that solutions to the three-dimensional quasi-static evolution problem of Prandtl-Reuss elastoplasticity converge to a quasi-static evolution of a suitable reduced model. In the limiting model, the admissible displacements are of Kirchhoff-Love type, and the stretching and bending components of the stress are coupled through a plastic flow rule. This limiting model has in general a genuinely three-dimensional nature and cannot be reduced to a two-dimensional setting, as usually done for classical elastoplastic plates. Similar reduced linearized plate models are studied in [25, 26], starting from three-dimensional finite plasticity with hardening.

The convergence of the elastoplasticity model to that of rigid-plasticity as the elasticity coefficients diverge has been established in [3]. Models of rigid-plasticity are particularly important and useful in a plane stress setting, where the slip-line method enables one to obtain explicit analytical solutions.

Dynamic models. The first well-posedness result in dynamical perfect plasticity has been obtained in [1] by means of a visco-elastic regularization of Kelvin-Voigt type for a Von Mises elasticity set. It also has been derived in various ways: first by means of vanishing isotropic hardening as in [11] and then by vanishing visco-plasticity of Perzyna type as in [12]. These results have been extended in [6] to any kind of elasticity sets, possibly depending on the hydrostatic pressure. The approximation of such models by means of cap models (where unbounded elasticity sets are cut in the unbounded directions) is also presented in [6]. The derivation of dynamical models of perfectly plastic plates has been studied in [55]. In [5], a hyperbolic approach to dynamical perfect plasticity has been proposed in the spirit of entropic solutions to systems of conservation laws. In [30], weak solutions to the dynamical perfect plasticity system are recovered by minimizing a suitable one-parameter family of functionals over entire trajectories and by passing to the parameter limit. In the paper [23], the question of convergence of a dynamical elasto-visco-plastic models toward a quasi-static limit is addressed when the data vary slowly.

Regularity of solutions. The question of regularity of solutions has been extensively developed in the monograph [42]. For dynamical problems, it has only recently been established in [62] (see also [5] in a simplified scalar setting), that for any elasticity set, the solutions are smooth for a short time, provided the data are smooth and compactly supported in space. Such a result does not hold in the static or quasi-static cases as noticed in [31]. However, partial regularity results are available for the stress tensor in some particular situations. In [9, 65, 31], the local H1 regularity of the stress tensor is established in the case of Von Mises yield criterion. This result has been adapted to the case of plates in [33, 29]. The extension to the case of a pressure dependent yield criterion of Mohr-Coulomb or Drucker-Prager type has been performed in [66], in the static case, and in [7], in the quasi-static case. A nice consequence of this regularity for the stress is that it enables one to use the quasi-continuous representative of the stress to give a pointwise meaning to the flow rule (see [38, 7]), which usually holds in a weak measure theoretic sense due to the fact that the energy spaces for the stress and the strain are not in duality. As far as the displacement field is concerned, it is known that it can develop singularities of both jump or Cantor types since it is a vector field with bounded deformation. In [38], using a Von Mises yield criterion, a simple sufficient condition on the Cauchy stress is introduced, which prevents the onset of plastic slip, i.e., no additional jump of velocity can occur. Thanks to a recent result [34] concerning the rank-one symmetric structure of the singular part of symmetric gradients of BD functions, it turns out that no additional Cantor part of the velocity can be created.

Concerning global regularity results for both the displacement field and the stress tensor, the only available results are obtained in [47, 48, 49], using some power law regularization of Norton-Hoff type or linear hardening.

Finite plasticity. Besides small strain elastoplasticity, let us mention a few contributions concerning finite plasticity at large strain. The main difficulty is related to the (nonlinear) multiplicative decomposition of the strain into a product of the elastic strain and of the plastic strain, which replaces the (linear) additive decomposition in the small strain setting. Let us quote the works [57, 58] where the energetic formulation of [56] is adapted to this setting in the presence of hardening, and [59] in the presence of viscosity. The case of gradient plasticity at large strains with kinematic hardening is analyzed in [60]. Another related work is contained in [27], where the usual ordered multiplicative decomposition is reversed and a well-posedness result is established in the presence of kinematic hardening.

The problem of relaxation of an energy functional associated to a rigid-plastic body at finite strain endowing one or two slip systems is the object of the papers [15, 13, 14], where tools of quasi-convexity are used to evidence fine scale microstructures.

Finally the question of convergence of finite plasticity to small strain plasticity is studied in [44, 61].

Other developments. Closer to the content of this book we would like to mention [2], where the author revisited the proof of the trace theorem extensively used in this book for functions with bounded deformation [space BD(Ω)] after [69, 70], [71], and [63], where a relaxation of the Hencky model for perfect plasticity is performed. In another direction, the papers [52, 53] study evolution equations for a minimal surface operator. This was meant as a step toward the study of the Prandtl-Reuss equations of plasticity. Although the equations studied in [52, 53] are of the first order in time, whereas the Prandtl-Reuss equations are of the second order in time, the two equations share the fact that they are derived from a functional with linear growth at infinity for the space variable. Hence the authors of [52, 53] use many of the concepts developed in this book; see also [10]. Finally, a study of the Prandtl-Reuss equations appears in [72, 73].

References

[1] G. Anzellotti, S. Luckhaus: Dynamical evolution in elasto-perfectly plastic bodies, Appl. Math. Opt., 15 (1987), 121–140.

[2] J.-F. Babadjian: *Traces of functions of bounded deformation*, Indiana University Math. Journal, 64 (2015), 1271–1290.

[3] J.-F. Babadjian, G. A. Francfort: A note on the derivation of rigid-plastic models, NoDEA Nonlinear Differential Equations Appl., 23 (2016), 23-37.

[4] J.-F. Babadjian, G. A. Francfort, M.G. Mora: Quasistatic evolution in non-associative plasticity - the cap model, SIAM J. Math. Anal., 44 (2012), 245–292.

[5] J.-F. Babadjian, C. Mifsud: Hyperbolic structure for a simplified model of dynamical perfect plasticity, Arch. Rational Mech. Anal., 223 (2017), 761–815.

[6] J.-F. Babadjian, M. G. Mora: Approximation of dynamic and quasi-static evolution problems in elasto-plasticity by cap models, Quart. Appl. Math., 73 (2015), 265–316.

[7] J.-F. Babadjian, M. G. Mora: Stress regularity in quasi-static perfect plasticity with a pressure dependent yield criterion, to appear in Journal Diff. Equations.

[8] S. Bartels, A. Mielke, T. Roubicek: Quasistatic small-strain plasticity in the limit of vanishing hardening and its numerical approximation, SIAM J. Numer. Anal., 50 (2012), 951–976.

[9] A. Bensoussan, J. Frehse: Asymptotic behaviour of the time dependent Norton-Hoff law in plasticity theory and H^1 regularity, Comment. Math. Univ. Carolin., 37 (1996), 285–304.

[10] Verena Bögelein, Frank Duzaar and Paolo Marcellini, Existence of evolutionary variational solutions via the calculus of variations, J. Differential Equations, 256, (2014), 12, 3912–3942.
[11] K. Chełmiński: Perfect plasticity as a zero relaxation limit of plasticity with isotropic hardening, Math. Meth. Appl. Sci., 24 (2001), 117–136.
[12] K. Chełmiński: Coercive approximation of viscoplasticity and plasticity, Asymptot. Anal., 26 (2001), 105–133.
[13] S. Conti, G. Dolzmann, C. Kreisbeck: Relaxation and microstructure in a model for finite crystal plasticity with one slip system in three dimensions, Discrete Contin. Dyn. Syst. Ser. S, 6 (2013), 1–16.
[14] S. Conti, G. Dolzmann, C. Kreisbeck: Asymptotic behavior of crystal plasticity with one slip system in the limit of rigid elasticity, SIAM J. Math. Anal.,, 43 (2011), 2337–2353.
[15] S. Conti, F. Theil: Single-slip elastoplastic microstructure, Arch. Rational Mech. Anal., 178 (2005), 125–148.
[16] F. Christowiak, C. Kreisbeck: Homogenization of layered materials with rigid components in single-slip finite crystal plasticity, Calc. Var. PDEs., 56 (2017), Art. 75, 28 pp.
[17] G. Dal Maso, A. De Simone, M.G. Mora: Quasistatic evolution problems for linearly elastic–perfectly plastic materials, Arch. Rational Mech. Anal., 180 (2006), 237–291.
[18] G. Dal Maso, A. De Simone, M. G. Mora, M. Morini: A vanishing viscosity approach to quasistatic evolution in plasticity with softening, Arch. Rational Mech. Anal., 189 (2008), 469–544.
[19] G. Dal Maso, A. De Simone, M. G. Mora, M. Morini: Time-dependent systems of generalized Young measures, Netw. Heterog. Media, 2 (2007); 1–36.
[20] G. Dal Maso, A. De Simone, M. G. Mora, M. Morini: Globally stable quasistatic evolution in plasticity with softening, Netw. Heterog. Media, 3 (2008), 567–614.
[21] G. Dal Maso, A. De Simone, F. Solombrino: Quasistatic evolution for Cam-Clay plasticity: a weak formulation via viscoplastic regularization and time rescaling, Calc. Var. PDEs, 40 (2011), 125–181.
[22] G. Dal Maso, A. De Simone, F. Solombrino: Quasistatic evolution for Cam-Clay plasticity: properties of the viscosity solutions, Calc. Var. PDEs., 44 (2012), 495–541.
[23] G. Dal Maso, R. Scala: Quasistatic evolution in perfect plasticity as limit of dynamic processes, Journal of Dynamics Diff. Eq., 26 (2014), 915–954.
[24] G. Dal Maso, F. Solombrino: Quasistatic evolution for Cam-Clay plasticity: the spatially homogeneous case, Netw. Heterog. Media, 5 (2010) 97–132.
[25] E. Davoli: Linearized plastic plate models as Γ-limits of 3D finite elastoplasticity, ESAIM: Cont. Optim. Calc. Var., 20 (2014), 725–747.
[26] E. Davoli: Quasistatic evolution models for thin plates arising as low energy ?-limits of finite plasticity, Math. Models Methods Appl. Sci. 24 (2014), 2085–2153.
[27] E. Davoli, G. A. Francfort: A critical revisiting of finite elasto-plasticity, SIAM J. Math Anal., 47 (2015), 526–565.
[28] E. Davoli, M.G. Mora: A quasistatic evolution model for perfectly plastic plates derived by Gamma-convergence, Ann. Inst. H. Poincar Anal. Nonlin., 30 (2013), 615–660.
[29] E. Davoli, M.G. Mora: Stress regularity for a new quasistatic evolution model of perfectly plastic plates,Partial Differential Equations, 54 (2015), 2581–2614.
[30] E. Davoli, U. Stefanelli: Dynamic perfect plasticity as convex minimization, Preprint (2016).
[31] A. Demyanov: Regularity of stresses in Prandtl-Reuss perfect plasticity, Partial Differential Equations, 34 (2009), 23–72.
[32] A. Demyanov: Quasistatic evolution in the theory of perfectly elasto-plastic plates. Part I: existence of a weak solution, Math. Models Methods Appl. Sci., 19 (2009), 229–256.
[33] A. Demyanov: Quasistatic evolution in the theory of perfect elasto-plastic plates. Part II: Regularity of bending moments, Ann. Inst. H. Poincar Anal. Nonlin., 26 (2009), 2137–2163.
[34] G. De Philippis, F. Rindler: On the structure of $\mathcal{A}$-free measures and applications, Ann. of Math., 184 (2016),1017–1039.
[35] G. A. Francfort, A. Giacomini: Small strain heterogeneous elasto-plasticity revisited, Comm. Pure Appl. Math., 65 (2012), 1185–1241.

[36] G. A. Francfort, A. Giacomini: On periodic homogenization in perfect elasto-plasticity, J. Eur. Math. Soc., 16(2014), 409–461.
[37] G. A. Francfort, A. Giacomini: The role of a vanishing interfacial layer in perfect elasto-plasticity, Chin. Ann. Math. Ser. B, 36 (2015), 813–828.
[38] G.A. Francfort, A. Giacomini, J.-J. Marigo: The taming of plastic slips in von Mises elasto-plasticity, Interfaces Free Bound., 17 (2015), 497–516.
[39] G. A. Francfort, A. Giacomini, A. Musesti: On the Fleck and Willis homogenization procedure in strain gradient plasticity, Discrete Contin. Dyn. Syst. Ser. S, 6 (2013) 43–62.
[40] G. A. Francfort, M. G. Mora: Quasistatic evolution in non-associative plasticity revisited, Calc. Var. Partial Differential Equations, 57 (2018),
[41] G. A. Francfort, U. Stefanelli: Quasistatic Evolution for the Armstrong-Frederick Hardening-Plasticity Model, Applied Maths. Res. Express, 2 (2013), 297–344.
[42] M. Fuchs, G. Seregin: Variational Methods for problems from plasticity theory and for generalized Newtonian fluids, Lecture notes in Mathematics, Springer (2000).
[43] A. Giacomini, A. Musesti: Two-scale homogenization for a model in strain gradient plasticity, ESAIM Control Optim. Calc. Var., 17 (2011), 1035–1065.
[44] A. Giacomini, A. Musesti: Quasi-static evolutions in linear perfect plasticity as a variational limit of finite plasticity: a one-dimensional case, Math. Models and Methods Appl. Sci., 23 (2013), 1275–1308.
[45] M. Heida, B. Schweizer: Stochastic homogenization of plasticity equations, ESAIM: Cont. Optim. Calc. Var., 24 (2018), 153–176.
[46] M. Jesenko, B. Schmidt: Homogenization and the limit of vanishing hardening in Hencky plasticity with non-convex potentials, Calc. Var. Partial Differential Equations, 57 (2018), Art. 2, 43 pp.
[47] D. Knees: Global regularity of the elastic fields of a power-law model on Lipschitz domains, Math. Models and Methods Appl. Sci., 29 (2006), 1363–1391.
[48] D. Knees: On global spatial regularity in elasto-plasticity with linear hardening, Calc. Var. Partial Differential Equations, 36 (2009), 611–625.
[49] D. Knees: On global spatial regularity and convergence rates for time dependent elasto-plasticity, Math. Models and Methods Appl. Sci., 20 (2010), 1823–1858.
[50] P. Laborde: A nonmonotone differential inclusion, Nonlin. Anal. Theory, Math. & Appl., 11 (1986), 757–767.
[51] P. Laborde: Analysis of the strain-stress relation in plasticity with non-associative laws, Int. J. Engrg Sci., 25 (1987), 655–666.
[52] A. Lichnewsky and R. Temam, Surfaces minimales d'évolution: le concept de pseudo-solution, C.R. Acad. Sc. Paris, Série A 284, (1977), 853–856.
[53] A. Lichnewsky and R. Temam: Pseudo-solutions of time dependent minimal surfaces, J. Diff. Equ., 30, (1978), 340–364.
[54] M. Liero, A. Mielke: An evolutionary elastoplastic plate model derived via Γ-convergence. Math. Models Meth. Appl. Sci., 21 (2001), 1961–1986.
[55] G. B. Maggiani, M. G. Mora: A dynamic evolution model for perfectly plastic plates, Math. Models Methods Appl. Sci., 26 (2016), 1825–1864.
[56] A. Mainik, A. Mielke: Existence results for energetic models for rate-independent systems, Calc. Var. Partial Differential Equations, 22 (2005), 73–99.
[57] A. Mielke: Energetic formulation of multiplicative elasto-plasticity using dissipation distances, Cont. Mech. Thermodynamics, 15 (2003), 351–382.
[58] A. Mielke: Existence of minimizers in incremental elasto-plasticity with finite strains, SIAM J. Math. Analysis, 36 (2004), 384–404.
[59] A. Mielke, R. Rossi, G. Savaré. Global existence results for viscoplasticity at finite strain, Archive Rational Mech. Anal., 227 (2018), 423–475.
[60] A. Mielke, T. Roubcek. Rate-independent elastoplasticity at finite strains and its numerical approximation. Math. Models Meth. Appl. Sciences, 26 (2016), 2203–2236.
[61] A. Mielke, U. Stefanelli. Linearized plasticity is the evolutionary Γ-limit of finite plasticity, J. Eur. Math. Soc., 15 (2013), 923-948.

[62] C. Mifsud: Short-time regularity for dynamic evolution problems in perfect plasticity, (2016) hal-01370797.
[63] M.G. Mora: Relaxation of the Hencky model in perfect plasticity, J. Math. Pures Appl. 106 (2016), 725–743.
[64] S. Repin, G. Seregin: Existence of weak solution of the minimax problem arising in Coulomb-Mohr plasticity, Amer. Math. Soc. Transl., 164 (1995), 189–220.
[65] G. Seregin: On the differentiability of extremals of variational problems of the mechanics of ideally elastoplastic media. (Russian) Differentsial'nye Uravneniya, 23 (1987), 1981–1991.
[66] G. Seregin: On the differentiability of the stress tensor in the Coulomb-Mohr theory of plasticity, St. Petersburg Math. J., 4 (1993), 1257–1271.
[67] F. Solombrino: Quasistatic evolution problems for nonhomogeneous elastic plastic materials, J. Convex Anal., 16 (2009), 89–119.
[68] F. Solombrino: Quasistatic evolution in perfect plasticity for general heterogeneous materials, Archive Rat. Mech. Anal., 212 (2014), 283–330.
[69] G. Strang and R. Temam: Existence de solutions relaxées pour les équations de la plasticité: étude d'un espace fonctionnel, C.R. Acad. Sc. Paris, Série A, 287, (1978), 515–518.
[70] G. Strang and R. Temam: Functions of bounded deformations, Arch. Rational Mech. Anal., 75, (1980), 7–21.
[71] P. Suquet: Un espace fonctionnel pour les équations de la plasticité, (*A functional space for the plasticity equations)*, Ann. Fac. Sci. Toulouse, 1, (1979), 77–87.
[72] R. Temam, Principe de dissipation maximale pour la loi de Prandtl-Reuss en plasticité, C. R. Acad. Sc. Paris, Série I, 302, (1986), 79–82.
[73] R. Temam, A generalized Norton-Hoff model and the Prandtl-Reuss law of plasticity, Arch. Rational Mech. Anal., 95, (1986), 137–183.

A CATALOG OF SELECTED

DOVER BOOKS

IN SCIENCE AND MATHEMATICS

Physics

THEORETICAL NUCLEAR PHYSICS, John M. Blatt and Victor F. Weisskopf. An uncommonly clear and cogent investigation and correlation of key aspects of theoretical nuclear physics by leading experts: the nucleus, nuclear forces, nuclear spectroscopy, two-, three- and four-body problems, nuclear reactions, beta-decay and nuclear shell structure. 896pp. 5 3/8 x 8 1/2. 0-486-66827-4

QUANTUM THEORY, David Bohm. This advanced undergraduate-level text presents the quantum theory in terms of qualitative and imaginative concepts, followed by specific applications worked out in mathematical detail. 655pp. 5 3/8 x 8 1/2. 0-486-65969-0

ATOMIC PHYSICS AND HUMAN KNOWLEDGE, Niels Bohr. Articles and speeches by the Nobel Prize–winning physicist, dating from 1934 to 1958, offer philosophical explorations of the relevance of atomic physics to many areas of human endeavor. 1961 edition. 112pp. 5 3/8 x 8 1/2. 0-486-47928-5

COSMOLOGY, Hermann Bondi. A co-developer of the steady-state theory explores his conception of the expanding universe. This historic book was among the first to present cosmology as a separate branch of physics. 1961 edition. 192pp. 5 3/8 x 8 1/2. 0-486-47483-6

LECTURES ON QUANTUM MECHANICS, Paul A. M. Dirac. Four concise, brilliant lectures on mathematical methods in quantum mechanics from Nobel Prize-winning quantum pioneer build on idea of visualizing quantum theory through the use of classical mechanics. 96pp. 5 3/8 x 8 1/2. 0-486-41713-1

THE PRINCIPLE OF RELATIVITY, Albert Einstein and Frances A. Davis. Eleven papers that forged the general and special theories of relativity include seven papers by Einstein, two by Lorentz, and one each by Minkowski and Weyl. 1923 edition. 240pp. 5 3/8 x 8 1/2. 0-486-60081-5

PHYSICS OF WAVES, William C. Elmore and Mark A. Heald. Ideal as a classroom text or for individual study, this unique one-volume overview of classical wave theory covers wave phenomena of acoustics, optics, electromagnetic radiations, and more. 477pp. 5 3/8 x 8 1/2. 0-486-64926-1

THERMODYNAMICS, Enrico Fermi. In this classic of modern science, the Nobel Laureate presents a clear treatment of systems, the First and Second Laws of Thermodynamics, entropy, thermodynamic potentials, and much more. Calculus required. 160pp. 5 3/8 x 8 1/2. 0-486-60361-X

QUANTUM THEORY OF MANY-PARTICLE SYSTEMS, Alexander L. Fetter and John Dirk Walecka. Self-contained treatment of nonrelativistic many-particle systems discusses both formalism and applications in terms of ground-state (zero-temperature) formalism, finite-temperature formalism, canonical transformations, and applications to physical systems. 1971 edition. 640pp. 5 3/8 x 8 1/2. 0-486-42827-3

QUANTUM MECHANICS AND PATH INTEGRALS: Emended Edition, Richard P. Feynman and Albert R. Hibbs. Emended by Daniel F. Styer. The Nobel Prize–winning physicist presents unique insights into his theory and its applications. Feynman starts with fundamentals and advances to the perturbation method, quantum electrodynamics, and statistical mechanics. 1965 edition, emended in 2005. 384pp. 6 1/8 x 9 1/4. 0-486-47722-3

Physics

INTRODUCTION TO MODERN OPTICS, Grant R. Fowles. A complete basic undergraduate course in modern optics for students in physics, technology, and engineering. The first half deals with classical physical optics; the second, quantum nature of light. Solutions. 336pp. 5 3/8 x 8 1/2. 0-486-65957-7

THE QUANTUM THEORY OF RADIATION: Third Edition, W. Heitler. The first comprehensive treatment of quantum physics in any language, this classic introduction to basic theory remains highly recommended and widely used, both as a text and as a reference. 1954 edition. 464pp. 5 3/8 x 8 1/2. 0-486-64558-4

QUANTUM FIELD THEORY, Claude Itzykson and Jean-Bernard Zuber. This comprehensive text begins with the standard quantization of electrodynamics and perturbative renormalization, advancing to functional methods, relativistic bound states, broken symmetries, nonabelian gauge fields, and asymptotic behavior. 1980 edition. 752pp. 6 1/2 x 9 1/4. 0-486-44568-2

FOUNDATIONS OF POTENTIAL THERY, Oliver D. Kellogg. Introduction to fundamentals of potential functions covers the force of gravity, fields of force, potentials, harmonic functions, electric images and Green's function, sequences of harmonic functions, fundamental existence theorems, and much more. 400pp. 5 3/8 x 8 1/2. 0-486-60144-7

FUNDAMENTALS OF MATHEMATICAL PHYSICS, Edgar A. Kraut. Indispensable for students of modern physics, this text provides the necessary background in mathematics to study the concepts of electromagnetic theory and quantum mechanics. 1967 edition. 480pp. 6 1/2 x 9 1/4. 0-486-45809-1

GEOMETRY AND LIGHT: The Science of Invisibility, Ulf Leonhardt and Thomas Philbin. Suitable for advanced undergraduate and graduate students of engineering, physics, and mathematics and scientific researchers of all types, this is the first authoritative text on invisibility and the science behind it. More than 100 full-color illustrations, plus exercises with solutions. 2010 edition. 288pp. 7 x 9 1/4. 0-486-47693-6

QUANTUM MECHANICS: New Approaches to Selected Topics, Harry J. Lipkin. Acclaimed as "excellent" (*Nature*) and "very original and refreshing" (*Physics Today*), these studies examine the Mössbauer effect, many-body quantum mechanics, scattering theory, Feynman diagrams, and relativistic quantum mechanics. 1973 edition. 480pp. 5 3/8 x 8 1/2. 0-486-45893-8

THEORY OF HEAT, James Clerk Maxwell. This classic sets forth the fundamentals of thermodynamics and kinetic theory simply enough to be understood by beginners, yet with enough subtlety to appeal to more advanced readers, too. 352pp. 5 3/8 x 8 1/2. 0-486-41735-2

QUANTUM MECHANICS, Albert Messiah. Subjects include formalism and its interpretation, analysis of simple systems, symmetries and invariance, methods of approximation, elements of relativistic quantum mechanics, much more. "Strongly recommended." – *American Journal of Physics.* 1152pp. 5 3/8 x 8 1/2. 0-486-40924-4

RELATIVISTIC QUANTUM FIELDS, Charles Nash. This graduate-level text contains techniques for performing calculations in quantum field theory. It focuses chiefly on the dimensional method and the renormalization group methods. Additional topics include functional integration and differentiation. 1978 edition. 240pp. 5 3/8 x 8 1/2. 0-486-47752-5

Physics

MATHEMATICAL TOOLS FOR PHYSICS, James Nearing. Encouraging students' development of intuition, this original work begins with a review of basic mathematics and advances to infinite series, complex algebra, differential equations, Fourier series, and more. 2010 edition. 496pp. 6 1/8 x 9 1/4. 0-486-48212-X

TREATISE ON THERMODYNAMICS, Max Planck. Great classic, still one of the best introductions to thermodynamics. Fundamentals, first and second principles of thermodynamics, applications to special states of equilibrium, more. Numerous worked examples. 1917 edition. 297pp. 5 3/8 x 8. 0-486-66371-X

AN INTRODUCTION TO RELATIVISTIC QUANTUM FIELD THEORY, Silvan S. Schweber. Complete, systematic, and self-contained, this text introduces modern quantum field theory. "Combines thorough knowledge with a high degree of didactic ability and a delightful style." – *Mathematical Reviews.* 1961 edition. 928pp. 5 3/8 x 8 1/2. 0-486-44228-4

THE ELECTROMAGNETIC FIELD, Albert Shadowitz. Comprehensive undergraduate text covers basics of electric and magnetic fields, building up to electromagnetic theory. Related topics include relativity theory. Over 900 problems, some with solutions. 1975 edition. 768pp. 5 5/8 x 8 1/4. 0-486-65660-8

THE PRINCIPLES OF STATISTICAL MECHANICS, Richard C. Tolman. Definitive treatise offers a concise exposition of classical statistical mechanics and a thorough elucidation of quantum statistical mechanics, plus applications of statistical mechanics to thermodynamic behavior. 1930 edition. 704pp. 5 5/8 x 8 1/4. 0-486-63896-0

INTRODUCTION TO THE PHYSICS OF FLUIDS AND SOLIDS, James S. Trefil. This interesting, informative survey by a well-known science author ranges from classical physics and geophysical topics, from the rings of Saturn and the rotation of the galaxy to underground nuclear tests. 1975 edition. 320pp. 5 3/8 x 8 1/2. 0-486-47437-2

STATISTICAL PHYSICS, Gregory H. Wannier. Classic text combines thermodynamics, statistical mechanics, and kinetic theory in one unified presentation. Topics include equilibrium statistics of special systems, kinetic theory, transport coefficients, and fluctuations. Problems with solutions. 1966 edition. 532pp. 5 3/8 x 8 1/2. 0-486-65401-X

SPACE, TIME, MATTER, Hermann Weyl. Excellent introduction probes deeply into Euclidean space, Riemann's space, Einstein's general relativity, gravitational waves and energy, and laws of conservation. "A classic of physics." – *British Journal for Philosophy and Science.* 330pp. 5 3/8 x 8 1/2. 0-486-60267-2

RANDOM VIBRATIONS: Theory and Practice, Paul H. Wirsching, Thomas L. Paez and Keith Ortiz. Comprehensive text and reference covers topics in probability, statistics, and random processes, plus methods for analyzing and controlling random vibrations. Suitable for graduate students and mechanical, structural, and aerospace engineers. 1995 edition. 464pp. 5 3/8 x 8 1/2. 0-486-45015-5

PHYSICS OF SHOCK WAVES AND HIGH-TEMPERATURE HYDRO DYNAMIC PHENOMENA, Ya B. Zel'dovich and Yu P. Raizer. Physical, chemical processes in gases at high temperatures are focus of outstanding text, which combines material from gas dynamics, shock-wave theory, thermodynamics and statistical physics, other fields. 284 illustrations. 1966–1967 edition. 944pp. 6 1/8 x 9 1/4. 0-486-42002-7

Engineering

FUNDAMENTALS OF ASTRODYNAMICS, Roger R. Bate, Donald D. Mueller, and Jerry E. White. Teaching text developed by U.S. Air Force Academy develops the basic two-body and n-body equations of motion; orbit determination; classical orbital elements, coordinate transformations; differential correction; more. 1971 edition. 455pp. 5 3/8 x 8 1/2. 0-486-60061-0

INTRODUCTION TO CONTINUUM MECHANICS FOR ENGINEERS: Revised Edition, Ray M. Bowen. This self-contained text introduces classical continuum models within a modern framework. Its numerous exercises illustrate the governing principles, linearizations, and other approximations that constitute classical continuum models. 2007 edition. 320pp. 6 1/8 x 9 1/4. 0-486-47460-7

ENGINEERING MECHANICS FOR STRUCTURES, Louis L. Bucciarelli. This text explores the mechanics of solids and statics as well as the strength of materials and elasticity theory. Its many design exercises encourage creative initiative and systems thinking. 2009 edition. 320pp. 6 1/8 x 9 1/4. 0-486-46855-0

FEEDBACK CONTROL THEORY, John C. Doyle, Bruce A. Francis and Allen R. Tannenbaum. This excellent introduction to feedback control system design offers a theoretical approach that captures the essential issues and can be applied to a wide range of practical problems. 1992 edition. 224pp. 6 1/2 x 9 1/4. 0-486-46933-6

THE FORCES OF MATTER, Michael Faraday. These lectures by a famous inventor offer an easy-to-understand introduction to the interactions of the universe's physical forces. Six essays explore gravitation, cohesion, chemical affinity, heat, magnetism, and electricity. 1993 edition. 96pp. 5 3/8 x 8 1/2. 0-486-47482-8

DYNAMICS, Lawrence E. Goodman and William H. Warner. Beginning engineering text introduces calculus of vectors, particle motion, dynamics of particle systems and plane rigid bodies, technical applications in plane motions, and more. Exercises and answers in every chapter. 619pp. 5 3/8 x 8 1/2. 0-486-42006-X

ADAPTIVE FILTERING PREDICTION AND CONTROL, Graham C. Goodwin and Kwai Sang Sin. This unified survey focuses on linear discrete-time systems and explores natural extensions to nonlinear systems. It emphasizes discrete-time systems, summarizing theoretical and practical aspects of a large class of adaptive algorithms. 1984 edition. 560pp. 6 1/2 x 9 1/4. 0-486-46932-8

INDUCTANCE CALCULATIONS, Frederick W. Grover. This authoritative reference enables the design of virtually every type of inductor. It features a single simple formula for each type of inductor, together with tables containing essential numerical factors. 1946 edition. 304pp. 5 3/8 x 8 1/2. 0-486-47440-2

THERMODYNAMICS: Foundations and Applications, Elias P. Gyftopoulos and Gian Paolo Beretta. Designed by two MIT professors, this authoritative text discusses basic concepts and applications in detail, emphasizing generality, definitions, and logical consistency. More than 300 solved problems cover realistic energy systems and processes. 800pp. 6 1/8 x 9 1/4. 0-486-43932-1

THE FINITE ELEMENT METHOD: Linear Static and Dynamic Finite Element Analysis, Thomas J. R. Hughes. Text for students without in-depth mathematical training, this text includes a comprehensive presentation and analysis of algorithms of time-dependent phenomena plus beam, plate, and shell theories. Solution guide available upon request. 672pp. 6 1/2 x 9 1/4. 0-486-41181-8

HELICOPTER THEORY, Wayne Johnson. Monumental engineering text covers vertical flight, forward flight, performance, mathematics of rotating systems, rotary wing dynamics and aerodynamics, aeroelasticity, stability and control, stall, noise, and more. 189 illustrations. 1980 edition. 1089pp. 5 5/8 x 8 1/4. 0-486-68230-7

MATHEMATICAL HANDBOOK FOR SCIENTISTS AND ENGINEERS: Definitions, Theorems, and Formulas for Reference and Review, Granino A. Korn and Theresa M. Korn. Convenient access to information from every area of mathematics: Fourier transforms, Z transforms, linear and nonlinear programming, calculus of variations, random-process theory, special functions, combinatorial analysis, game theory, much more. 1152pp. 5 3/8 x 8 1/2. 0-486-41147-8

A HEAT TRANSFER TEXTBOOK: Fourth Edition, John H. Lienhard V and John H. Lienhard IV. This introduction to heat and mass transfer for engineering students features worked examples and end-of-chapter exercises. Worked examples and end-of-chapter exercises appear throughout the book, along with well-drawn, illuminating figures. 768pp. 7 x 9 1/4. 0-486-47931-5

BASIC ELECTRICITY, U.S. Bureau of Naval Personnel. Originally a training course; best nontechnical coverage. Topics include batteries, circuits, conductors, AC and DC, inductance and capacitance, generators, motors, transformers, amplifiers, etc. Many questions with answers. 349 illustrations. 1969 edition. 448pp. 6 1/2 x 9 1/4. 0-486-20973-3

BASIC ELECTRONICS, U.S. Bureau of Naval Personnel. Clear, well-illustrated introduction to electronic equipment covers numerous essential topics: electron tubes, semiconductors, electronic power supplies, tuned circuits, amplifiers, receivers, ranging and navigation systems, computers, antennas, more. 560 illustrations. 567pp. 6 1/2 x 9 1/4. 0-486-21076-6

BASIC WING AND AIRFOIL THEORY, Alan Pope. This self-contained treatment by a pioneer in the study of wind effects covers flow functions, airfoil construction and pressure distribution, finite and monoplane wings, and many other subjects. 1951 edition. 320pp. 5 3/8 x 8 1/2. 0-486-47188-8

SYNTHETIC FUELS, Ronald F. Probstein and R. Edwin Hicks. This unified presentation examines the methods and processes for converting coal, oil, shale, tar sands, and various forms of biomass into liquid, gaseous, and clean solid fuels. 1982 edition. 512pp. 6 1/8 x 9 1/4. 0-486-44977-7

THEORY OF ELASTIC STABILITY, Stephen P. Timoshenko and James M. Gere. Written by world-renowned authorities on mechanics, this classic ranges from theoretical explanations of 2- and 3-D stress and strain to practical applications such as torsion, bending, and thermal stress. 1961 edition. 560pp. 5 3/8 x 8 1/2. 0-486-47207-8

PRINCIPLES OF DIGITAL COMMUNICATION AND CODING, Andrew J. Viterbi and Jim K. Omura. This classic by two digital communications experts is geared toward students of communications theory and to designers of channels, links, terminals, modems, or networks used to transmit and receive digital messages. 1979 edition. 576pp. 6 1/8 x 9 1/4. 0-486-46901-8

LINEAR SYSTEM THEORY: The State Space Approach, Lotfi A. Zadeh and Charles A. Desoer. Written by two pioneers in the field, this exploration of the state space approach focuses on problems of stability and control, plus connections between this approach and classical techniques. 1963 edition. 656pp. 6 1/8 x 9 1/4. 0-486-46663-9

Browse over 9,000 books at www.doverpublications.com